T0309115

Agricultural Water Management: Theories and Practices

Agricultural Water Management: Theories and Practices

Editor: Keith Wheatley

www.callistoreference.com

Callisto Reference,
118-35 Queens Blvd., Suite 400,
Forest Hills, NY 11375, USA

Visit us on the World Wide Web at:
www.callistoreference.com

ISBN: 978-1-64116-766-6 (Hardback)

Cataloging-in-Publication Data

Agricultural water management : theories and practices / edited by Keith Wheatley.
p. cm.
Includes bibliographical references and index.
ISBN 978-1-64116-766-6
1. Water in agriculture. 2. Water-supply, Agricultural. 3. Water resources development.
4. Irrigation. I. Wheatley, Keith.

S494.5.W3 W38 2023
631.7--dc23

Table of Contents

Preface

Water is an important input in agricultural production and contributes towards food security. Agricultural water management (AWM) aims to utilize water in such a way that the water requirement of crops and animals are met, while enhancing productivity and conserving natural resources. In addition to irrigation, AWM also involves soil, land and ecosystem conservation strategies, including fisheries management, drainage and watershed management along with technologies for lifting, storing and transferring water. Improvements in the management of water resources are based on the integrated management of nutrients, plants, water and soil. This can be achieved by optimization of irrigation scheduling and utilizing irrigation systems such as drip irrigation. This book contains some path-breaking studies on agricultural water management. It includes contributions of experts and scientists, which will provide innovative insights into this area of study.

The researches compiled throughout the book are authentic and of high quality, combining several disciplines and from very diverse regions from around the world. Drawing on the contributions of many researchers from diverse countries, the book's objective is to provide the readers with the latest achievements in the area of research. This book will surely be a source of knowledge to all interested and researching the field.

In the end, I would like to express my deep sense of gratitude to all the authors for meeting the set deadlines in completing and submitting their research chapters. I would also like to thank the publisher for the support offered to us throughout the course of the book. Finally, I extend my sincere thanks to my family for being a constant source of inspiration and encouragement.

Editor

1

Deficit Drip Irrigation in Processing Tomato Production in the Mediterranean Basin

Rosa Francaviglia and Claudia Di Bene *

Council for Agricultural Research and Economics, Research Centre for Agriculture and Environment (CREA-AA), 00184 Rome, Italy; rosa.francaviglia@crea.gov.it
* Correspondence: claudia.dibene@crea.gov.it

Abstract: In this study, the effects of deficit irrigation (DI) on crop yields and irrigation water utilization efficiency (IWUE) of processing tomato are contrasting. This study aimed at analyzing a set of field experiments with drip irrigation available for Mediterranean Italy in terms of marketable yields and IWUE under DI. Both yields and IWUE were compared with the control treatment under full irrigation, receiving the maximum water restoration (MWR) in each experiment. The study also aimed at testing the effect of climate (aridity index) and soil parameters (texture). Main results indicated that yields would marginally decrease at 70–80% of MWR and variable irrigation regimes during the crop cycle resulted in higher crop yields. However, results were quite variable and site-dependent. In fact, DI proved more effective in fine textured soils and semiarid climates. We recommend that further research should address variable irrigation regimes and soil and climate conditions that proved more unfavorable in terms of crop response to DI.

Keywords: deficit irrigation; Mediterranean region; tomato fruit yield; irrigation water use efficiency

---

1. Introduction

Water resources are extremely scarce in many areas of the world, and water saving has become a priority due to the increase in population and global climate change [1,2]. Agriculture is a major water consumer in regions where irrigation is required for profitable yields, and strategies to reduce water use have the potential to increase sustainability of production. Globally, agricultural irrigation is responsible for 70–80% of freshwater consumption [3,4]. Increased water savings and optimization of irrigation management as much as possible are, thus, urgently needed. A small amount of water saved can be used for other purposes. Therefore, in recent decades, agricultural water use efficiency has been improved by innovations in technology and plant breeding. Irrigation systems and scheduling mainly affect crop yields. Thus, the knowledge of crop water requirement, the reference crop evapotranspiration, and the rainfall of the target region is recommended [5]. While full irrigation (FI) aims to meet crop water requirements to maximize crop yield, in deficit irrigation (DI) water use is optimized in relation to crop yield per volume of water consumed. Modest yield reductions can be acceptable if connected to a significant reduction in water use [6]. DI has been found suitable for grapevine and fruit crops, but vegetables might suffer from losses in yield and quality. DI effects on crop yield and water efficiency have been studied on several crops, including tomato (*Solanum lycopersicum* L.), though with contrasting results that can be ascribed to the different cultivars cultivated or the period of DI application during the crop cycle [6]. Although farm income is higher with increasing yields when more water is supplied with irrigation, water availability is continuously decreasing due to the competing requirements of agriculture, industry, recreation, and the environment. In addition, DI provides an effective adaptive response to water scarcity within a climate change perspective [7], and lower yields might be compensated by the increased production and farm income from additional lands irrigated with the water saved by DI [8,9].

Most of horticultural production areas are in hot and dry climates due to favorable weather conditions (high light, high temperature), as in Mediterranean regions, however, soil water deficit is rather frequent. Tomato is considered one of the most commonly consumed vegetables and economically important crops in the world, and has the highest planted areas of all vegetables worldwide. It is characterized by high-water needs [10] due to the high temperatures and the large gap between rainfall and evapotranspiration (ET) during the long spring-summer growing season [2]. Tomato is a drought sensitive plant because its yield decreases considerably after short periods of water deficiency [5]. The application of DI strategies during tomato growing season may greatly contribute to saving irrigation water [11,12] without affecting tomato yield, compared with (FI) receiving the maximum water restoration (MWR), at a rate of 100% ET [13]. Nevertheless, water deficits at different growth stages can differentially affect tomato yield. Results of crop simulation models showed that certain tomato life stages, such as the flowering and fruiting stages, were more susceptible to water stress than the seedling stage [14].

According to the World Processing Tomato Council [15] (an international non-profit organization representing the tomato processing industry), Italy is ranked as the second producer of processing tomato worldwide, after California and followed by China. Italy is the leading country in Europe, contributing to 44% of the total amount, followed by Spain (27%), Turkey (12%) and Portugal (11%). The total national production in 2018 was 4,811,955 t, cultivated on a surface of 72,504 ha (average marketable yield is 66 t ha^{-1}). The two most important production regions are Emilia-Romagna in the north and Apulia in the south (concentrated in the Capitanata plain, in Foggia province). The two regions contributed in 2018 to 35 and 32% respectively of the national production of processing tomato, as reported by official statistics [16]. Average yields were 69 and 85 t ha^{-1}, respectively. In these areas processing tomato cultivation is highly intensive due to large and regular application of irrigation water and nutrient inputs during flowering and fruit formation [17,18], which may create the potential for negative side-effects on the environment [19]. Thus, the application of water saving strategies is of particular interest where water availability is limited and to save water while maximizing tomato yields under water deficit conditions [20,21]. Since the price of water is increasing, DI is an effective strategy to provide an adequate economic profit for farmers in Mediterranean environments [22]. Moreover, results of crop simulation models in southern Italy have shown that climate change would decrease tomato yields due to the shorter crop cycle induced by the temperature increase [23,24]. Besides water savings, gains in fruit quality (higher soluble solid contents and fruit color intensity) can often compensate for the losses in fruit yields [11,21]. However, the contrasting results available in the scientific literature suggest the need to better understand site-dependent plant responses to water deficit with DI [25,26].

A quantitative analysis is important to provide suggestions for improving crop yield and recommendations of irrigation water inputs in processing tomato cultivation under Mediterranean conditions. The present study is aimed at evaluating the effect of DI irrigation on processing tomato in field experiments derived from a literature search. Data were analyzed in terms of marketable yields, water restoration, and irrigation water use efficiency (IWUE) under DI compared with the control treatment under full irrigation, receiving the MWR in each experiment. The study also aimed at testing the effect of climate (aridity index) and soil parameters (texture).

2. Materials and Methods

2.1. Data Collection and Case Studies

To assess the effect of deficit drip irrigation on tomato yields, a data search of existing field studies was performed. The literature search was performed with SCOPUS with no source limitations (all years, article types, and access types). Literature was screened by searching three fields in the title, abstract, and keywords of the source reference: "Mediterranean" AND "Italy" AND "tomato". Results referring to greenhouse studies, pot experiments, Life Cycle Assessment, and simulation studies addressing crop development and water dynamics were excluded from the data analysis.

Information derived from field experiments included: region, province, altitude (m above sea level), long-term mean annual temperature (MAT °C) and total rainfall (MAP mm), aridity index, rainfall and irrigation during the growth cycle, marketable yield as fresh and dry matter, irrigation treatments, and soil texture group (Table S1). Crop evapotranspiration (ETc) in the different field experiments was estimated by the authors as the product of reference evapotranspiration (ETo), calculated with the FAO Penman–Monteith equation [27] or using Class A pan evaporation and Kpan [28], and tomato-specific crop coefficient (Kc). All experiments preformed were under drip irrigation.

Using these criteria 10 studies, totaling 54 yield observations, were found in four regions of Italy: Apulia (3), Basilicata (3), Latium (2), and Sicily (2).

2.1.1. Apulia

A two-year field research (2011–2012) was carried out at Valenzano (41°03′N, 16°52′E, altitude 72 m a.s.l) in Bari Province [29]. MAT and MAP were 16.2°C and 523 mm, respectively. Tomato (cv. Tomato F1) was grown under three irrigation regimes: full recover of crop evapotranspiration (I100), 50% of full irrigation supply (I50), and rainfed (I0). Tomato was transplanted in mid-April, and fertilized with 100, 120, and 150 kg ha^{-1} of N, P_2O_5, and K_2O, respectively.

A field research was carried out in 2011 in Foggia province (41°45′N, 15°50′E, altitude 90 m a.s.l), with MAT and MAP of 15.8°C and 526 mm, respectively [22]. Four irrigation regimes re-establishing 125% (ET125), 100% (ET100), 75% (ET75), and 50% (ET50) of ETc were considered. Tomato (cv. Defender F1) was transplanted in mid-May and fertilized with 133, 75, and 90 kg ha^{-1}, respectively, of N, P_2O_5, and K_2O.

A two-year experiment (2009–2010) was carried out in Foggia province (41°24′N, 15°45′E., altitude 30 m a.s.l) with MAT and MAP of 15.8°C and 526 mm respectively [30]. Tomato was cultivated under four irrigation regimes: DI, constant regime with restoration of 60% of maximum ETc during the crop cycle; RDI, variable irrigation regime with 60%, 80%, and 60% of maximum ETc through the three main phenological stages of the crop cycle; FI, full irrigation regime with the restoration of 100% ETc; FaI, farmer irrigation regime based on usual farming routine. Tomato (cv. Genius F1) was transplanted in the first decade of May and fertilized with 154 and 56 kg ha^{-1} of N and P_2O_5, respectively.

2.1.2. Basilicata

A two-year experiment (2002–2003) was carried out in Lavello (41°03′N, 15°42′E, altitude 180 m a.s.l) in Potenza province [31]. MAT and MAP were 14.5°C and 518 mm, respectively. Tomato was cultivated under six irrigation regimes: (i) four constant irrigation regimes with restoration of 0 (T0), 50 (T1), 75 (T2), and 100% (T3) of ETc during the whole crop cycle; (ii) two variable irrigation regimes with 100% restoration of ETc during the first period of the crop growth, followed by 75 or 50% restoration of ETc in the second part of the cycle (T4 and T5 treatments respectively). Tomato (cv. Pullrex) was transplanted after mid-May, and fertilized with 182, 214, and 160 kg ha^{-1} of N, P_2O_5, and K_2O, respectively.

A first field experiment carried out at Metaponto (40°24′N, 16°48′E, altitude 10 m a.s.l) in Matera province [32] reported the results related to 2007 and 2009 growing cycles. MAT and MAP were 16.5°C and 493 mm, respectively. Three irrigation treatments were compared: re-establishing 50 (I1), 75 (I2), and 100% (I3) of the crop evapotranspiration (ETc). Tomato (cv Tomito) was transplanted in mid-May and fertilized with 180 kg ha^{-1} of N.

A second two-year experiment (2008–2009) in the same area compared three irrigation regimes: V100, full restoration (100%) of ETc, V50, 50% restoration of ETc, and V0, no water restoration [33]. Tomato (cv. Faino F1) was transplanted in late May and fertilized with 160 kg ha^{-1} of N.

2.1.3. Latium

Two field experiments were carried out in Viterbo province (42°43′N, 12°07′E, altitude 310 m a.s.l). MAT and MAP were 14.4°C and 746 mm respectively. The first research [34] was conducted in 1997, and compared four irrigation regimes: 50–75, 50–100, 75–50, and 100–75 % restitution of ETc in the first (from planting to fruit set) and in the second (from fruit set to harvest) growth period. Tomato (hybrid PS 1296) was transplanted at the end of May with three fertilization treatments: control (no fertilization), D1 with 79, 68, and 107 kg ha^{-1} of N, P_2O_5, and K_2O respectively, and D2 (double the doses of D1).

The second experiment [35] was carried out in 2006–2007 and compared two irrigation treatments: full irrigation (FULL) restoring 100% of ETc, and deficit irrigation (DI), restoring 50% of ETc. Tomato (cv. Carioca). which were transplanted in mid-May and fertilized with 152, 200, and 150 kg ha^{-1} of N, P_2O_5, and K_2O, respectively.

2.1.4. Sicily

A two-year field experiment [12] was carried out in 2001–2002 in Enna province (37°27′N, 14°14′E, altitude 550 m a.s.l.). MAT and MAP were 15.4°C and 514 mm respectively. Four irrigation treatments were compared: no irrigation after plant establishment (V0), 100% (V100) or 50% (V50) ETc restoration up to fruit maturity, 100% ETc restoration up to flowering, then 50% ETc restoration (V100-50). Tomato (cv. Brigade) was transplanted in early May and fertilized with 150, 229, and 120 kg ha^{-1} of N, P_2O_5, and K_2O, respectively.

A field experiment [26] was conducted in 2002 in Siracusa province (37°03′N, 15°18′E, altitude 10 m a.s.l.). MAT and MAP were 17.8°C and 504 mm, respectively. Five irrigation regimes were compared: no irrigation after plant establishment (NI), long-season full irrigation with 100% ETc restoration (LF), long-season deficit irrigation with 50% ETc restoration (LD), short-season full irrigation up to first fruit set with 100% ETc restoration (SF), and short-season deficit irrigation up to first fruit set with 50% ETc restoration (SD). Tomato (cv. Brigade) was transplanted in early May and fertilized with 150, 229, and 120 kg ha^{-1} of N, P_2O_5, and K_2O respectively.

2.2. Data Evaluation

Marketable fruit yields (Mg ha^{-1} fresh weight) under deficit irrigation (DI) were compared with yields of the control treatment with the maximum water restoration (MWR) of each experiment, including rainfall:

$$\text{Yield } (\%) = \text{Yield}_{DI}/\text{Yield}_{MWR} \times 100 \quad (1)$$

Irrigation water use efficiency (IWUE) of the different treatments was calculated according to [36]:

$$\text{IWUE} = \text{Yield}/\text{TWS} \quad (2)$$

where Yield is the fruit dry biomass at harvest (kg ha^{-1}), and TWS is the total water supply including irrigation and rainfall from planting to harvest (m^3 ha^{-1}).

The Aridity index [37] was calculated with the formula Aridity index = MAP/(MAT+10) that defines aridity classes as humid (30–60), sub-humid (20–30), semi-arid (15–20), arid (5–15), and strongly-arid (< 5). Total water supplies with deficit irrigation during the crop cycle were divided in five classes based on the % of maximum water restoration (MWR): 0–20, 20–40, 40–60, 60–80, and 80–100%. Soil texture group was evaluated according to Soil Taxonomy [38] as (C) coarse (sandy loam, sandy clay loam, loamy sand), (M) medium (clay loam, loam, silty clay loam, silt, silt loam), and (F) fine (clay, silt clay, sandy clay).

Statistical analyses were performed using Statistica 7.0 (Statsoft, Tulsa, OK, USA). Significant differences among means were evaluated through the Fisher's protected least significant difference test (LSD post hoc test).

3. Results

A summary of maximum marketable fruit yields (Mg fresh weight ha^{-1}) under full irrigation and the related total water supply (mm) by rainfall and irrigation during the crop cycle in the different provinces are shown in Figures 1 and 2. Marketable yields ranged from 114.2 to 51.0 Mg ha^{-1}, respectively, at Matera and Siracusa. Total water supplies ranged from 768 to 395 mm at Foggia and Enna, respectively. The result for Matera, coupling a high marketable fruit yield (114.2 Mg ha^{-1}) and a low water supply (517 mm), are an indication of a proper irrigation schedule when fully restoring crop evapotranspiration. Conversely, at Foggia a slightly lower marketable fruit yield (95.2 Mg ha^{-1}) was coupled with a higher water supply (768 mm), indicating the ineffectiveness in productive terms of water supplies following the farmer routine, using more water than the full irrigation regime, restoring 100% of ETc [30].

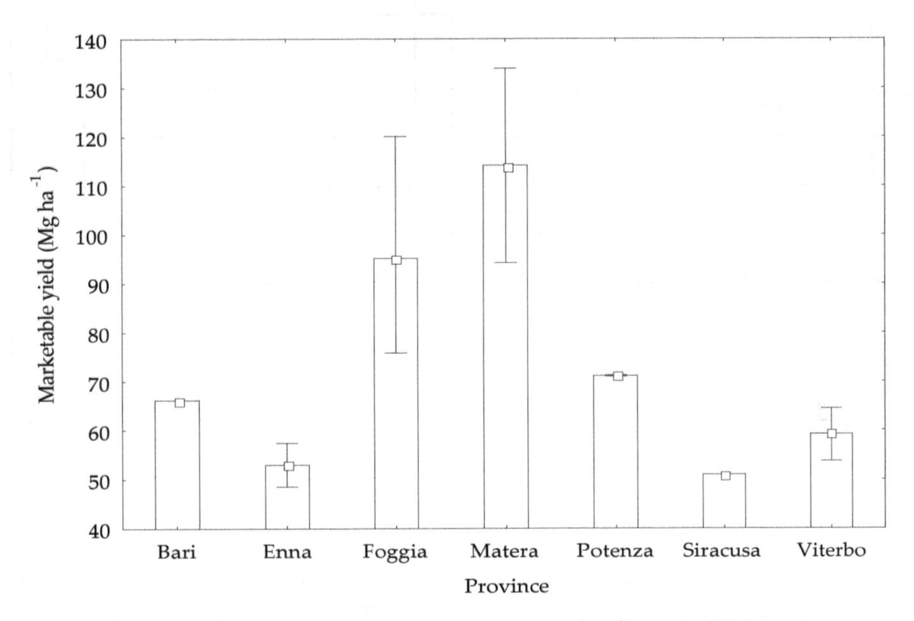

Figure 1. Maximum marketable fruit yield (Mg fresh weight ha^{-1}) in the different provinces under full irrigation. Boxes represent mean values, whiskers represent Min–Max interval.

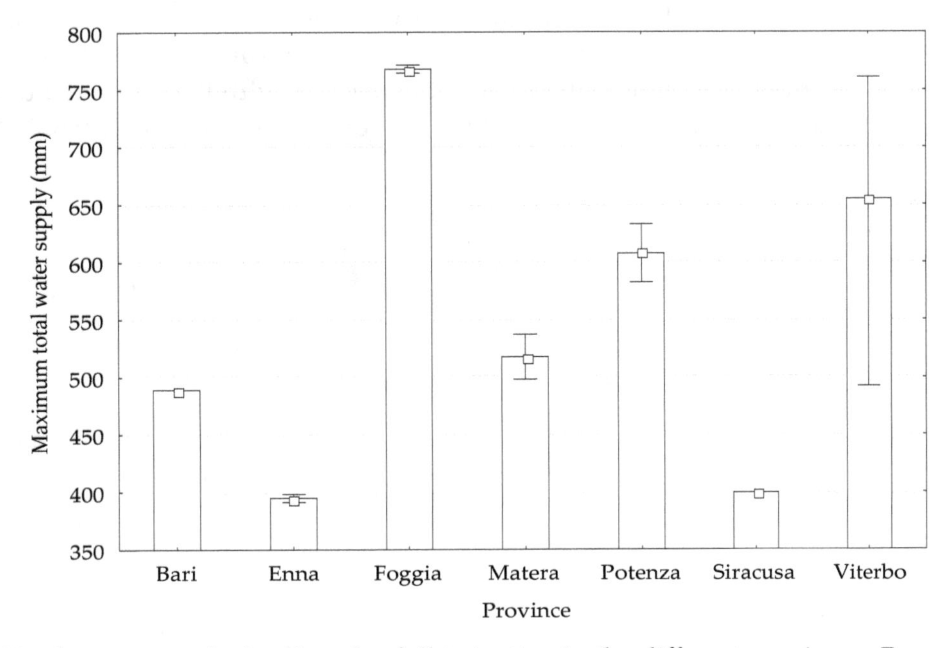

Figure 2. Total water supply (mm) under full irrigation in the different provinces. Boxes represent mean values, whiskers represent Min–Max interval.

3.1. *Marketable Yield and Water Restoration*

When comparing marketable fruit yields under DI with the control treatment (Equation (1)), average yields (%) significantly differed among irrigation classes ($p = 0.0000$). In detail (Figure 3), yields were significantly lower in 0–20 and 20–40 irrigation classes (31.5 and 27.3 % respectively) and higher in 40–60, 60–80, and 80–100 classes (74.9, 72.6, and 87.4% respectively).

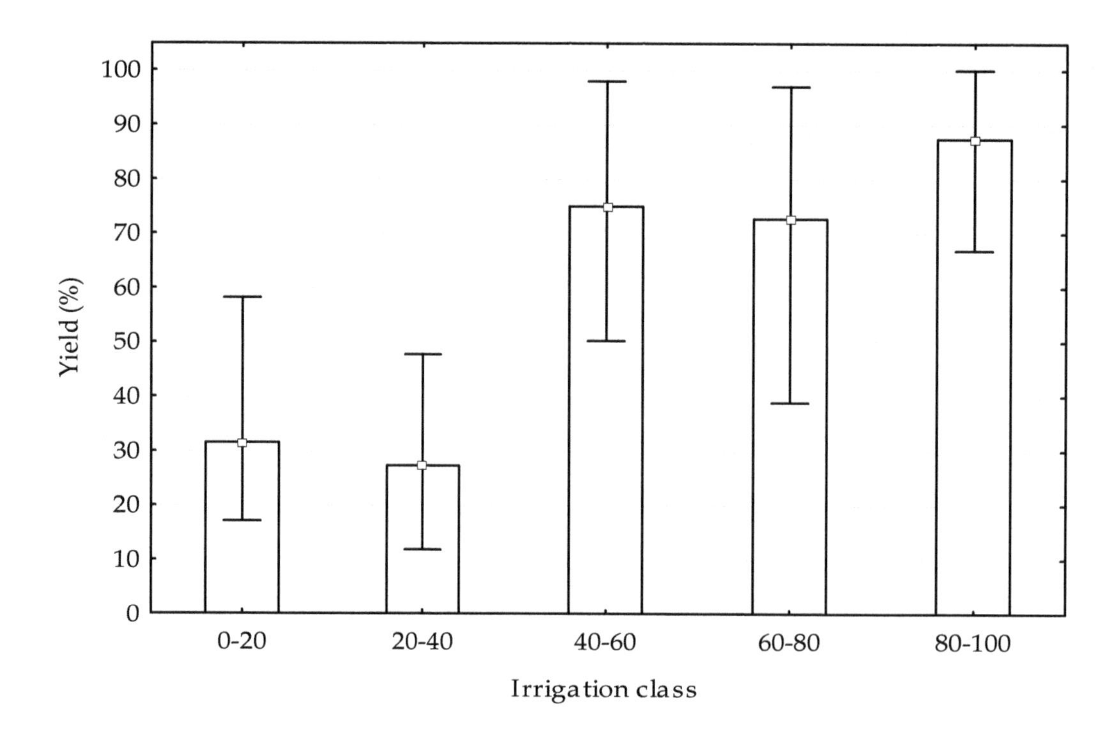

Figure 3. Tomato fresh fruit yield (%) compared with the control treatment based on the irrigation classes as % of maximum water restoration (MWR), F (4;35) = 14.1531; $p = 0.00000$. Boxes represent mean values, whiskers Min–Max interval.

Average yields (%) significantly differed among irrigation regimes ($p = 0.0000$). As expected, yields (Figure 4) were significantly lower with none irrigation excluding rainfall during the crop cycle (26.8%), and did not differ between constant and variable regimes with the regulated deficit irrigation (RDI) but were lower under constant irrigation (74.0%) in comparison with RDI (85.7%).

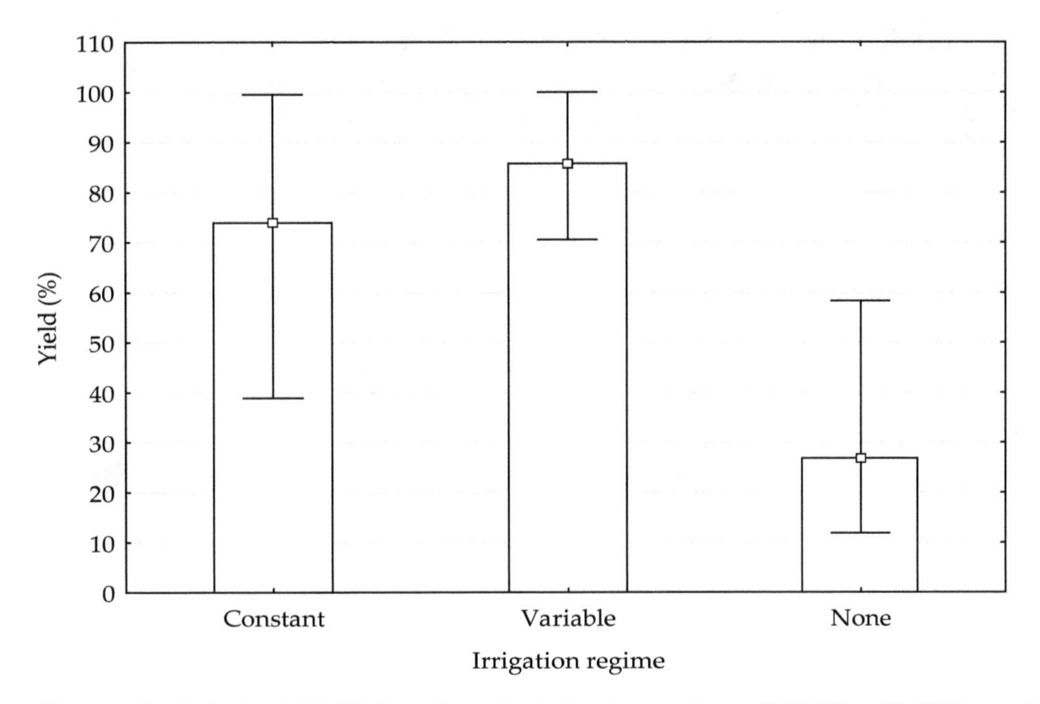

Figure 4. Tomato fresh fruit yield (%) based on the irrigation regimes, F (2;37) = 25.6319; $p = 0.00000$. Boxes represent mean values, whiskers represent Min–Max interval.

Fresh fruit yield (%) and maximum water restoration (%) supplied with DI were interpolated (Figure 5) with a polynomial equation: $y = -0.0039x^2 + 1.2053x + 16.8326$ ($R^2 = 0.7045$). The interpolating function indicated that yields would decrease by 6.3, 8.9, and 11.7% at 90, 85, and 80% of maximum water restoration (MWR), but would still be acceptable at 75 and 70% of MWR with decreases of 14.7 and 17.9%.

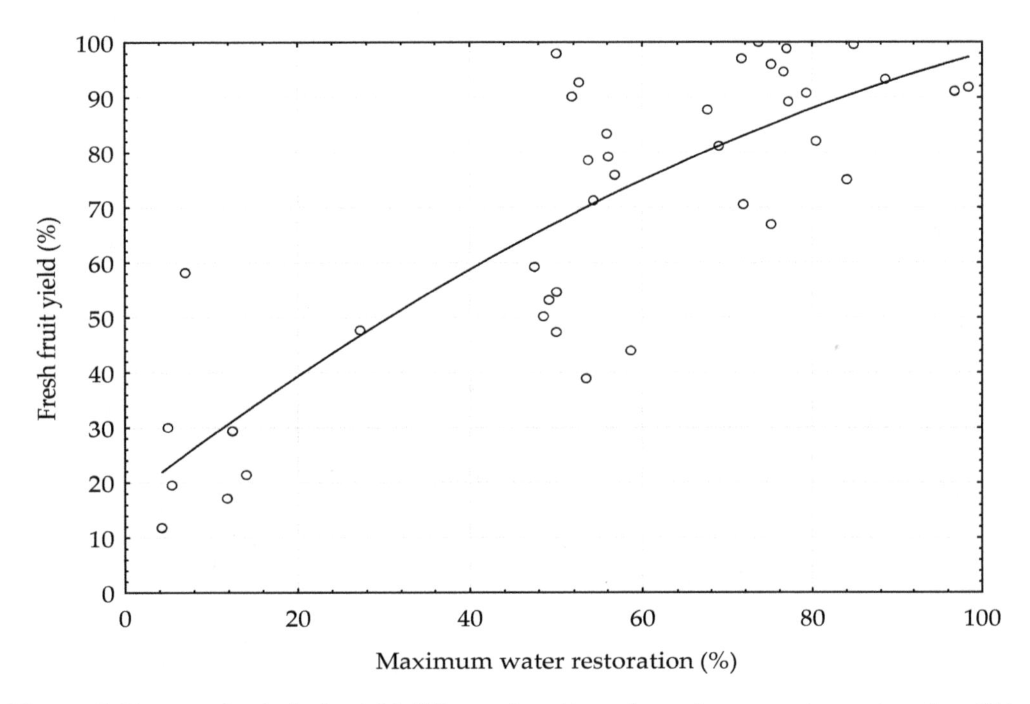

Figure 5. Tomato fresh fruit yield (%) as a function of maximum water restoration (%).

Marketable fruit yields (Mg fresh weight ha^{-1}) under DI significantly differed among the provinces ($p = 0.00006$) where the field experiments were conducted (Figure 6). Average yields were significantly lower at Bari, Siracusa, Enna, Viterbo, and Potenza (22.2, 27.4, 36.0, 41.1, and 50.3 Mg ha^{-1}, respectively) compared to Foggia and Matera (75.4 and 92.7 Mg ha^{-1}, respectively). Compared with marketable yields under full irrigation (Figure 1), decreases were as follows: Bari (67%), Siracusa (46%), Enna (32%), Viterbo (31%), Potenza (29%), Foggia (21%), and Matera (19%).

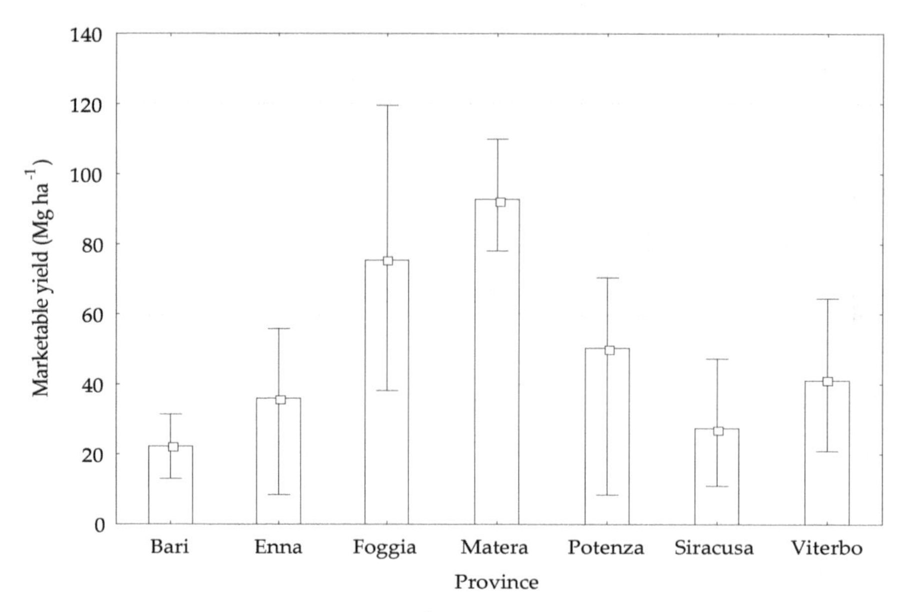

Figure 6. Marketable yield (Mg fresh weight ha^{-1}) in the different provinces under DI, F (6;33) = 7.1298; p = 0.00006. Boxes represent mean values, whiskers represent Min–Max interval.

Marketable fruit yields (Mg fresh weight ha^{-1}) under DI also significantly differed among soil texture groups (p = 0.0049) and average yields (Figure 7) were significantly lower in field experiments with coarse and medium texture (32.5 and 51.7 Mg fresh weight ha^{-1}, respectively) compared to fine textured soils (75.4 Mg fresh weight ha^{-1}). This result is coherent with the soil textures of the field experiments, which were fine and medium at Foggia and Matera, respectively, and also showed the lowest yield decreases; conversely, at Bari soils were coarse textured and presented the highest yield decrease under DI (Figure 6).

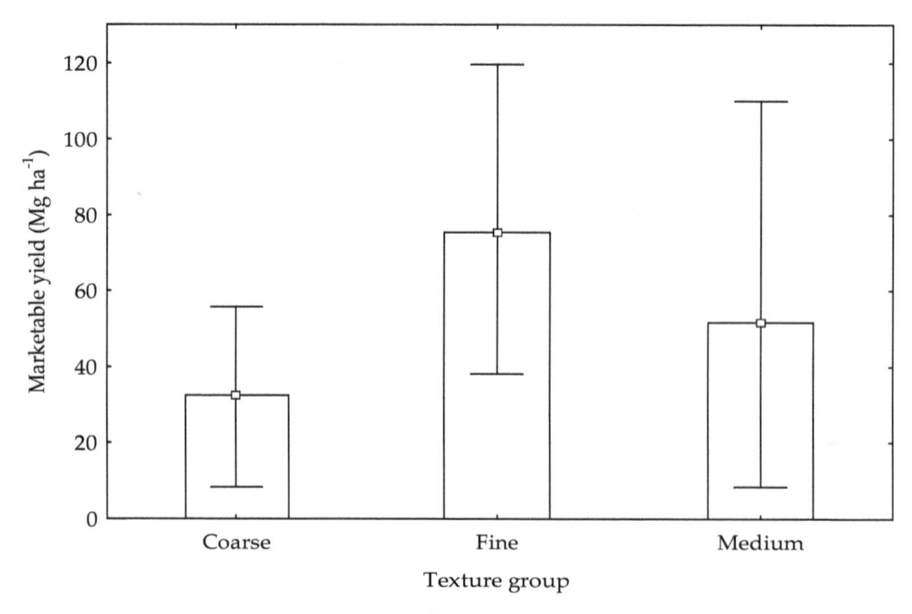

Figure 7. Marketable yield (Mg fresh weight ha^{-1}) under DI based on soil texture groups according to United States Department of Agriculture (USDA), F (2;37) = 6.1615; p = 0.0049. Boxes represent mean values, whiskers represent Min–Max interval.

3.2. *Irrigation Water Use Efficiency and Water Restoration*

Irrigation water use efficiency (IWUE in kg dry weight m^{-3}) was weakly significantly different ($p = 0.0683$) among irrigation classes (Figure 8). Average IWUE was significantly higher only in 0–20 irrigation class (1.67 kg m^{-3}) compared to the other classes.

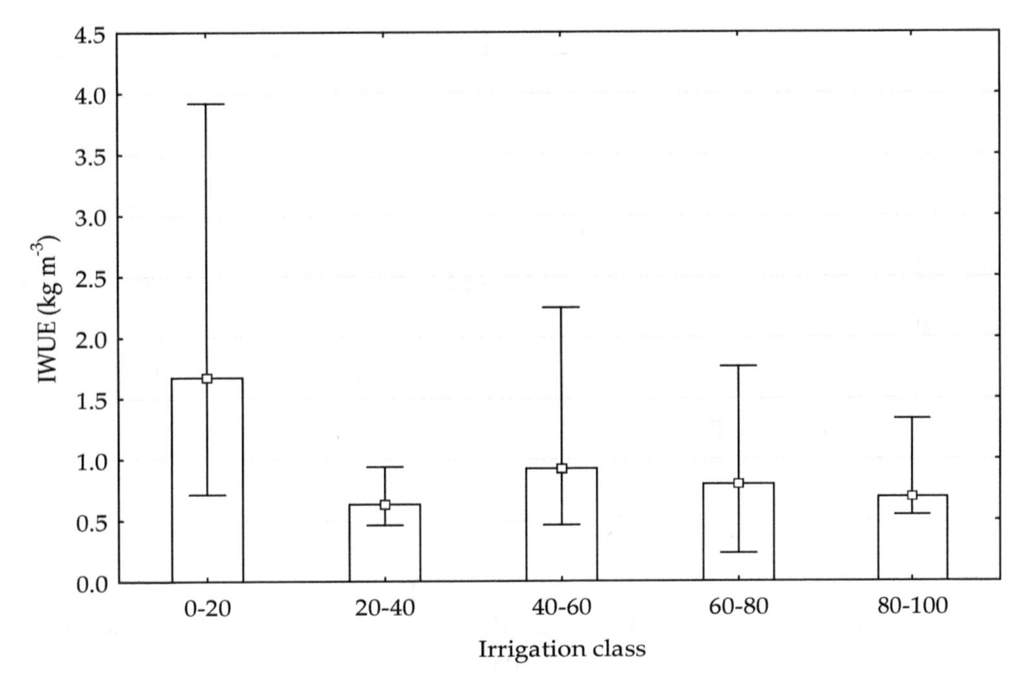

Figure 8. Irrigation Water Use Efficiency (IWUE) (kg dry weight m^{-3}) based on the irrigation classes as % of maximum water restoration (MWR), F (4;35) = 2.403; $p = 0.0683$. Boxes represent mean values, whiskers represent Min–Max interval.

In relation to irrigation regimes (Figure 9), IWUE was not significantly different ($p = 0.2366$). IWUE was higher with no irrigation (1.18 kg m^{-3}) and decreased with constant and variable irrigation regimes (0.86 and 0.66 kg m^{-3}, respectively).

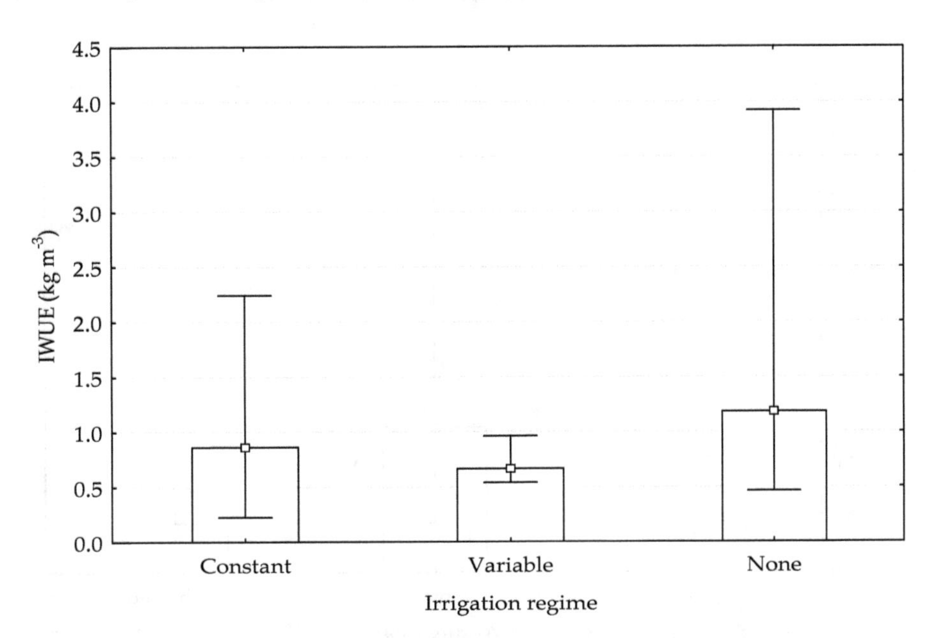

Figure 9. Irrigation Water Use Efficiency (IWUE) (kg dry weight m^{-3}) based on irrigation regimes, F (2;37) = 1.4992; $p = 0.2366$. Boxes represent mean values, whiskers represent Min–Max interval.

IWUE significantly differed among the provinces ($p = 0.0000$) where the field experiments were conducted (Figure 10). In detail, average IWUE was significantly lower at Viterbo, Bari, Potenza, Foggia Siracusa, and Enna (0.47, 0.49, 0.61, 0.75, 0.80, and 0.98 kg m^{-3}, respectively) compared to Matera (2.31 kg m^{-3}). Results for Matera are in agreement with the low decreases observed in marketable yields.

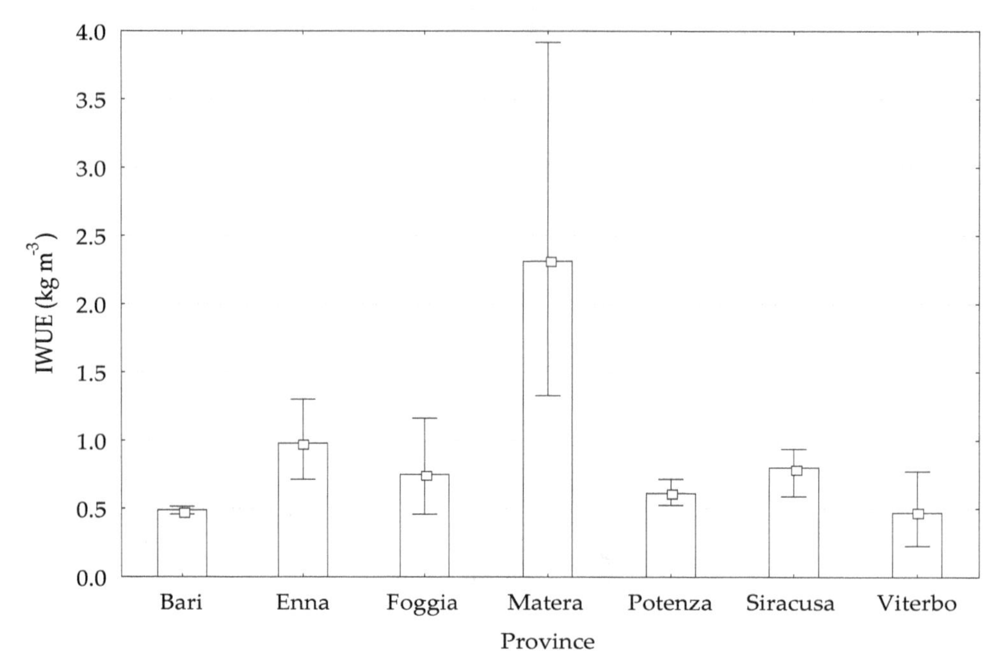

Figure 10. Irrigation Water Use Efficiency (IWUE) (kg dry weight m^{-3}) in the different provinces, $F_{(6;33)} = 11.5912$; $p = 0.0000$. Boxes represent mean values, whiskers represent Min–Max interval.

IWUE also significantly differed among aridity classes ($p = 0.0092$) and on average was significantly lower (Figure 11) in field experiments under humid and sub-humid climates (0.47 and 0.75 kg m^{-3}, respectively) compared to semiarid conditions (1.34 kg m^{-3}). Considering the location of field experiments, humid climate conditions are related to Viterbo in Latium, and semiarid conditions to Matera in Basilicata.

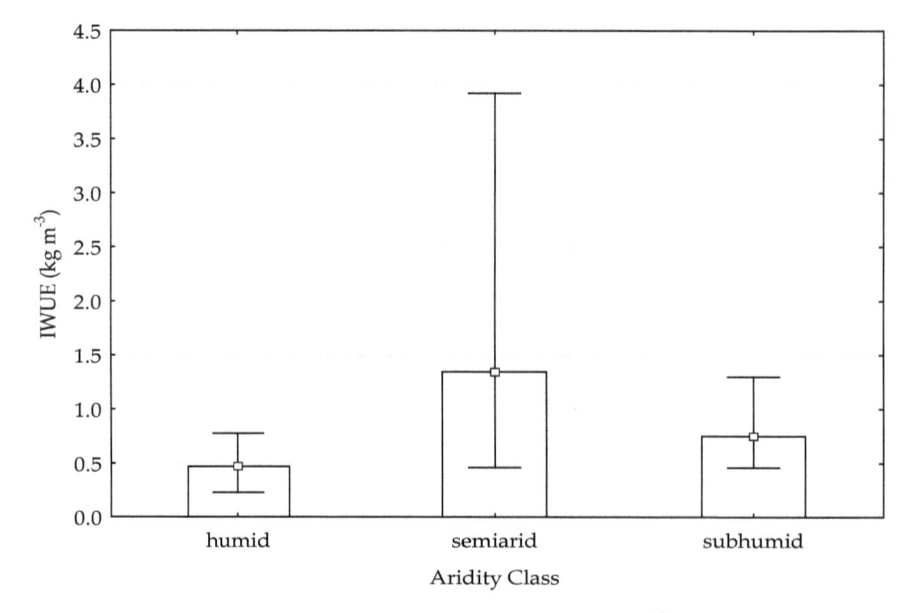

Figure 11. Irrigation Water Use Efficiency (IWUE) (kg dry weight m^{-3}) based on the aridity classes [37], $F_{(2;37)} = 5.3391$; $p = 0.0092$. Boxes represent mean values, whiskers represent Min–Max interval.

4. Discussion

In the environments typical of the Mediterranean area the use of water resources for irrigation is a priority to be managed through sustainable regulation of water supplies, contributing to water savings with environmental and economic benefits [39] but avoiding high productivity losses to maintain profit for farmers [6,30]. Research also indicated that DI may have positive side effects, such as contributing to decreased soil CO_2 emissions and enhanced C sequestration in soils, by decreasing microbial activity in response to decreased soil moisture levels [39]. In addition, nitrogen fertilization also results in lower N_2O emissions in Mediterranean regions with drip irrigation systems that are commonly used in tomato cultivation compared with sprinkler irrigation methods [40].

Results from the different field experiments examined in this study are in contrast with each other. Data obtained in a study at Foggia in Apulia [30] indicated that farmers tend to over irrigate tomato crops, with no significant increase in the marketable fruit yield and quality, as reported in other research [20]. Moreover, the same authors [30] indicated that the adoption of variable irrigation regimes as RDI restoration of 60%, 80%, and 60% of the maximum ETc during the three main tomato phenological stages (i.e., from plant establishment to flowering of the first truss, from flowering of the first truss to fruit breaking colors of the first truss, and from fruit breaking colors of the first truss to harvest), was effective to save water, as shown by other authors [31,34] at Viterbo (Latium) and Lavello (Basilicata). A study at Matera in Basilicata [32] indicated that water restoration of 50, 75, and 100% of crop evapotranspiration showed no statistical differences among the irrigation volumes in relation to tomato yield and quality. Conversely, in the same environment another study [33] reported statistically significant differences in both marketable yields and fruit quality when restoring 0, 50, and 100% of ETc.

The study conducted at Enna in Sicily [12] showed that marketable yields were strongly decreased by early soil water deficit following plant establishment, while a reduced irrigation rate after the initial stages or after flowering did not induce any significant loss. The study also indicated that DI has beneficial effects on fruit quality. In particular, a high total solids content of the fruit improves the efficiency of the industrial process due to the lower energy required to evaporate water from fruit. Tomato yield also proved more sensitive to the length of the irrigation period rather than to the total water supplied during DI experiments in Sicily [26]. In fact, the long-season deficit irrigation (LD) with 50% ETc restoration and the short-season full irrigation (SF) with 100% ETc restoration received about the same amount of water, but yields decreased by 46% in SF. In addition, irrigation cut-off during the ripening period did not significantly affect marketable yields and enhanced fruit quality [31].

In the case of Foggia and Viterbo, the average irrigation supplied (Figure 2) was the highest (768 and 655 mm, respectively). However, marketable yields with full irrigation (Figure 1) were higher in Foggia than in Viterbo (95 and 59 Mg fresh weight ha^{-1}). This result can be ascribed firstly to the lower fertilization supplied at Viterbo (average N fertilization was about two thirds compared to the amount supplied at Foggia), and secondly to other environmental conditions that can positively or negatively affect crop yields (e.g., air temperature). In fact, average temperature is 15.8 °C at Foggia and climate is sub-humid; at Viterbo temperature is 14.4 °C and climate is humid. The same consideration is valid in relation to IWUE. In fact, average IWUE was higher at Matera (Figure 10), where temperature is 16.5 °C and climate is semi-arid (Figure 11).

Generally, our data-analysis has confirmed that results are quite variable and strongly site-dependent due to different climate and soil conditions that may mask the actual effect of the irrigation regime, and consequently cannot be generalized. Based on the field experiments considered, a limited decrease in water restoration according to the calculated interpolating function (Figure 5) would marginally decrease yield by 17.9% and 11.7%, at 70 and 80% of maximum water restoration, respectively. Marketable yields did not differ significantly at 40–60% and 60–80% of maximum water restoration (Figure 3) but were higher when 80–100% of maximum water restoration was supplied with DI, in agreement with previous research [20]. In addition, variable irrigation regimes during the crop cycle showed a higher and significant response to crop yields (Figure 4). Yield responses to DI were significantly lower in soils with coarse and medium textures (Figure 7). Irrigation water use efficiency was weakly significantly different among irrigation classes and water regimes but was significantly higher in the experiment conducted at Matera (Figure 10), coupled with semiarid climate conditions (Figure 11).

5. Conclusions and Recommendations

Under Mediterranean conditions, water management is a crucial factor for tomato crops, due to the limited availability of water resources during the growing season, when evapotranspiration is not balanced by the moderate amount of rainfall. Therefore, in this environment, the sustainable use of water resources is a priority. A proper application of DI can save huge amounts of water, particularly in semi-arid environments where water scarcity is an increasing concern and water costs are continuously rising.

Our results provide practical guidelines for irrigation water use in processing tomato cultivation that can be easily addressed by farmers to avoid over-irrigation and to adopt reduced irrigation rates during the less sensitive growth stages. Our recommendation is that further research should address the response of crop yield under variable irrigation regimes adopting RDI, and in relation either to coarse and medium soil textures and sub-humid climate conditions that are very frequent in Mediterranean Italy.

Alternative strategies to reduce irrigation water use can be recommended, namely sensor-based irrigation scheduling [41] or partial root-zone drying [35]. However, their implementation involves higher costs for farmers in terms of irrigation equipment and management compared to deficit or regulated deficit irrigation.

Author Contributions: R.F. and C.D.B. made substantial contributions to the manuscript. R.F. performed data curation and writing—review and editing; C.D.B. performed writing—review and editing.

References

1. FAO (Food and Agriculture Organization of the United Nations). FAO Statistical Yearbook 2012. Available online: http://www.fao.org/docrep/015/i2490e/i2490e00.htm (accessed on 7 April 2019).
2. Du, Y.D.; Niu, W.Q.; Gu, X.B.; Zhang, Q.; Cui, B.J. Water- and nitrogen-saving potentials in tomato production: A meta-analysis. *Agr. Water Manag.* **2018**, *210*, 296–303. [CrossRef]
3. Molden, D. *Water for Food, Water for Life: A Comprehensive Assessment of Water Management in Agriculture*; Earthscan and Colombo, International Water Management Institute: London, UK, 2007; 688p.
4. FAO (Food and Agriculture Organization of the United Nations). AQUASTAT Main Database 2016. Available online: http://www.fao.org/nr/water/aquastat/data/query/index.html (accessed on 7 April 2019).
5. Yahyaoui, I.; Tadeo, F.; Vieira, M. Energy and water management for drip-irrigation of tomatoes in a semi-arid district. *Agr. Water Manag.* **2017**, *183*, 4–15. [CrossRef]
6. Costa, J.M.; Ortuno, M.F.; Chaves, M.M. Deficit irrigation as a strategy to save water: Physiology and potential application to horticulture. *J. Integr. Plant Biol.* **2007**, *49*, 1421–1434. [CrossRef]
7. Mushtaq, S.; Moghaddasi, M. Evaluating the potentials of deficit irrigation as an adaptive response to climate change and environmental demand. *Environ. Sci. Policy* **2011**, *14*, 1139–1150. [CrossRef]
8. Ali, M.H.; Hoque, M.R.; Hassan, A.A.; Khair, A. Effects of deficit irrigation on yield, water productivity, and economic returns of wheat. *Agr. Water Manag.* **2007**, *92*, 151–161. [CrossRef]
9. Vazifedoust, M.; Van Dam, J.C.; Feddes, R.A.; Feizi, M. Increasing water productivity of irrigated crops under limited water supply at field scale. *Agr. Water Manag.* **2008**, *95*, 89–102. [CrossRef]
10. Rodriguez-Ortega, W.; Martinez, V.; Rivero, R.; Camara-Zapata, J.; Mestre, T.; Garcia-Sanchez, F. Use of a smart irrigation system to study the effects of irrigation management on the agronomic and physiological responses of tomato plants grown under different temperatures regimes. *Agr. Water Manag.* **2017**, *183*, 158–168. [CrossRef]

11. Zegbe-Domínguez, J.A.; Behboudian, M.H.; Lang, A.; Clothier, B.E. Deficit irrigation and partial rootzone drying maintain fruit dry mass and enhance fruit quality in 'Petopride' processing tomato (*Lycopersicon esculentum*, Mill.). *Sci. Hortic.* **2003**, *98*, 505–510. [CrossRef]
12. Patanè, C.; Tringali, S.; Sortino, O. Effects of deficit irrigation on biomass, yield, water productivity and fruit quality of processing tomato under semi-arid Mediterranean climate conditions. *Sci. Hortic.* **2011**, *129*, 590–596. [CrossRef]
13. Zhang, H.M.; Xiong, Y.W.; Huang, G.H.; Xu, X.; Huang, Q.Z. Effects of water stress on processing tomatoes yield, quality and water use efficiency with plastic mulched drip irrigation in sandy soil of the Hetao Irrigation District. *Agric. Water Manag.* **2017**, *179*, 205–214. [CrossRef]
14. Chen, J.L.; Kang, S.Z.; Du, T.S.; Guo, P.; Qiu, R.Q.; Chen, R.Q.; Gu, F. Modeling relations of tomato yield and fruit quality with water deficit at different growth stages under greenhouse condition. *Agr. Water Manag.* **2014**, *146*, 131–148. [CrossRef]
15. WPTC (World Processing Tomato Council). Available online: http://www.wptc.to/ (accessed on 22 March 2019).
16. ISTAT (Istituto Nazionale di Statistica). Available online: http://agri.istat.it/ (accessed on 22 March 2019).
17. Battilani, A. Processing tomato water and nutrient integrated crop management: State of art and future horizons. *Acta Hortic.* **2003**, *613*, 63–73. [CrossRef]
18. Rinaldi, M.M.; Thebaldi, M.S.; da Rocha, M.S.; Sandri, D.; Felisberto, A.B. Postharvest quality of the tomato irrigated by different irrigation systems and water qualities. *IRRIGA* **2013**, *18*, 59–72. [CrossRef]
19. Benincasa, P.; Guiducci, M.; Tei, F. The nitrogen use efficiency: Meaning and sources of variation – case studies on three vegetable crops in Central Italy. *Hort. Technol.* **2011**, *21*, 266–273. [CrossRef]
20. Fereres, E.; Soriano, M.A. Deficit irrigation for reducing agricultural water use. *J. Exp. Bot.* **2007**, *58*, 147–159. [CrossRef] [PubMed]
21. Giuliani, M.; Nardella, E.; Gagliardi, A.; Gatta, G. Deficit irrigation and partial root-zone drying techniques in processing tomato cultivated under Mediterranean climate conditions. *Sustainability* **2017**, *9*, 2197. [CrossRef]
22. Rinaldi, M.; Garofalo, P.; Vonella, A.V. Productivity and water use efficiency in processing tomato under deficit irrigation in Southern Italy. In XIII International Symposium on Processing Tomato, Sirmione, Italy. *Acta Hort.* **2015**, *1081*, 97–104. [CrossRef]
23. Ventrella, D.; Giglio, L.; Charfeddine, M.; Lopez, R.; Castellini, M.; Sollitto, D.; Castrignanò, A.; Fornaro, F. Climate change impact on crop rotations of winter durum wheat and tomato in southern Italy: Yield analysis and soil fertility. *Ital. J. Agron.* **2012**, *7*, 15. [CrossRef]
24. Ventrella, D.; Giglio, L.; Garofalo, P.; Dalla Marta, A. Regional assessment of green and blue water consumption for tomato cultivated in Southern Italy. *J. Agr. Sci.* **2018**, *156*, 689–701. [CrossRef]
25. Marouelli, W.A.; Silva, W.L.C. Water tension thresholds for processing tomatoes under drip irrigation in Central Brazil. *Irrig. Sci.* **2007**, *25*, 411–418. [CrossRef]
26. Patanè, C.; Cosentino, S.L. Effects of soil water deficit on yield and quality of processing tomato under a Mediterranean climate. *Agr. Water Manag.* **2010**, *97*, 131–138. [CrossRef]
27. Allen, R.G.; Pereira, L.S.; Raes, D.; Smith, M. Crop evapotranspiration. In *Guidelines for Computing Crop Water Requirements*; FAO: Rome, Italy, 1998; Irrigation and Drainage Paper Volume 56, p. 300.
28. Doorenbos, J.; Pruitt, W.O. *Crop Water Requirements*; FAO Irrigation and Drainage Paper 24; FAO: Rome, Italy, 1977; p. 144.
29. Cantore, V.; Lechkar, O.; Karabulut, E.; Sellami, M.H.; Albrizio, R.; Boari, F.; Stellacci, A.M.; Todorovic, M. Combined effect of deficit irrigation and strobilurin application on yield, fruit quality and water use efficiency of "cherry" tomato (*Solanum lycopersicum* L.). *Agr. Water Manag.* **2016**, *167*, 53–61. [CrossRef]
30. Giuliani, M.M.; Gatta, G.; Nardella, E.; Tarantino, E. Water saving strategies assessment on processing tomato cultivated in Mediterranean region. *Ital. J. Agron.* **2016**, *11*, 69–76. [CrossRef]
31. Lovelli, S.; Potenza, G.; Castronuovo, D.; Perniola, M.; Candido, V. Yield, quality and water use efficiency of processing tomatoes produced under different irrigation regimes in Mediterranean environment. *Ital. J. Agron.* **2017**, *12*, 17–24. [CrossRef]
32. Leogrande, R.; Lopedota, O.; Montemurro, F.; Vitti, C.; Ventrella, D. Effects of irrigation regime and salinity on soil characteristics and yield of tomato. *Ital. J. Agron.* **2012**, *7*, 50–57. [CrossRef]
33. Candido, V.; Campanelli, G.; D'Addabbo, T.; Castronuovo, D.; Perniola, M.; Camele, I. Growth and yield

promoting effect of artificial mycorrhization on field tomato at different irrigation regimes. *Sci. Hortic.* **2015**, *187*, 35–43. [CrossRef]

34. Colla, G.; Casa, R.; Lo Cascio, B.; Saccardo, F.; Temperini, O.; Leoni, C. Responses of processing tomato to water regime and fertilization in Central Italy. In Proceedings of the VI International Symposium on Processing Tomato and Workshop on Irrigation and Fertigation of Processing Tomato, Pamplona, Spain. *Acta Hort.* **1999**, *487*, 531–536. [CrossRef]
35. Casa, R.; Rouphael, Y. Effects of partial root-zone drying irrigation on yield, fruit quality, and water-use efficiency in processing tomato. *J. Hortic. Sci. Biotech.* **2014**, *89*, 389–396. [CrossRef]
36. Howell, T.A.; Steiner, J.E.; Schneider, A.D.; Evertt, S.R.; Tolk, J.A. Seasonal and maximum daily evapotranspiration of irrigated winter wheat, sorghum and corn: Southern high plains. *Trans. Asae* **1997**, *40*, 623–634. [CrossRef]
37. De Martonne, E. Une nouvelle fonction climatologique: l'indice d'aridite. *Meteorologie* **1926**, *2*, 449–458.
38. Soil Survey Staff. *Keys to Soil Taxonomy*, 12th ed.; USDA-Natural Resources Conservation Service: Washington, DC, USA, 2014; p. 360.
39. Zornoza, R.; Rosales, R.M.; Acosta, J.A.; de la Rosa, J.M.; Arcenegui, V.; Faz, Á.; Pérez-Pastor, A. Efficient irrigation management can contribute to reduce soil CO_2 emissions in agriculture. *Geoderma* **2016**, *263*, 70–77. [CrossRef]
40. Cayuela, M.L.; Aguilera, E.; Sanz-Cobena, A.; Adams, D.C.; Abalos, D.; Barton, L.; Ryals, R.; Silver, W.L.; Alfaro, M.A.; Pappa, V.A.; et al. Direct nitrous oxide emissions in Mediterranean climate cropping systems: Emission factors based on a meta-analysis of available measurement data. *Agr. Ecosyst. Environ.* **2017**, *238*, 25–35. [CrossRef]
41. Zotarelli, L.; Dukes, M.D.; Scholberg, J.M.S.; Munoz-Carpena, R.; Icerman, J. Tomato nitrogen accumulation and fertilizer use efficiency on a sandy soil, as affected by nitrogen rate and irrigation scheduling. *Agr. Water Manag.* **2009**, *96*, 1247–1258. [CrossRef]

2

Feeding Emitters for Microirrigation with a Digestate Liquid Fraction up to 25% Dilution did not Reduce their Performance

Simone Bergonzoli [1], Massimo Brambilla [1], Elio Romano [1], Sergio Saia [2,*], Paola Cetera [2], Maurizio Cutini [1], Pietro Toscano [1], Carlo Bisaglia [1] and Luigi Pari [2]

1 Council for Agricultural Research and Economics-Research Centre for Engineering and Agro-Food Processing, CREA-IT, Treviglio, 24047 Bergamo, Italy; simone.bergonzoli@crea.gov.it (S.B.); massimo.brambilla@crea.gov.it (M.B.); elio.romano@crea.gov.it (E.R.); maurizio.cutini@crea.gov.it (M.C.); pietro.toscano@crea.gov.it (P.T.); carlo.bisaglia@crea.gov.it (C.B.)

2 Council for Agricultural Research and Economics—Research Centre for Engineering and Agro-Food Processing (CREA-IT), Via della Pascolare, 16, Monterotondo, 00015 Roma, Italy; paola.cetera@crea.gov.it (P.C.); luigi.pari@crea.gov.it (L.P.)

* Correspondence: sergio.saia@crea.gov.it

Abstract: Irrigation with wastewater can strongly contribute to the reduction of water abstraction in agriculture with an especial interest in arid and semiarid areas. However, its use can have drawbacks to both soil and micro-irrigation systems, especially when the total solids in the wastewater are high, such as in digestate liquid fractions (DLF) from plant material. The aim of this study was thus to evaluate the performances of a serpentine shaped micro-emitter injected with a hydrocyclone filtered DLF (HF-DLF) from corn + barley biomass and evaluate the traits of the liquid released within a 8-h irrigation cycle. HF-DLF was injected at 10%, 25%, and 50% dilution compared to tap water (at pH = 7.84) and the system performances were measured. No clogging was found, which likely depended on both the shape of the emitter and the high-pressure head (200 kPa). HF-DLF dilution at 10%, 25%, and 50% consisted in +1.9%, +3.5, and −4.9% amount of liquid released compared to the control. Fluid temperature during irrigation (from 9:00 to 17:00) did not explain the difference in the released amounts of liquid. In 10% HF-DLF % and 25% HF-DLF, a pH difference of + 0.321 ± 0.014 pH units compared to the control was found, and such difference was constant for both dilutions and at increasing the time. In contrast, 50% HF-DLF increased pH by around a half point and such difference increased with time. Similar differences among treatments were found for the total solids in the liquid. These results indicate that 50% HF-DLF was accumulating materials in the serpentine. These results suggest that a low diluted HF-DLF could directly be injected in irrigation systems with few drawbacks for the irrigation system and contribute to water conservation since such wastewater are available from the late spring to the early fall, when water requirements are high.

Keywords: clogging; drip irrigation; emitter; hydrocyclone; digestate liquid fraction; wastewater

---

1. Introduction

Water availability for crops in various areas of the world is reducing because of climate change and the use of fresh water for other human uses. Climate change is increasing the demand for water in agriculture through both a general increase of temperatures, and thus of the evapotranspiration demand, and the increase of their variability [1,2].

Irrigation with low-quality water, and especially wastewater, was thus proposed a long time ago as a suitable measure to mitigate the shortage of high quality water [3,4]. The use of wastewater or filtrate of the liquid fractions of various wastes can increase the availability of water for agriculture. However, its use may result in a wealth of problems following the effects wastewater has on soil and irrigation systems. These include salinization or pH variation [5], a reduction of other soil fertility properties [6], and an increase of soil hydrophobicity [7], the clogging of the emitters [3], as well as the deposition of solid materials in the tanks or other parts of the irrigation system [8]. Besides, wastewaters may contain pollutants and pathogens which harm plants and animals, albeit the treatments they undergo are meant to prevent health risk following their use or disposal [9,10].

Drip systems allow the achievement of high irrigation efficiency in areas with high water demand and low water availability. In these systems, water pressure is usually below 200 kPa, and emitters have internal serpentine to compensate pressure loss along the line and potential fluctuations in the water pressure.

The success of the use of wastewater in the irrigation depends on a wealth of factors. These include the amount of solids in the wastewater or its filtrate, the ability of the suspended material to form biofilms, the pressure of the water in the system, the type of filters and emitters, and age of the systems [8,11–14]. In case of low pressure (60 kPa) and low rate emitters (0.9 and 1.4 L h^{-1} $emitter^{-1}$), high quality drip tapes showed a reduction of uniformity of distribution by 5.2% on average depending on the activated sludge used as secondary effluent [15]. Similar results were found by Puig-Bargués et al. [14], who also reported that pressure compensating emitters performed better than non-pressure compensating emitters. Chlorination or flushing of the pipes at the end of each irrigation cycle proved to reduce the impact on clogging [14,16]. However, such treatments imply an additional cost, and application of chlorine to the soil may have harmful effects both on the soil and on plants. In the latter work [16], application of compressed air cleaning at a pressure of 1.96 kPa did not mitigate the incidence of drippers clogging.

Digestate from crop biomass and manure is increasingly being used, and its liquid fraction was indicated as a potential source for a wastewater irrigation [17]. When used for irrigation purposes, information on the solid particles fractions, mostly salts, of these liquids in the irrigation systems are scarce. Such salts are likely to precipitate and, together with other suspended solids, can easily clog the emitters of a drip irrigation system by a fouling accumulation [18]. In turn, digestate filtrates used for irrigation can increase plant yield [6], even when compared to an irrigated + fertilized treatment [19]. However, little information is available about the efficiency of many emitters when subjected to wastewater, especially when using the liquid fraction of the biomass-based digestate, as the solid fraction can contain high amounts of organic material [20].

The aim of the present study was thus to test the efficiency of a commercial emitter when injected with the liquid fraction of the effluent from an agricultural biogas unit previously treated with a hydrocyclone filtration system.

2. Materials and Methods

2.1. Experimental Setup

The study was conducted in February 2020 at the CREA-IT institute (45°31′21.9″ N 9°33′54.9″ E, Treviglio, Bergamo, Italy). The liquid digestate used was gathered from an anaerobic digestion plant, stored in a tank and filtered using a hydrocyclone filter (Alfaturbo Hydrocyclone Sand separator 2″). The hydrocyclone filter was placed between a storage tank and the operating tank. Both tanks had a 1 m^3 maximum volume. After the filtration, the filtered liquid was shaken once per day before the beginning of the tests (see below) by gently shaking the tank using a forklift truck. The storage of the liquid digestate before using it at the irrigation setup lasted for 8 days.

The characteristic of the liquid digestate before the filtration are depicted in Table 1. The digestate liquid fraction was kindly provided by the Società Agricola Pallavicina S.R.L. (Via Fara—24047 Treviglio, IT), which also undergoes quality analyses, and did not display the presence of any pathogen nor pollutants according to the Italian laws.

Table 1. Composition of the digestate liquid fraction used for the test. Values are means ± s.d. (3 subsamples) provided by the management office of the digester. The analyses were made on the digestate liquid fraction hydrocyclone filtering.

Trait	Value	Unit	Method
Dry matter content at 105 °C	4.91 ± 0.03	%	IRSA CNR Q 64 Vol 2 1985
Ashes (dry matter content at 600 °C)	1.42 ± 0.02	%	
Chemical Oxigen demand (COD)	51.0 ± 1.8	g O_2 kg^{-1}	APAT IRSA CNR 5130 Man 29 2003
Total N	3.55 ± 0.02	g N kg^{-1}	IRSA CNR Q 64 Vol 3 1985
NH_4^+-N	1.848 ± 0.025	g N kg^{-1}	
Phosphorus	0.843 ± 0.051	g P kg^{-1}	UNI EN 16174:2012 and 16170:2015
Potassium	4.96 ± 0.25	g K kg^{-1}	

A water pump of 0.75 kW power (Pedrollo company, model: JSWm 2CX, San Bonifacio, Verona, Italy) was used to pump the digestate liquid fraction in the irrigation system. The operating pressure was set to 0.2 MPa. Samples were taken from the non-filtered digestate liquid fraction (DLF), from the filter outlet and from the filtered DLF to determine dry matter content and pH.

The experimental design consisted of two factors: filtrate dilution (FD) × time of the sampling within the irrigation cycle (TIME). The FD factor had four treatments: One control using freshwater and three filtrate dilutions (10%, 25%, and 50% of filtrate in freshwater). Each irrigation cycle lasted 8 h and samples were taken once per hour.

One irrigation cycle per day was performed, the irrigation cycles began with the tap water and continued with each increasing concentration of the digestate to avoid contamination. Each water-HF-DLF mixture was prepared mixing the relevant amount of tap water and HF-DLF in an operating tank and reflushing it several times with the pump. The water pump used was set at 0.2 MPa operating pressure, and the irrigation tank was filled with 400 L of DLF. Before starting each test, three samples were collected from the irrigation tank to measure the dry matter and the pH of the solution, following the methodology described above. In addition, flushing with tap water was performed by 15 min at the end of each cycle to allow for the cleaning the system. A pre-flushing was also made before the beginning of the first experiment with tap water. Within each irrigation cycle, the pump recycled part of the water or diluted HF-DLF into the tank to keep it mixed and avoid particle deposition. The irrigation system was organized by three polyethylene 1-m long dripper tubes (Stocker company N°26085, Bozen, Italy), as replicates, with three emitters each. The dripper tubes used had 0.016 m of diameter (maximum design pressure 0.4 MPa) and were spaced 0.33 m each other. Water flow declared by the manufacturer was 2 L h^{-1}. Emitters were not changed from each cycle to the following one.

2.2. *Measurements and Analyses*

During the tests, the water or filtrate dropping from the tubes was collected in plastic flagons placed underneath each emitter (Figure 1). The flagons were weighted once per hour with a portable scale (RADWAG WLC6/C1/R, Radom, Poland, used with 0.1 g sensitivity) to calculate the water flow (g h^{-1}). At the time of weighting the turbidity and the temperature of the liquid were measured. The temperature of the water or DLF were measured in °C using the DS18B20 digital thermometer (Maxim IC, San Jose, CA, USA). The turbidity of water or DLF were measured by using the turbidity sensor SKU:SEN0189 (Arduino, Ivrea, Italia), which was used as an indirect measurement of filtrate and water quality. The turbidity sensor SKU:SEN0189 uses light to detect suspended particles in water by measuring the light transmittance and scattering rate, which depends on the concentration of the total suspended solids (TSS) in the solution/dispersion. In particular, the sensor provide an output expressed in mV, which should be calibrated to the corresponding Nephelometric Turbidity Units (NTU). The output slightly and linearly decreases at increasing temperatures. In addition, the relationship between output and NTU value is not linear (for a brief description see https://wiki.dfrobot.com/Turbidity_sensor_SKU__SEN0189). In particular, the higher the sensor output, the lower the liquid NTU value. The manufacturer provide an output for pure water of 4.1 ± 0.3*V* when temperature span from 10 °C to 50 °C. Integration of the temperature and turbidity systems was made according to [21]. In the present work, the output of sensor was provided along with a direct measurement of the total suspended solids.

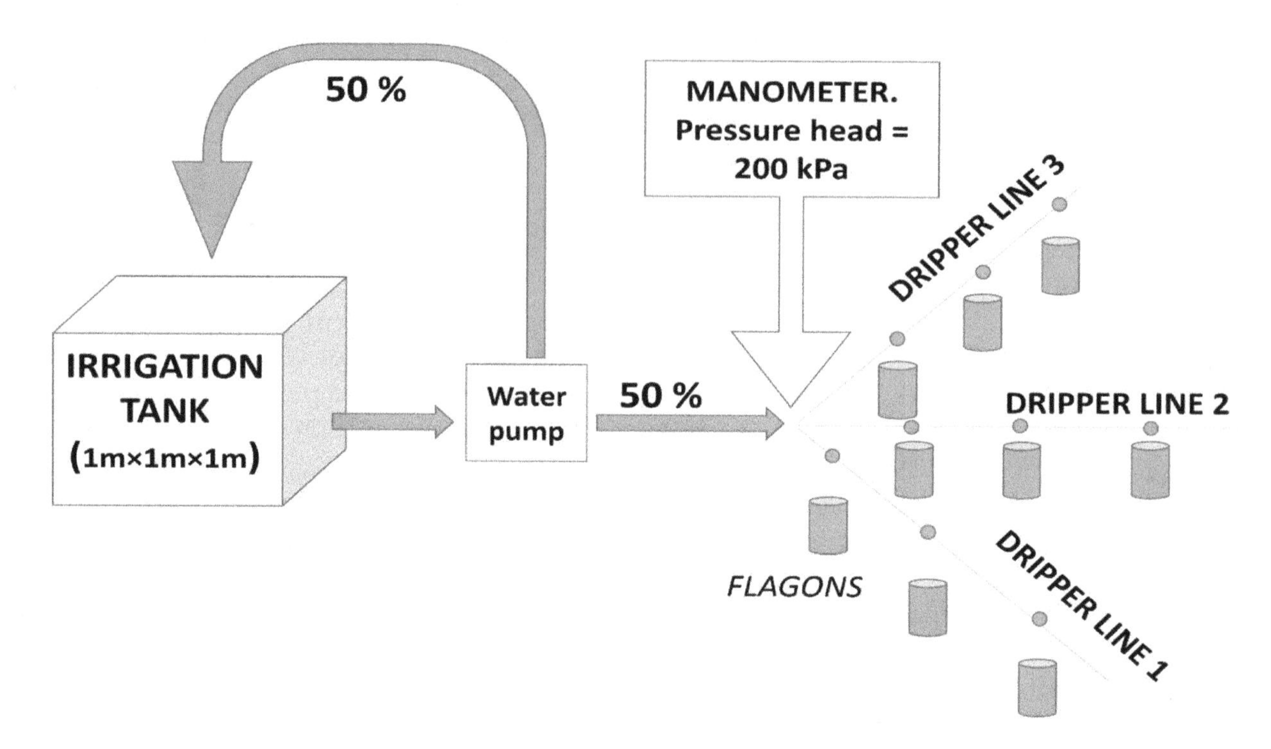

Figure 1. Design of the irrigation tests.

Dry matter and moisture content of the samples were assessed by oven drying at 105 °C until constant weight [22]. The pH of the samples was measured with no dilution by using a CRISON GLP21 pH-meter (Hach Lange Spain, S.L.U., Barcelona, Spain).

Then, the samples of each emitter per line were mixed and a random composite subsample of 500 mL of liquid was taken. In total, 24 sample per irrigation test were obtained (8 sampling moments × 3 irrigation lines). Each irrigation test consisted of the injection of a DLF dilution in an 8-h irrigation cycle. Thus 72 total samples of DLF released by each line were obtained. Dry matter of each subsample and its pH, following the methodology described above, were measured. For the control test (100% water), turbidity and pH were measured only before starting the test and no samples were collected during the test. This was done since these variables did not change by the time in the control from previous tests (data not shown). To monitor the air temperature and humidity of the indoor environment where the test was performed a sensor DHT22 (Guangzhou Aosong Electronics Co., Ltd., Guangzhou, China) was used. The Waterproof DS18B20 Digital Temperature Sensor was used to read the liquid temperature.

2.3. *Computations and Statistical Analysis*

The amounts of OH^- ions per ton of solution released by the emitters were computed by using the pH and used as a proxy of the potential of the irrigation with the DLF to increase the pH of a soil compared to the tap water used as a control. The analysis of variance was performed according to the statistical design by means of a general linear mixed model (Glimmix procedure in SAS/STAT 9.2 statistical package; SAS Institute Inc., Cary, NC, USA). The model used was similar to that shown in Saia et al. [23] (see the supplementary material in [23] for both a description of the procedure and the SAS package model applied) in which TIME was modelled as a repeated measurement [24]. Unbiased estimates of variance and covariance parameters were estimated by restricted maximum likelihood (REML). Repeated measurement analysis was modelled by applying a random statement with a first-order autoregressive covariance structure. In particular, the subject of reference was the emitter for the data related to the amount of liquid released, its turbidity and temperature, and it was the line for date on the pH. Denominator degrees of freedom of each error were estimated by Kenward–Roger approximation and least square means (LSmeans, see below for a definition) of the treatment distributions were computed. Data were provided both as LSmeans in figures and arithmetic means in supplemental materials, along with each standard error estimation or computation, respectively. Differences among means were compared by applying t-grouping at the 5% probability level to the LSMEANS p-differences. Time-sliced significance was also computed.

When the effect of time was significant, variation per unit time was modelled. Variation by time per each variable and treatment significantly varying by time was fitted to a linear distribution function and significance of the regression models were computed using the Slide Write Plus for Windows version 7.01 (Advanced Graphics Software, Inc., Encinitas, CA, USA).

3. Results

3.1. *Effects of the Hydrocyclone Filtration and Resting Time on the Traits of the Digestate Liquid Fraction*

Filtration of the digestate liquid fraction (DLF) influenced the pH of the different resulting fractions (Figure 2) pointing out that the native fraction before filtering had a pH value higher than 8.2 and not statistically appreciable differences were found after the filtration (hereafter referred as hydrocyclone filtered DLF, or HF-DLF). When the HF-DLF was allowed to rest for eight days before the beginning of the experiment, the solution at 50% dilution showed a lower pH compared to both the freshly made native DLF and the freshly made HF-DLF soon after the filtration. However, such latter difference was not statistically appreciable according to the conservative post-hoc test used. The pH value of the native DLF was higher than the fraction discarded from the filter and the tap water.

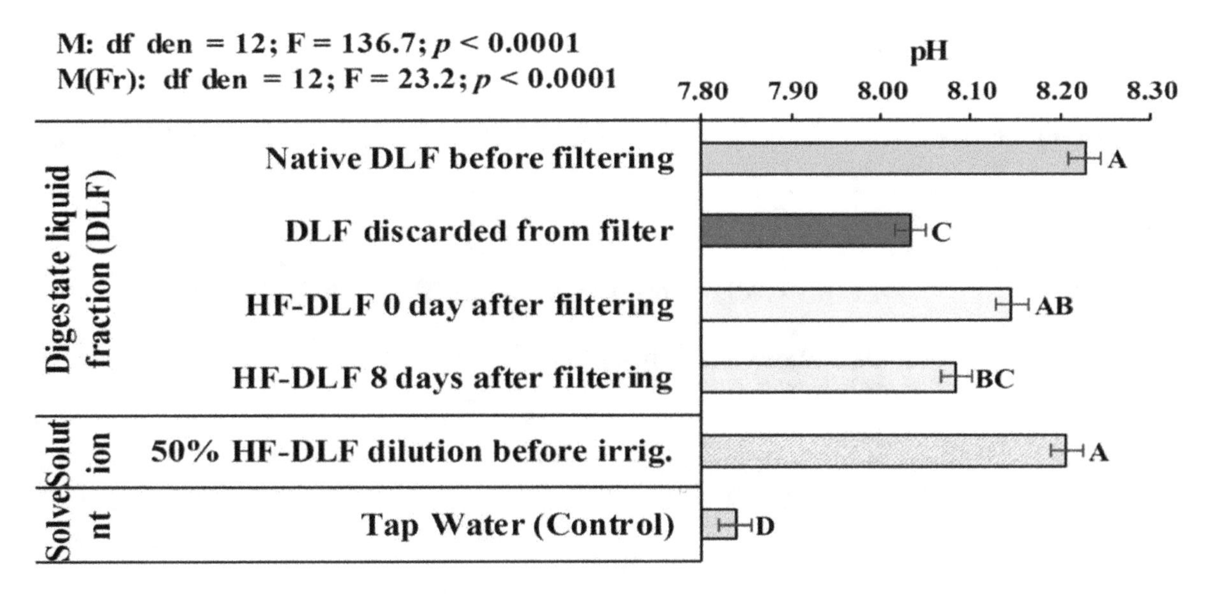

Figure 2. Values of pH of the digestate liquid fraction (DLF), before and after hydrocyclone filtering (HF), the 50% dilution of the HF-DLF, and the tap water used for the experiment. Bares are least square means (LSmeans) ± Lsmeans standard error estimates. Bars with a letter in common cannot be considered different according to a conservative Tukey-grouping applied to the p-differences of the LSmeans. Results of the statistical analysis are embedded.

The output of the turbidity sensor did not change by the filtration among the DLFs (Figure 3), which recorded value closed to 14.2 mV (please mind that, in theory, the lower the conductibility, the higher the turbidity). The tap water showed a turbidity sensor output close to 800 mV, 56-fold than any of DLFs or HF-DLF, on average.

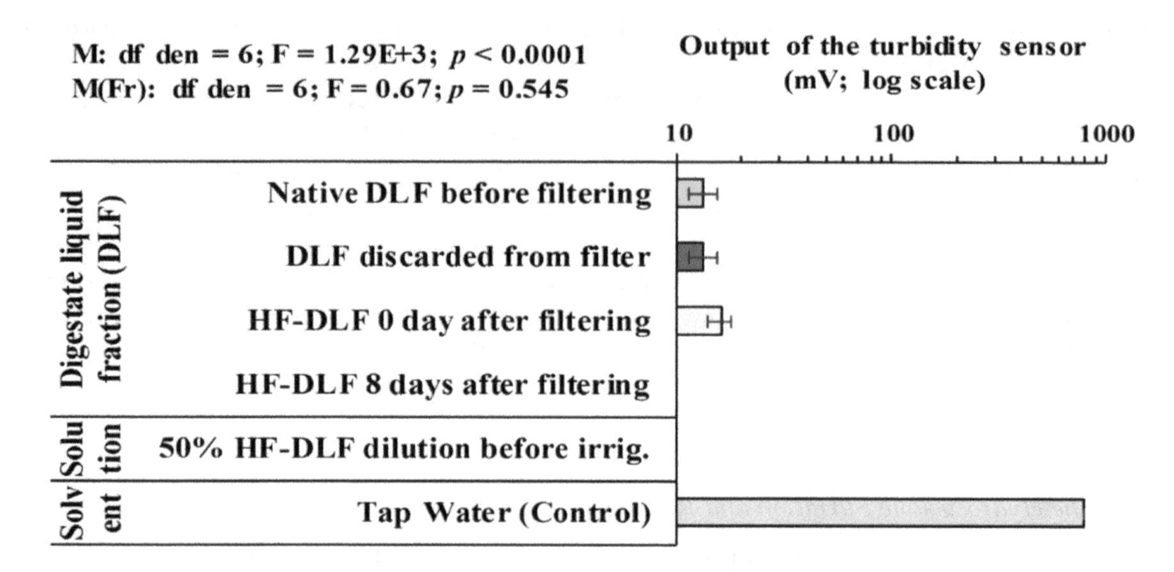

Figure 3. Output of the turbidity sensor in the digestate liquid fraction (DLF), before and after hydrocyclone filtering (HF), the 50% dilution of the HF-DLF, and the tap water used for the experiment. Bars are LSmeans ± Lsmeans standard error estimates. Bars with a letter in common cannot be considered different according to a conservative Tukey-grouping applied to the p-differences of the LSmeans. Results of the statistical analysis are embedded.

The dry matter concentration varied among the DLFs (Figure 4), with the fraction discarded from the filter showing a relative concentration compared to the native DLF 45% higher. The native DLF also showed a marginally, albeit significantly, lower dry matter concentration than the HF-DLF (−1.1% relative difference, corresponding to −0.02%).

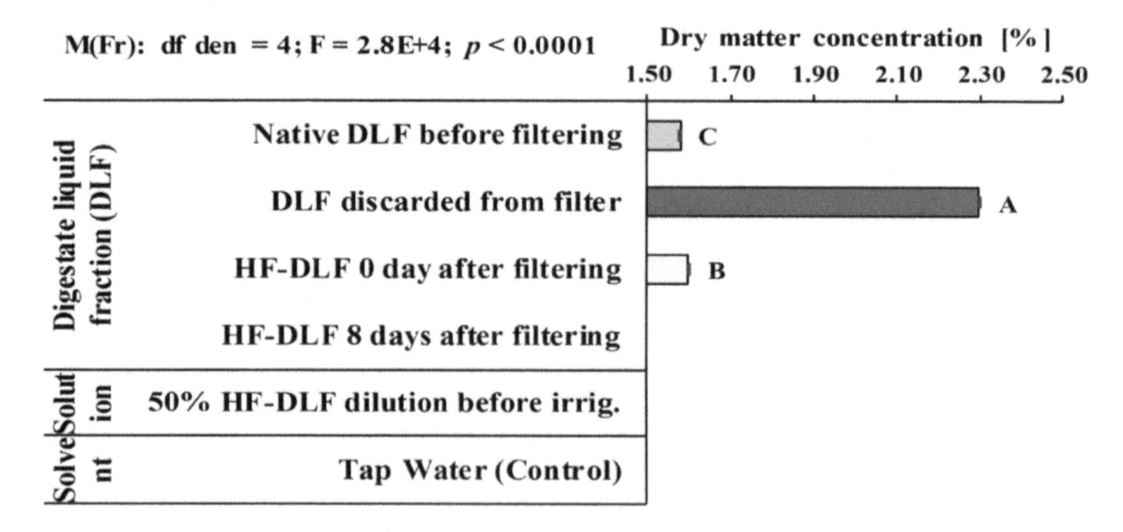

Figure 4. Dry matter concentration of the digestate liquid fraction (DLF), before and after hydrocyclone filtering (HF), the 50% dilution of the HF-DLF, and the tap water used for the experiment. Bars are LSmeans ± Lsmeans standard error estimates. Bars with a letter in common cannot be considered different according to a conservative Tukey-grouping applied to the p-differences of the LSmeans. Results of the statistical analysis are embedded.

3.2. *Effects of the Digestate Liquid Fraction Dilution on the Emitter Performances and Solution Traits*

Results of the statistical analyses of the irrigation test are shown in Table 2. Both the amount of water in the control and HF-DLF and its turbidity varied by the treatment at increasing time, with differences more marked among treatments in the early stages of the irrigation. The dilution of the DLF influenced the quantity of the solution released by the emitters during the test depending on the percentage of the dilution (Figure 5, Supplementary Material Table S1). In particular, water release increased almost constantly, whereas 10% and 25% dilution during the first 3 h. The 50% showed milder increases, and a total amount of water released slightly lower than the other treatments. The coefficient of variation of the system was in general lower than 5%, with some outlier only in the HF-DLF 25% dilution (Table 3).

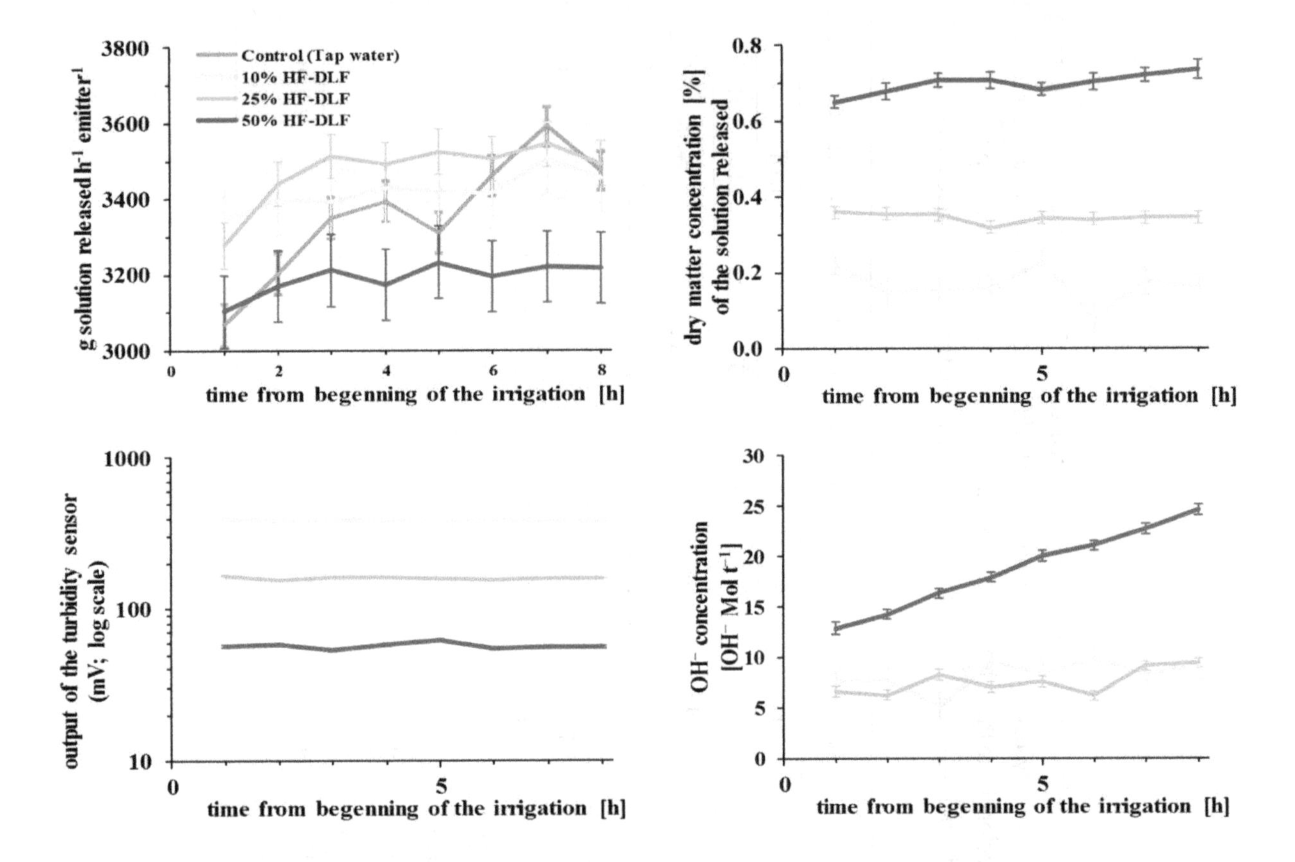

Figure 5. Amount of solution (tap water in the control and digestate liquid fractions (HF-DLF) diluted at the 10%, 25%, and 50%) released by the emitter each hour (upper left panel); dry matter concentration of the DLF released (upper right panel); turbidity of the HF-DLF released (lower left panel); and amount of OH^- ions released per ton of water released each hour. Data are LSmeans ± LSmeans standard error estimates. For post-hoc comparisons and raw data see Supplementary Material Table S3.

Table 2. Results of the statistical analysis (degrees of freedom estimate of the error (DF den); F statistics; and *p* values) of the general linear mixed model applied to the amount of solution released each hour, its turbidity, dry matter concentration, pH difference compared to the control, and the amount of OH^- released with the solution per hour (expressed as mol ton^{-1}). Factors were the solution dilution (FD) and variation in time (1-h step of a 8-h irrigation cycle). Data were analyzed with a repeated treatment option and after which differences by each time-step were sliced. When the *p*-values were lower than 0.05, F and *p* were shown in bold. See the Supplementary Material for the post-hoc test.

	Amount of Solution Released Each Hour			**Turbidity**			**Dry Matter Concentration**			**[OH⁻] Concentration**			**pH Difference Compared to Control**		
Effect	Den DF	F	*p*	Den DF	F	*p*	Den DF	F	p	Den DF	F	*p*	Den DF	F	*p*
FD	101.8	3.2	0.0284	52.03	3.6×10^4	<0.0001	7.395	540.3	<0.0001	11.12	912.2	<0.0001	15.5	188.2	<0.0001
Time	83.06	47.3	<0.0001	139.6	9.1	<0.0001	30.7	1.2	0.3452	25.62	24.0	<0.0001	20.06	12.7	<0.0001
FD × Time	149.6	52.9	<0.0001	149.8	7.4	<0.0001	28.01	2.2	0.0389	29.78	12.9	<0.0001	21.15	7.4	<0.0001
Sliced By Time [h]															
1	97.0	3.9	0.011	64.9	9.7×10^3	<0.0001	23.6	121.5	<0.0001	18.5	34.4	<0.0001	13.0	41.4	<0.0001
2	109.4	4.7	0.004	91.6	9.7×10^3	<0.0001	11.4	97.8	<0.0001	24.5	72.8	<0.0001	20.0	64.2	<0.0001
3	111.0	3.2	0.025	95.0	1.0×10^4	<0.0001	24.1	165.7	<0.0001	25.5	74.4	<0.0001	20.5	65.4	<0.0001
4	111.4	3.1	0.031	95.4	9.8×10^3	<0.0001	12.7	128.3	<0.0001	25.5	111.4	<0.0001	20.5	85.9	<0.0001
5	111.5	4.1	0.008	95.4	9.1×10^3	<0.0001	22.3	129.1	<0.0001	25.5	149.7	<0.0001	20.5	101.7	<0.0001
6	111.6	3.0	0.034	95.4	9.8×10^3	<0.0001	13.2	142.3	<0.0001	25.5	205.9	<0.0001	20.5	140.0	<0.0001
7	111.7	4.4	0.006	95.4	9.5×10^3	<0.0001	20.3	159.0	<0.0001	25.5	183.6	<0.0001	20.5	110.3	<0.0001
8	111.8	2.5	0.065	95.4	9.5×10^3	<0.0001	15.5	112.9	<0.0001	25.5	226.2	<0.0001	20.5	119.0	<0.0001

Table 3. Coefficient of variation of each treatment (n = 9, consisting of 3 lines with 3 emitters each) and relative distribution percentiles at 0.025, 0.25, 0.5 (median), 0.75, and 0.975. Data of the three dilutions of the hydrocyclone filtered digestate liquid fraction (HF-DLF) were showed singly and bulked.

	Control (Tap Water)	**10% Dil. HF-DLF**	**25% Dil. HF-DLF**	**50% Dil. HF-DLF**	**10% + 25% + 50% Dil. HF-DLF Bulked Data**
Mean of coefficient of variation	1.65	3.41	3.69	3.66	3.59
Percentile 0.025	1.24	3.26	2.65	3.55	2.71
Percentile 0.250	1.45	3.35	2.79	3.59	2.89
Percentile 0.500 (median)	1.66	3.41	2.82	3.65	3.43
Percentile 0.750	1.86	3.46	2.86	3.68	3.61
Percentile 0.975	2.07	3.49	8.86	3.86	6.44

The dry matter content of the solutions (Figure 5) showed that the concentration strongly depended on the dilution rate, and to a lesser extent on time. In fact, higher values >0.6%) were recorded in the DLF 50%, compared to those in DLF 10% < 0.2%) and DLF 25% showed values ranging from 0.4% to 0.3%. The value of dry matter concentration (%) was quite stable all along the 8 h of irrigation test for all the treatments, except for DLF50% that showed a slight increase with time (Supplementary Material Table S1; Supplementary Material Table S2).

A similar trend, but more pronounced by the time, was found regarding turbidity (Figure 5; please mind that the higher is the turbidity value, the lower the liquid turbidity). For all the treatments the values recorded were stable over time and around 60 mV, 160 mV, and 380 mV for DLF 50%, 25% and 10% respectively. The analysis of the concentration of ions OH^- (Figure 5), calculated using the pH values, showed that even if the trend of treatments DLF 10% and 25% was slightly variable during the irrigation test, the values were included between 5 and 10 OH^- mol t^{-1} solution, with scarce differences by the time. Instead, treatment DLF 50% showed an increasing trend during the test with an initial value of 13 OH^- mol $t^{-1}$and a final value of 25 OH^- mol t^{-1}. Finally, we inspected a serpentine from the control and the DLF 50% (Figure 6) and found that no clogging occurred.

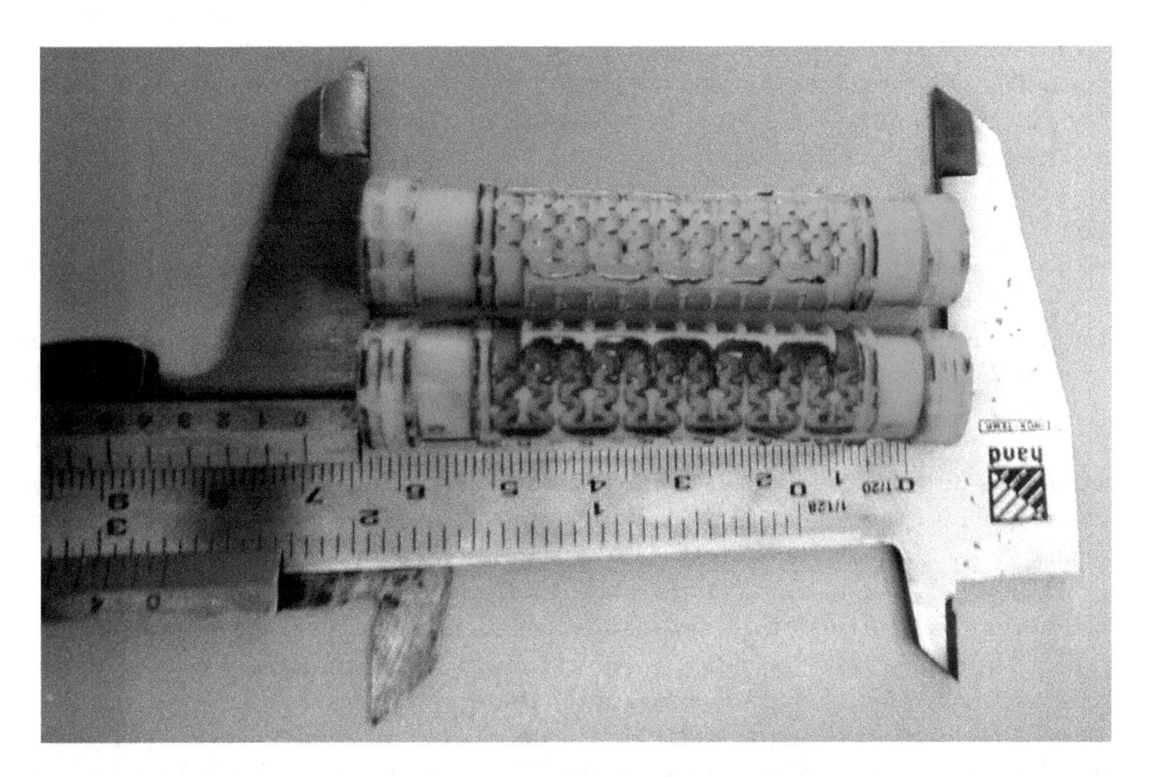

Figure 6. Serpentine inspection of the control (tap water; above) vs. 50% diluted hydrocyclone filtered digestate liquid fraction (below) emitters.

4. Discussion

In the present work, we studied the role of an increasing ratio between a hydrocyclone-filtered digestate liquid fraction (referred to as HF-DLF) and tap water on the performance of an irrigation system and water quality. Treatments included 3 HF-DLF ratios (10%, 25%, and 50% of total solution used for the irrigation) in contrast to tap water as control and measurements were taken at an hourly basis on an 8-h irrigation cycle, that simulates most of the irrigation cycles occurring in a broad range of crops.

The native digestate liquid fraction used before the hydrocyclone filtering had a higher pH than the tap water used (pH = 8.23 vs. 7.84, respectively) and such pH slightly reduced after the hydrocyclone filtering. Information on the effect that the hydrocyclone filtering has on the pH of digestate liquid fraction and its total solids are scarce, nonetheless, centrifuge filtering was shown to have relatively high efficiency [25]. Thus, the pH reduction after the hydrocyclone treatment may have been due to the fractionation of the calcium carbonates or other high-weight solids in the digestate liquid fraction, including cations. The digester diet of the material in the present study was mainly composed of corn and barley biomasses (residual and dedicated) and to a lesser extent of cow slurry. This kind of digestate has high contents in potassium, chloride, carbonates, and proteins [20]. It is thus likely that the high pH of the HF-DLF under study was due to a high content of basic, high molecular weight proteins, which can be removed by hydrocyclone filtering. Such a hypothesis is corroborated by the further reduction of pH of the DLF found eight days after filtering, before the irrigation experiment started, which may have been due to oxidation of the organic material in the HF-DLF that in such time-lapse was resting. Hydrocyclone filtering, however, did not result in a reduction of the turbidity and such results could be due to the high total solid concentration in the DLF following incomplete filtering, as pointed by Guilayn et al. [25]. Indeed, in our study, the HF-DLF showed a dry matter concentration of 16.0‰ (on a weight basis) and a pH = 8.15 soon after the filtration. These traits suggested a low quality fraction for drip irrigators according to the early classification by Nakayama and Bucks [3]. This likely was a main cause of the differences among the amount of HF-DLF released by the emitters at increasing time and varying the HF-DLF ratios within the irrigation system.

The manufacturer declared the used emitters as and releasing 2 L h^{-1} at 100 kPa. When subjected to the 200 kPa pressure of the present study, we found that the amount of water released in the control ranged from 3.07 L in the first hour and such an amount increased on average by the 1.9% h^{-1}. Such variation were higher than those found by Bodole et al. [26]. The variation of the amount of HF-DLF each emitter released per hour also increased with time in the three dilution treatments (0.5–0.9% h^{-1}), but to a lesser extent compared to the control. Such variations likely depended on the usury of the system. In particular, the dilution at 10% and 25% released 1.9% and 3.5%, respectively, more HF-DLF per cycle than the water released by the control, whereas the dilution at 50% released 4.9% less HF-DLF per cycle than the control. Besides, the differences between each HF-DLF dilution and control in the amount of water released per unit time declined with time. The temperature of the tap water or the HF-DLFs was similar among the treatments and increased linearly during each irrigation cycle, starting at 9.42 °C ± 0.27 °C at the 9:00 a.m. (moment of the beginning of each experiment) and increased at a rate of 0.94 °C h^{-1} ± 0.05 °C h^{-1} (data not shown). This implies that differences by time can only partly be explained by a heating of the emitters and thus the expansion of their pore size. We hypothesize that the emitters used in this experiment rapidly lost their ability to compensate for the irrigation rate at the pressure we used. In addition, the 10% and 25% diluted HF-DLF did not likely consist in a strong occlusion of the emitters. This is consistent with the constant rate of the dry matter content of 10% and 25% HF-DLF and increasing rate of the dry matter content of the 50% HF-DLF, which progressively increased at a rate of $27.98 \times 10^{-3} \pm 0.95 \times 10^{-3}$ pH units h^{-1}. When using the 50% dilution, 6.63% less HF-DLF was released, on average, if compared to the water released in the control or the HF-DLF in the 10% and 25% dilutions. Despite such difference, the potential effect on the pH (expressed as $[OH^-]$ amount of a putative medium receiving the HF-DLF at the 50% dilution) strongly increased over time, whereas it did not vary in the in the control or the HF-DLF in the 10% and 25% dilutions

1.68 ± 0.04 Mol OH^- (t HF-DLF)$^{-1}$ h^{-1}. Results from other experiments were variable and depended on the pressure, kind of emitter and kind of wastewater. In contrast to our study, Gamri et al. [27] found a strong reduction of the emitter performance with time and difference between the present and the one by Gamri et al. [27] experiment can be due to the higher pressure we used, which is two-fold compared to their study, and this occurred despite our HF-DLF had a solid concentration 2.3–9.8 fold higher than the synthetic wastewater composition used by Gamri et al. [27]. Nonetheless, these differences can be due to the high-frequency flushing in our experiment (one every 8 h). And indeed, Puig-Bargués et al. [14] showed that flushing every 540 h was sufficient to almost completely avoid the emitter clogging. Puig-Bargués et al. [14] also found, as in the present study, that dripline flow increased 8% and 25% over time when using a pressure compensating and a non-pressure compensating emitter, respectively, when used with a tertiary effluent from a wastewater treatment plant filtered with a 0.130 mm filtration level. Similarly, we found that coefficient of variation computed at an hourly basis was 1.653% ($CI_{95\%}$ 1.240–2.065%) in the control and 3.587% ($CI_{95\%}$ 2.713–6.444%) in the HF-DLF, with scarce differences among dilution, suggesting that accumulation of deposited material in the emitters affected the dripline flow performance. Such coefficient of variation was lower than those found in other similar studies [28,29] and can be marginally acceptable as indicated by Bodole et al. [26], according to which a test duration of more than 60 min is enough to minimize the uncertainty due to the initial fluctuation of the data. The lower variation of the HF-DLF is likely due to an anti-clogging shape of the present emitters if compared to other emitters [30].

5. Conclusions

In conclusions, hydrocyclone filtration scarcely affected the traits of the digestate liquid fraction used for the irrigation. Irrigation with hydrocyclone-filtered digestate liquid fraction (HF-DLF) injected in the system at 10% and 25% dilution did not affect the performance of the system nor the traits of the liquid fraction released by the emitters, whereas using 50% dilution of the HF-DLF consisted in a lower amount of liquid released at increasing pH. In particular, HF-DLF dilution at 10%, 25%, and 50% consisted in +1.9%, +3.5, and −4.9% amount of liquid released compared to the control. In 10% HF-DLF % and 25% HF-DLF, a constant pH difference of $+ 0.321 \pm 0.014$ pH units compared to the control was found, in the 50% HF-DLF pH increased by around a half point and such difference increased with time.

This implies that that highly concentrated digestate liquid fractions, i.e., low dilutions, can pose problems for the functioning of the system and may have potentially harmful effects on soils with high pH. Nonetheless, the use of digestate liquid fractions for irrigation purposes may be a valuable option in those areas with a high amount of biogas plants and digestate production, such as various nations in Europe, America and Asia including USA, China, Germany, United Kingdom, Italy, and France [31,32]. Results from the present study have beneficial implication on the on water conservation since digestate production by feeding the digester with barley and corn provide wastewater from the late spring to the early fall, when water requirements are high. At the one time, the ability to use such wastewaters with minimal impact on the irrigation system, and thus with reduced negative impacts due to the system maintenance and disposal.

The digestate liquid fraction used in the present study was previously subjected to an additional hydrocyclone filtering, that likely discarded the high-molecular weight fraction. However, since few differences were found between filtered and non-filtered liquid digestate fraction, it is likely that a dilution at least up to 25%, according to the present study, can allow for a direct use of the digestate liquid fraction in microirrigation system with a minimal harming of the system performances. However, since the present is a short-term experiment, these results would require additional experiments with unfiltered liquid fractions.

Supplementary Materials: Table S1: LSmeans estimates and relative standard errors of the traits under study [Solution released; output of the turbidity sensor; dry matter concentration; [OH^-] concentration; and pH difference than control]. LSmeans with a letter in common can't be considered different according to a conservative Tukey-grouping applied to the p-differences of the LSmeans.; Table S2: Linear models of the variation in time of the temperature in the whole experiment and of dry matter concentration, $[OH^-]$ concentration and pH, and in the 50% diluted hydrocyclone filtered digestate liqud fraction; Table S3: Raw data of the amount of water or diluted hydrocyclone filtered digestate liqud fraction, the output of the turbidity sensor (mV), dry matter content (%), pH of the solution, and the amount of $[OH^-]$ concentration.

Author Contributions: Conceptualization, S.B., M.B., E.R., S.S., L.P.; methodology, S.B., M.B., E.R., S.S.; software, S.B., E.R., S.S.; validation, S.B., S.S.; formal analysis, S.S.; investigation, S.B., M.B., E.R., S.S., P.C.; resources, M.C., P.T., C.B., L.P.; data curation, S.B., S.S., P.C.; writing—original draft, S.B., S.S.; writing—review and editing, S.B., S.S., P.C.; supervision, L.P.; project administration, L.P.; funding acquisition, L.P. All authors have read and agreed to the published version of the manuscript.

Acknowledgments: The authors wish to thank Ivan Carminati, Gianluigi Rozzoni, Alex Filisetti and Elia Premoli for their assistance in performing the tests and for their professionalism and availability.

References

1. Elliott, J.; Deryng, D.; Müller, C.; Frieler, K.; Konzmann, M.; Gerten, D.; Glotter, M.; Flörke, M.; Wada, Y.; Best, N.; et al. Constraints and potentials of future irrigation water availability on agricultural production under climate change. *Proc. Natl. Acad. Sci. USA* **2013**, *111*, 3239–3244. [CrossRef] [PubMed]
2. Döll, P. Impact of Climate Change and Variability on Irrigation Requirements: A Global Perspective. *Clim. Chang.* **2002**, *54*, 269–293. [CrossRef]
3. Nakayama, F.; Bucks, D. Water quality in drip/trickle irrigation: A review. *Irrig. Sci.* **1991**, *12*, 12. [CrossRef]
4. Pereira, L.S.; Oweis, T.; Zairi, A. Irrigation management under water scarcity. *Agric. Water Manag.* **2002**, *57*, 175–206. [CrossRef]
5. Saia, S.; Fragasso, M.; De Vita, P.; Beleggia, R. Metabolomics Provides Valuable Insight for the Study of Durum Wheat: A Review. *J. Agric. Food Chem.* **2019**, *67*, 3069–3085. [CrossRef]
6. Rusan, M.J.M.; Hinnawi, S.; Rousan, L. Long term effect of wastewater irrigation of forage crops on soil and plant quality parameters. *Desalination* **2007**, *215*, 143–152. [CrossRef]
7. Tarchitzky, J.; Lerner, O.; Shani, U.; Arye, G.; Brener, A.; Chen, Y.; Lowengart-Aycicegi, A. Water distribution pattern in treated wastewater irrigated soils: Hydrophobicity effect. *Eur. J. Soil Sci.* **2007**, *58*, 573–588. [CrossRef]
8. Adin, A.; Sacks, M. Dripper-Clogging Factors in Wastewater Irrigation. *J. Irrig. Drain. Eng.* **1991**, *117*, 813–826. [CrossRef]
9. Elgallal, M.; Fletcher, L.; Evans, B. Assessment of potential risks associated with chemicals in wastewater used for irrigation in arid and semiarid zones: A review. *Agric. Water Manag.* **2016**, *177*, 419–431. [CrossRef]
10. Drechsel, P.; Scott, C.A.; Raschid-Sally, L.; Redwood, M.; Bahri, A. *Wastewater Irrigation and Health*; Bahri, A., Drechsel, P., Raschid-Sally, L., Redwood, M., Eds.; Routledge: London, UK; Sterling, VA, USA, 2009; 432p, ISBN 9781849774666.
11. Capra, A.; Scicolone, B. Emitter and filter tests for wastewater reuse by drip irrigation. *Agric. Water Manag.* **2004**, *68*, 135–149. [CrossRef]
12. Ahmed, B.A.O.; Yamamoto, T.; Fujiyama, H.; Miyamoto, K. Assessment of emitter discharge in microirrigation system as affected by polluted water. *Irrig. Drain. Syst.* **2007**, *21*, 97–107. [CrossRef]
13. Goyal, M.R. *Wastewater Management for Irrigation*; Goyal, M.R., Triphati, V.K., Eds.; Apple Academic Press & CRC Press: Boca Raton, FL, USA, 2016; ISBN 9780429152498.
14. Puig-Bargués, J.; Arbat, G.; Elbana, M.; Duran-Ros, M.; Barragan, J.; De Cartagena, F.R.; Lamm, F. Effect of flushing frequency on emitter clogging in microirrigation with effluents. *Agric. Water Manag.* **2010**, *97*, 883–891. [CrossRef]
15. Hills, D.J.; Brenes, M.J. Microirrigation of Wastewater Effluent Using Drip Tape. *Appl. Eng. Agric.* **2001**, *17*, 17. [CrossRef]
16. Cararo, D.C.; Botrel, T.A.; Hills, D.J.; Leverenz, H.L. Analysis of Clogging in Drip Emitters during Wastewater Irrigation. *Appl. Eng. Agric.* **2006**, *22*, 251–257. [CrossRef]

17. Makádi, M.; Tomcsik, A.; Orosz, V. Digestate: A New Nutrient Source—Review. *Biogas* **2012**, 1–17. [CrossRef]
18. Green, O.; Katz, S.; Tarchitzky, J.; Chen, Y. Formation and prevention of biofilm and mineral precipitate clogging in drip irrigation systems applying treated wastewater. *Irrig. Sci.* **2018**, *36*, 257–270. [CrossRef]
19. Barzee, T.J.; Edalati, A.; El-Mashad, H.; Wang, D.; Scow, K.; Zhang, R. Digestate Biofertilizers Support Similar or Higher Tomato Yields and Quality Than Mineral Fertilizer in a Subsurface Drip Fertigation System. *Front. Sustain. Food Syst.* **2019**, *3*, 3. [CrossRef]
20. Akhiar, A.; Battimelli, A.; Torrijos, M.; Carrere, H. Comprehensive characterization of the liquid fraction of digestates from full-scale anaerobic co-digestion. *Waste Manag.* **2017**, *59*, 118–128. [CrossRef]
21. Tahir, M.U.; Ahsan, S.M.; Arif, S.M.; Abdullah, M. GSM Based Advanced Water Quality Monitoring System Powered by Solar Photovoltaic System. In Proceedings of the 2018 Australasian Universities Power Engineering Conference (AUPEC), Auckland, New Zealand, 27–30 November 2018; pp. 1–5.
22. Standard Methods for the Examination of Water and Wastewater. In Proceedings of the AWRA's 1999 Annual Water Resources Conference: Watershed Management to Protect Declining Species, Seattle, WA, USA, 5–9 December 1999; APHA: Washington, DC, USA, 1999.
23. Saia, S.; Aissa, E.; Luziatelli, F.; Ruzzi, M.; Colla, G.; Ficca, A.G.; Cardarelli, M.; Rouphael, Y. Growth-promoting bacteria and arbuscular mycorrhizal fungi differentially benefit tomato and corn depending upon the supplied form of phosphorus. *Mycorrhiza* **2019**, *30*, 133–147. [CrossRef]
24. Giovino, A.; Militello, M.; Gugliuzza, G.; Saia, S. Adaptation of the tropical hybrid Euphorbia×lomi Rauh to the exposure to the Mediterranean temperature extremes. *Urban For. Urban Green.* **2014**, *13*, 793–799. [CrossRef]
25. Guilayn, F.; Jimenez, J.; Rouez, M.; Crest, M.; Patureau, D. Digestate mechanical separation: Efficiency profiles based on anaerobic digestion feedstock and equipment choice. *Bioresour. Technol.* **2019**, *274*, 180–189. [CrossRef] [PubMed]
26. Bodole, C.; Koech, R.; Pezzaniti, D. Laboratory evaluation of dripper performance. *Flow Meas. Instrum.* **2016**, *50*, 261–268. [CrossRef]
27. Gamri, S.; Soric, A.; Tomas, S.; Molle, B.; Roche, N. Biofilm development in micro-irrigation emitters for wastewater reuse. *Irrig. Sci.* **2013**, *32*, 77–85. [CrossRef]
28. Pinto, M.F.; Molle, B.; Alves, D.G.; Ait-Mouheb, N.; De Camargo, A.P.; Frizzone, J.A. Flow rate dynamics of pressure-compensating drippers under clogging effect. *Rev. Bras. Eng. Agríc. Ambient.* **2017**, *21*, 304–309. [CrossRef]
29. Dalri, A.B.; Santos, G.O.; Dantas, G.D.F.; De Faria, R.T.; Zanini, J.R.; Palaretti, L.F. Performance of drippers in two filtering systems using sewage treatment effluent. *Rev. Bras. Eng. Agríc. Ambient.* **2017**, *21*, 363–368. [CrossRef]
30. Zhang, J.; Zhao, W.; Tang, Y.; Lu, B. Anti-clogging performance evaluation and parameterized design of emitters with labyrinth channels. *Comput. Electron. Agric.* **2010**, *74*, 59–65. [CrossRef]
31. Deng, Y.; Xu, J.; Liu, Y.; Mancl, K. Biogas as a sustainable energy source in China: Regional development strategy application and decision making. *Renew. Sustain. Energy Rev.* **2014**, *35*, 294–303. [CrossRef]
32. Akhiar, A. Characterization of Liquid Fraction of Digestates after Solid-Liquid Separation from Anaerobic Co-Digestion Plants. Ph.D. Thesis, Université Montpellier, Montpellier, France, 2017. Submitted 15/01/2018 (NNT: 2017MONTS004).

3

Effect of Irrigation Water Regimes on Yield of *Tetragonia Tetragonioides*

Gulom Bekmirzaev [1,*], Jose Beltrao [2] and Baghdad Ouddane [3]

[1] Tashkent Institute of Irrigation and Agricultural Mechanization Engineers, Department of Irrigation and Melioration, Kori-Niyoziy street 39, 100000 Tashkent, Uzbekistan
[2] Research Centre for Spatial and Organizational Dynamics, University of Algarve, Campus de Gambelas, 8005-139 Faro, Portugal; jbeltrao@ualg.pt
[3] Physico-Chemistry Team of the Environment, Sciences and Technologies, University of Lille, LASIR UMR-CNRS 8516, Building C8, 59655 Villeneuve d'Ascq, CEDEX, France; baghdad.ouddane@univ-lille1.fr
* Correspondence: gulombek@gmail.com

Abstract: The main purpose of this experiment was to study the effect of several irrigation water regimes on *Tetragonia tetragonioides* (Pall) O. Kuntze in semi-arid regions. During the experiment period, it was measured that several irrigation regimes were affected in terms of growth, biomass production, total yield, mineral composition, and photosynthetic pigments. The experiment was conducted in the greenhouse at the University of Algarve (Portugal). The study lasted from February to April in 2010. Three irrigation treatments were based on replenishing the 0.25-m-deep pots to field capacity when the soil water level was dropped to 70% (T1, wet treatment), 50% (T2, medium treatment), and 30% (T3, dry treatment) of the available water capacity. The obtained results showed that the leaf mineral compositions of chloride and sodium, the main responsible ions for soil salinization and alkalization in arid and semi-arid regions, enhanced with the decrease in soil water content. However, the minimum amounts of chlorophyll, carotenoids, and soluble carbohydrates in the leaf content were obtained in the medium and driest treatments. On the other hand, growth differences among the several irrigation regimes were very low, and the crop yield increased in the dry treatment compared to the medium treatment; thus, the high capacity of salt-removing species suggested an advantage of its cultivation under dry conditions.

Keywords: irrigation water regimes; leaf mineral composition; semi-arid regions; available water capacity; biomass production; total yield

---

1. Introduction

In arid and semi-arid regions, such as the Mediterranean, supplies of good-quality water allocated to agriculture are expected to decrease because most available fresh/potable water resources were already mobilized [1]. According to the Food and Agriculture Organization (FAO) [2], due to the shortage of water, there is an enlargement of saline land in agricultural areas in some developing countries. As a result, yield is decreasing, provoking an increasing cost of agricultural products [3].

Soil salinization is recognized worldwide as being among the most important problems for crop production in arid and semi-arid regions [4]. Water deficit and salinity are the major limiting factors for plant productivity, affecting more than 10% of arable land on our planet, resulting in a yield reduction of more than 50% for most major crop plants [5]. The usually noted abiotic stresses that include a component of cellular water deficit are salinity and low temperature; stresses can also severely limit crop production [6]. Abiotic stresses, such as drought, salinity, extreme temperatures, chemical toxicity, and oxidative stress are serious threats to agriculture, and result in the deterioration of the environment. Abiotic stress is the primary cause of crop loss [7,8]. This problem is intensified in coastal areas due to sea-water intrusion. This results from reduced ground-water levels as the water demand exceeds the annual groundwater recharge [9]. As reported above, some of the emerging regions in risk of increasing levels of salinization of their soils are located in the Mediterranean Basin [10,11], Australia [12], Central Asia [13], and Northern Africa [14]. Salinity is one of the rising problems causing tremendous yield losses in many regions of the world, especially in arid and semi-arid regions. The use of halophytes can be an effective way of accumulating the salt in soil [15].

Intensive irrigation of agricultural crops with a high level of water mineralization causes salts to accumulate in the root zones, which adversely affects the crop productivity. In order to reduce such negative impacts, a regulated deficit irrigation (RDI) technique was adopted to combat salinization in arid and semi-arid environments by reducing the water application during certain growth stages of the crops [16].

When RDI is not feasible, halophyte crops might be a solution for the salinization of agricultural land. These crops can be irrigated by, for example, seawater, salt-contaminated phreatic sheets, brackish water, wastewater, or drainage water from other plantations [17,18].

Hence, our aims were to choose a salt-removing crop, tolerant to salinity, along with interest as a food crop, and to test its drought tolerance through its response to several water regimes. *Tetragonia tetragonioides* was the selected crop. In a previous experiment, its capability as a high biomass horticultural leaf crop was demonstrated, producing a plant dry weight of 40,000–50,000 dry mass (DM) $kg \cdot ha^{-1}$ if the plant population density is around 75,000 $plants \cdot ha^{-1}$ [19].

2. Materials and Methods

2.1. Experimental Procedure

The experimental work was conducted in the greenhouse of Horto at the University of Algarve, Faro, Portugal (37°2′37.1 N 7°58′30.8 W), from February to April in 2010. The salt-removing species *T. tetrogonioides* was selected. Plants were transplanted to 7-L pots when they had four leaves (10 February). The number of plants per pot was three, with four replications. The species were irrigated with tap water every three days until the beginning of the treatments (1 February–8 March). A nitrogen fertigation treatment was started on 8 March, with daily applied concentrations of 2 mM NO_3^- and 2 mM NH_4^+ as the cumulative amount of NO_3NH_4 ($g \cdot plant^{-1}$) to the end of the experimental studies (22 April). The electrical conductivity (EC_w) of irrigation water was 0.6 $dS \cdot m^{-1}$ and pH 7.

The treatments consisted of three irrigation regimes in a randomized complete block design with three replicated treatments based on replenishing the 0.25-m-deep pots to field capacity when the soil water level dropped to 70% (T1, wet treatment), 50% (T2, medium treatment), and 30% (T3, dry treatment) of the available water capacity (aw). This concept was developed by Reference [20], where "aw" is the range of available water that can be stored in soil and is available for growing crops. It was assumed by the same authors that the soil water content readily available to plants (θ_{aw}) is the difference between the volume of water content at field capacity (θ_{fc}) and at the permanent wilting point (θ_{wp}), calculated as follows:

$$\theta_{aw} = \theta_{fc} - \theta_{wp} \tag{1}$$

The watering volume was estimated to replenish the soil profile to field capacity at a depth of 0.25 m. The volumetric soil water content (m^3 water/m^3 soil; $m^3 \cdot m^{-3}$) was determined just before the water application.

To control soil water along the soil profile, the irrigation frequency, and the water amount, the pots were weighed every day. The soil water content was monitored periodically, gravimetrically measured for a depth of 0.00–0.25 m.

The plants were harvested destructively (26 April), washed in water, and dried with paper towels. Then, the fresh weight (FW) was measured. The fresh samples were dried in a forced drought oven at 70 °C for 48 h, and the dry weight (DW) was measured. Plant materials were collected for chemical analyses. The electrical conductivity (EC_s) and pH of soil were measured before and after the experiment.

2.2. *Growth and Chemical Analysis*

During the vegetation period, the stem length was measured, as well as the number of nodes and number of leaves of *T. tetragonioides* every seven days.

The plants' leaves were analyzed on total growth and mineral compositions (Na, Cl, N, K, P, Ca, and Mg). Dried leaves and stems were finally grounded and analyzed using the dry-ash method. The levels of Na and K were determined using a flame photometer, and the remaining cations (Na, K, Ca and Mg) were assessed by atomic absorption spectrometry. Chloride ions were determined in the aqueous extract by titration with silver nitrate according to the method of Reference [21]. Plant nitrogen (N) content was determined using the Kjeldhal method. Phosphorus was determined using the colorimetric method according to the vanadate–molybdate method. All mineral analyses were only performed on the leaves.

The analysis of pigments was done with a disc size of 0.66 cm and a total area of 1.37 cm^2. For sugars. there were ten discs, with a disc size of 0.66 cm and a total area of 3.42 cm^2. The amount of photosynthetic pigments (chlorophyll a (Chla), b (Chlb), total (ChlT), and carotenoids) was determined according to the method of Reference [22]. Shoot samples (0.25 g) were homogenized in acetone (80%). The extract was centrifuged at $3000\times g$, and absorbance was recorded at wavelengths of 646.8 and 663.2 nm for the chlorophyll assay and at 470 nm for the carotenoid assay using a Varian Cary 50 ultraviolet–visible light (UV–Vis) spectrophotometer. The levels of Chla, Chlb, ChlT, and carotenoids were calculated. Soluble sugars (glucose) in leaves were extracted as described by Reference [23]. The change in absorbance was continuously followed at 340 nm using an Anthos hat II microtiter-plate reader (AnthosLabtec Instrument, Hanau, Germany).

2.3. *Statistical Analyses*

Data ($n = 4$) were examined by one-way ANOVA. Multiple comparisons of the means of data between different salinity treatments within the plants were performed using Duncan's test at the $p < 0.05$ significance level (all tests were performed with the SPSS program version 17.0 for Windows).

2.4. *Soil*

Table 1 shows the soil texture and soil parameters before the experiment. According to the FAO, based on the United States Department of Agriculture (USDA) particle-size classification, the soil texture was sandy clay loam. The soil parameters show that the range in the soil's pH value was slightly alkaline and that the electrical conductivity (EC_s) was 1.1 $dS{\cdot}m^{-1}$ (non-saline soil) at 25 °C.

Table 1. Soil parameters before the experiment.

Soil Texture		Soil Parameters			
Sand (%)	58.9	Field capacity θ_{fc} ($m^3{\cdot}m^{-3}$)	0.22	pH (H_2O)	7.7
Silt (%)	18	Wilting point θ_{wp} ($m^3{\cdot}m^{-3}$)	0.12	EC_e* ($dS{\cdot}m^{-1}$)	1.1
Clay (%)	24.1	Available soil water θ_{aw} ($m^3{\cdot}m^{-3}$)	0.12		
Classification: Sandy clay loam		Bulk density ($g{\cdot}cm^{-3}$)	1.41		

EC_e*—Electrical conductivity of the extract of a saturated soil paste ($dS{\cdot}m^{-1}$).

Table 2 shows the volumetric soil water content (m^3 water/m^3 soil; $m^3{\cdot}m^{-3}$) just before the water application. The volumetric soil water content in soil ranged between 0.20 and 0.15 $m^3{\cdot}m^{-3}$.

Table 2. Volumetric soil water content (m^3 water/m^3 soil; $m^3 \cdot m^{-3}$) just before the water application.

Treatment	Determination	Θ ($m^3 \cdot m^{-3}$)
T1	$\theta_{wp} + 0.70 \times \theta_{aw}$	$\Theta_1 = 0.20$
T2	$\theta_{wp} + 0.50 \times \theta_{aw}$	$\Theta_2 = 0.17$
T3	$\theta_{wp} + 0.30 \times \theta_{aw}$	$\Theta_3 = 0.15$

2.5. Climate Condition in Greenhouse

The average climatic data during the experimental period in the greenhouse were as follows: maximal relative humidity, 88.4%; minimal relative humidity, 11.3%; maximal temperature, 45.8 °C; minimal temperature, 11.4 °C.

During the experimental period, the relative humidity of the greenhouse was increased, and the maximal temperature decreased.

3. Results and Discussion

3.1. Effect of Irrigation Water Regimes on Plant Growth

Table 3 shows the irrigation water regimes' effects on the *T. tetragonioides* growth (stem length, number of nodes, and number of leaves). A significant effect on the stem length can be seen. In the beginning of the experiment, the stem length of the crop showed very low variations between T1 and T2 treatments. During the last three weeks of the experimental period, the stem length increased showing equal differences between each treatment—T1 and T2, and T2 and T3 (Δ stem length ~0.5 cm). The number of nodes and number of leaves were also higher in treatment T1.

Table 3. Effect of irrigation water regimes on stem length, number nodes, and number of leaves of the species. Different letters within a column represent significant differences ($p \leq 0.05$).

Treatment	*Tetragonia tetragonioides*		
	Stem Length (cm)	Number of Nodes	Number of Leaves
T1	38.8 ± 1.9 a	22.5 ± 0.6 a	9.9 ± 0.58 a
T2	34.2 ± 0.5 b	18.1 ± 0.6 b	8.1 ± 0.37 b
T3	29.3 ± 1.5 c	19.2 ± 0.8 b	9.5 ± 0.22 b

3.2. Fresh (FW) and Dry (DW) Weight of Crop

The fresh weight (FW) of *T. tetragonioides* species showed low variation among treatments (Figure 1). There was a low increase of the fresh weight of stem, leaves, and seeds in treatment T1. Surprisingly, the obtained results in treatment T3 were slightly higher than in treatment T2.

The obtained results of dry matter show that the stem, leaves, and seeds of treatment T1 were slightly higher than other treatments. There was very low variation of dry matter between T2 and T3 treatments (Figure 2).

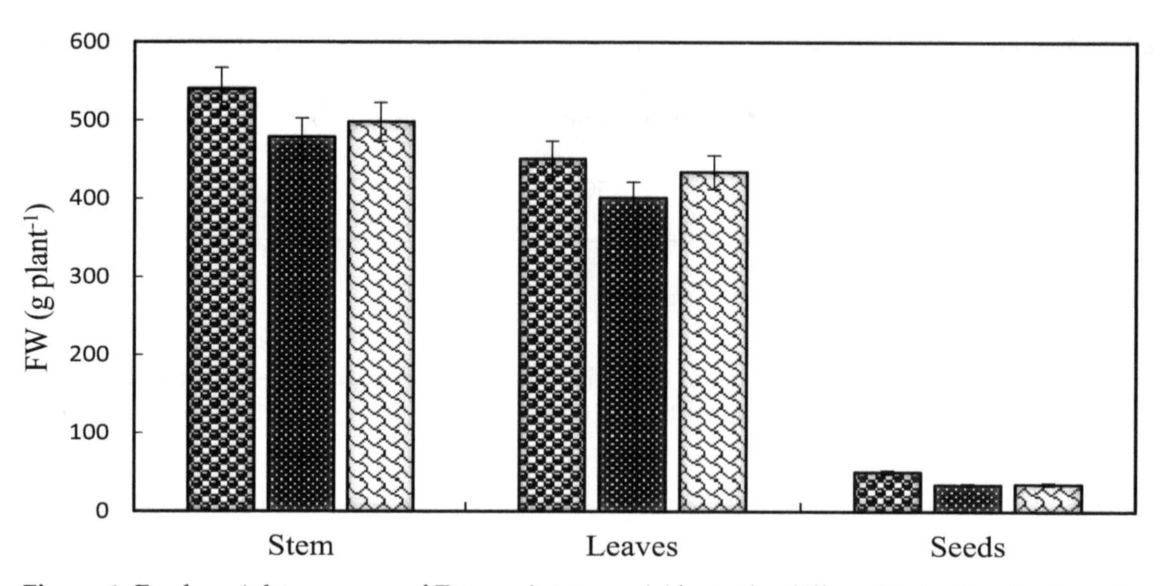

Figure 1. Fresh weight response of *Tetragonia tetragonioides* to the different irrigation treatments.

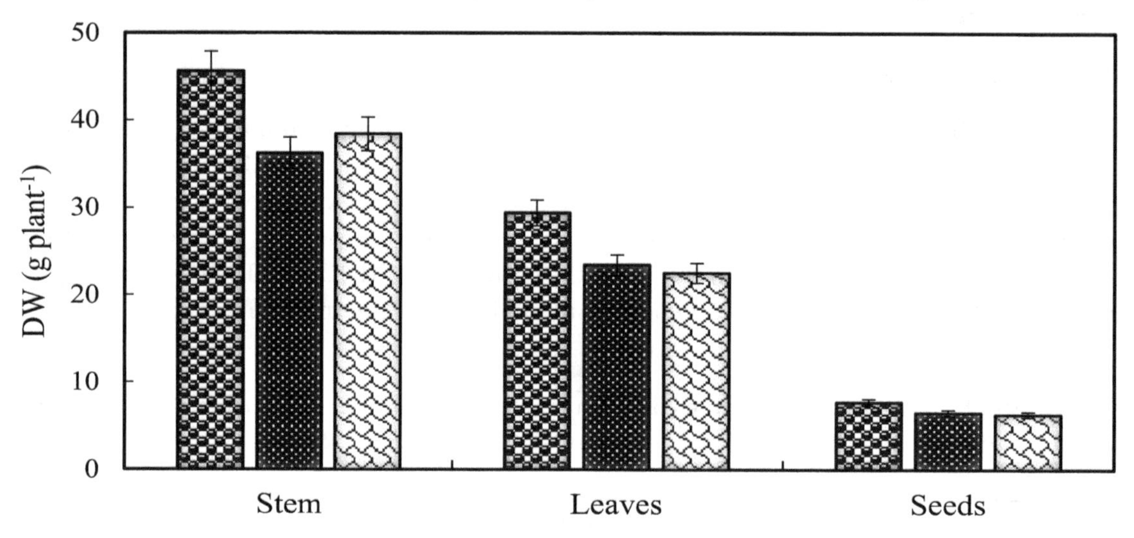

Figure 2. Dry weight response of *Tetragonia tetragonioides* to the different irrigation treatments.

3.3. Effect of Irrigation Water Regimes on Mineral Composition in Plant Leaves

Table 4 shows the effects of water application treatments on the mineral composition of *T. tetragonioides* leaves. In summary, the total nitrogen leaf content of the species showed low variation among treatments. There was an enhancement of chloride and sodium concentration with the decrease in water content. There was a general decrease in phosphorus, calcium, potassium, iron, and magnesium in the leaf content under drought conditions.

Table 4. Effect of irrigation water regimes on the leaf mineral composition.

Treatment	Leaf Mineral Composition (%)			
	Na	**Cl**	**Mg**	**Ca**
T1	3.4 ± 0.19	1.3 ± 0.07	0.43 ± 0.03	0.0006 ± 0.04
T2	4.3 ± 0.25	3.0 ± 0.22	0.36 ± 0.02	0.0004 ± 0.02
T3	4.4 ± 0.42	3.5 ± 0.09	0.35 ± 0.01	0.0004 ± 0.03
	N	K	P	Fe
T1	0.34 ± 0.02	4.2 ± 0.23	3.1 ± 0.02	0.0001 ± 0.02
T2	0.34 ± 0.01	3.7 ± 0.17	2.7 ± 0.04	0.0002 ± 0.03
T3	0.37 ± 0.01	4.1 ± 0.14	2.7 ± 0.02	0.0001 ± 0.01

3.4. Effect of Irrigation Water Regimes on Chlorophyll Content

The reaction of chlorophyll content of leaves of *T. tetragonioides* to the different water regimes is shown in Table 5. The results show that the chlorophyll content was higher in treatment T2 and lower in treatments T1 and T3. These results are in agreement with the findings obtained by Reference [24], where the minimum amounts of chlorophyll a, chlorophyll b, and total chlorophyll were obtained from the wettest and driest treatment in *Matricariachamomilla* L. potted plants. Similar results were obtained by References [25,26].

Table 5. Leaf chlorophyll content in leaf. DM—dry matter.

Treatment	Chlorophyll Content					
	C_a ($mg \cdot m^{-2}$)	C_b ($mg \cdot m^{-2}$)	C_{a+b} ($mg \cdot m^{-2}$)	C_a ($mg \cdot g^{-1}$; DM)	C_b ($mg \cdot g^{-1}$; DM)	C_{a+b} ($mg \cdot g^{-1}$; DM)
T1	232 ± 12.1	81 ± 5.8	313 ± 17.6	27 ± 1.4	9.2 ± 0.7	36 ± 1.9
T2	289 ± 7.3	110 ± 4.5	400 ± 11.4	32 ± 1.3	12.2 ± 0.5	45 ± 1.7
T3	254 ± 13	92 ± 5.6	346 ± 18	30 ± 2	10.7 ± 0.6	41 ± 2.1

3.5. Effect of Irrigation Water Regimes on Carotenoid Content in Leaves

Carotenoids which exist in all higher plants are synthesized and located in the chloroplast along with the chlorophyll. Table 6 shows the carotenoid content of the leaves of *T. tetragonioides* under different irrigation water regimes. The maximum leaf carotenoid content was 8.44 $mg \cdot g^{-1}$ DW in treatment T2. In wetter and drier treatments (T1 and T3), the carotenoid content was lower, with values of 7.2 and 7.9 $mg \cdot g^{-1}$ DW, respectively. Lower carotenoid content was also obtained for stress water regimes of some fenugreek varieties [27]. Moreover, the leaf carotenoid levels of green beans decreased, which was attributed to water stress; the vegetation index (NDVI) then showed the highest correlations with the chlorophyll (a, b, and total) and carotene content of leaves [28].

Table 6. Carotenoid leaf content of the species.

Treatment	Leaf Carotenoid Content	
	Car ($mg \cdot m^{-2}$)	Car ($mg \cdot g^{-1}$; DW)
T1	62.9 ± 3.2	7.2 ± 0.3
T2	75.8 ± 2.2	8.4 ± 0.4
T3	67.6 ± 4.1	7.9 ± 0.5

3.6. Effect of Irrigation Water Regimes on Soluble Carbohydrates Content

The irrigation water regimes had a slight effect on the soluble carbohydrate content on leaves of the species *T. tetragonioides*. The glucose and soluble carbohydrate content in leaves increased in the wet (T1) and dry (T3) treatments: glucose, 0.58 and 0.57 $mg \cdot mL^{-1}$, respectively; soluble carbohydrates, 1.71 and 1.67 mg, respectively. These results are confirmed by Reference [29]. There was a decrease in glucose (0.54 $mg \cdot mL^{-1}$) and soluble carbohydrate (1.59 g) content in leaves in the medium (T2) treatment (Table 7).

Table 7. Soluble carbohydrates content of leaves.

Treatment	Soluble Carbohydrates				
	Glucose ($mg \cdot mL^{-1}$)	Area (cm^2)	Soluble Carbohydrates (mg)	DW (cm^2)	Soluble Carbohydrates (g)
T1	0.57 ± 0.05	3.42	1.67 ± 0.14	0.001	1.9 ± 0.15
T2	0.54 ± 0.03	3.42	1.59 ± 0.08	0.001	1.8 ± 0.09
T3	0.58 ± 0.03	3.42	1.71 ± 0.09	0.001	2.0 ± 0.12

3.7. Yield of Species

T. tetragonioides produced significant amounts of dry matter, which ranged from 82.7 to 66.1 g·plant^{-1}. The partition of the plant dry matter to plant organs was changed by the effect of the irrigation water regimes (Table 8). The fact that the species was irrigated during the vegetation period T1 (70%, wet treatment) significantly increased the dry biomass of the species at the harvest time, averaging 6616 kg·ha^{-1}. The dry matter of the species decreased when the soil water decreased in treatments T2 (50%, medium treatment) and T3 (30%, dry treatment). There was no significant difference between treatments. The obtained results confirmed that the species *T. tetragonioides* is tolerant to drought conditions. The yield of the crop shows that the drought had less effect than the salinity (6616–5288 kg DM·ha^{-1}). These results are confirmed by the previous study of Reference [3].

Table 8. Effect of irrigation water regimes on yield of the species. FW—fresh weight; DW—dry weight; FM—fresh matter.

Treatment	*Tetragonia tetragonioides*				
	FW (g·plant^{-1})	**DW (g·plant^{-1})**	**Yield (%)**	**FM (kg·ha^{-1})**	**DM (kg·ha^{-1})**
T1	1041.2 ± 12	82.7 ± 4	7.8 ± 0.3	83,284 ± 967	6609 ± 329
T2	913.7 ± 23	66 ± 3	7.2 ± 0.5	73,094 ± 1805	5289 ± 248
T3	966.2 ± 22	67.3 ± 3	6.8 ± 0.3	77,300 ± 1787	5377 ± 242

4. Conclusions

The experimental results showed several effects of the water irrigation regimes on the growth, mineral composition, and photosynthetic pigments of *T. tetragonioides*, as listed below.

- Plant growth (stem, leaves, and seeds) increased slightly with an enhancement of the water level (near the field capacity), and the growth difference between the drier water regimes was very low. This increase was probably due to the increase of stomatal conductance and, consequently, transpiration and CO_2 fixation were higher. Hence, it is not surprising that experimental results, in which the only variable was water application, agree quite well with this supposed theory.

- Leaf mineral composition of chloride and sodium are the main responsible ions for soil salinization and alkalization, respectively, in arid and semi-arid regions, enhanced by the decrease in soil water content. The content was very high in relation to other plants, showing its high capacity as a salt-removing species.

- There was a generally low decrease in phosphorus, calcium, potassium, iron, and magnesium in leaf content under drought conditions, probably due to the chloride and potassium competition.

- The total nitrogen leaf content of species showed very low variation, probably due to the same fertigation for all irrigation treatments.

-The minimum carotenoid amounts of chlorophyll a, chlorophyll b, and total chlorophyll were obtained from the wettest and the driest treatment in *T. tetragonioides* plants, probably due to higher plant senescence provoked by these regimes.

- The glucose and soluble carbohydrate contents of leaves increased in the driest treatments and had enhanced tolerance to drought conditions.

- The yield of the species increased in the wettest and the driest treatments.

In conclusion, it can be suggested that *T. tetragonioides* is a species tolerant to drought conditions. Its capacity as a halophyte and salt-removing species when the soil water content decreases was shown, suggesting its use in arid and semi-arid regions. Moreover, growth and yield differences in the various irrigation regimes were very low, which suggests another important advantage of these species—its cultivation under dry conditions, when used as a leafy vegetable for human consumption or for animal feeding. Nevertheless, more research is needed in order to test plant development under drier conditions in arid and semi-arid climates.

Author Contributions: The paper is the result of the collaboration among all authors; however, G.B. and J.B. contributed to the all sections. B.O. contributed to the sections on Effect of Irrigation Water Regimes on Mineral Composition in Plant Leaves.

References

1. Costa, M.; Beltrao, J.; De Brito, J.C.; Guerrero, C. Turfgrass plant quality response to different water regimes. *WSEAS Trans. Environ. Dev.* **2011**, *7*, 167–176.
2. FAO. *The State of the World's Land and Water Resources for Food and Agriculture (SOLAW)—Managing Systems at Risk*; Food and Agriculture Organization of the United Nations and Earthscan: London, UK, 2011.
3. Bekmirzaev, G.; Beltrao, J.; Neves, M.A.; Costa, C. Climatical changes effects on the potential capacity of salt removing species. *Int. J. Geol.* **2011**, *5*, 79–85.
4. Szabolcs, I. Salt affected soils as the ecosystem for halophytes. In *Halophytes as a Resource for Livestock and for Rehabilitation of Degraded Lands*; Squires, V.R., Ayoub, A.T., Eds.; Kluwer Academic Publisher: London, UK, 1992; pp. 19–24.
5. Bartels, D.; Sunkar, R. Drought and salt tolerance in plants. *Crit. Rev. Plant. Sci.* **2005**, *24*, 23–58. [CrossRef]
6. Ansari, M.I.; Lin, T.P. Molecular analysis of dehydration in plants. *Int. Res. J. Plant. Sci.* **2010**, *1*, 21–25.
7. Boyer, J.S. Plant Productivity and Environment. *Science* **1982**, *218*, 443–448. [CrossRef] [PubMed]
8. Bray, E.A.; Bailey-Serres, J.; Weretilnyk, E. Responses to abiotic stresses. In *Biochemistry and Molecular Biology of Plants*; Buchanan, B.B., Gruissem, W., Jones, R.L., Eds.; American Society of Plant Physiologists: Rockville, MD, USA, 2000; pp. 1158–1203.
9. Ben-Asher, J.; Beltrao, J.; Costa, M.; Anaç, S.; Cuartero, J.; Soria, T. Modelling the effect of sea water intrusion on ground water salinity in agricultural areas in Israel, Portugal, Spain and Turkey. *Acta Hortic.* **2000**, *573*, 119–128. [CrossRef]
10. Nedjimi, B.; Daoud, Y.; Touati, M. Growth, water relations, proline and ion content of in vitro cultured *Atriplexhalimus* subsp. *schweinfurthii* as affected by $CaCl_2$. *Commun. Biom. Crop. Sci.* **2006**, *1*, 79–89.
11. Beltrao, J.; Correia, P.J.; Costa, M.; Gamito, P.; Santos, R.; Seita, J. The influence of nutrients on turfgrass response to treated wastewater application, under several saline conditions and irrigation regimes. *Environ. Proc.* **2014**, *1*, 105–113. [CrossRef]
12. FAO. Land and Plant Nutrition Management Service. Available online: http://www.fao.org/ag/agl/agll/spush/ (accessed on 15 May 2008).
13. Hamidov, A.; Khamidov, M.; Beltrão, J. Application of surface and groundwater to produce cotton in semi-arid Uzbekistan. *Asian Australas. J. Plant. Sci. Biotechnol.* **2013**, *7*, 67–71.
14. Yensen, N.P. Halophyte uses for the twenty-first century. In *Ecophysiology of High Salinity Tolerant Plants*; Springer: Dordrecht, The Netherlands, 2008; pp. 367–396.
15. Hasanuzzaman, M.; Nahar, K.; Alam, M.; Bhowmik, P.C.; Hossain, M.; Rahman, M.M.; Prasad, M.N.V.; Ozturk, M.; Fujita, M. Potential use of halophytes to remediate saline soils. *BioMed Res. Int.* **2014**. [CrossRef]
16. Cameron, R.W.F.; Harrison-Murray, R.S.; Atkinson, C.J.; Judd, H.L. Regulated deficit irrigation—A means to control growth in woody ornamentals. *J. Hortic. Sci. Biotechnol.* **2006**, *81*, 435–443. [CrossRef]
17. Grieve, C.M.; Suarez, D.L. Purslane (*Portulacaoleracea* L.): A halophytic crop for drainage water reuse systems. *Plant. Soil* **1997**, *192*, 277–283. [CrossRef]
18. Asher, J.B.; Beltrao, J.; Aksoy, U.; Anac, D.; Anac, S. Controlling and simulating the use of salt removing species. *Int. J. Energy Environ.* **2012**, *6*, 360–369.
19. Neves, A.; Miguel, M.G.; Marques, C.; Panagopoulos, T.; Beltrão, J. The combined effects of salts and calcium on growth and mineral accumulation of *Tetragoniate tragonioides*—A salt removing species. *WSEAS Trans. Environ. Dev.* **2008**, *4*, 1–5.
20. Veihmeyer, F.J.; Hendrickson, A.H. The moisture equivalent as a measure of the field capacity of soils. *Soil Sci.* **1931**, *32*, 181–193. [CrossRef]
21. Radojevic, M.; Bashkin, V.N. *Practical Environmental Analysis*; The Royal Society of Chemistry: Cambridge, UK, 1999.
22. Lichtenthaler, H.K. Chlorophylls and carotenoids: Pigments of photosynthetic biomembranes. *Meth. Enzymol.* **1987**, *148*, 350–382.

23. Dubois, M.; Gilles, K.A.; Hamilton, J.K.; Rebers, P.T.; Smith, F. Calorimetric method for determination of sugars and related substances. *Anal. Chem.* **1956**, *28*, 350–356. [CrossRef]
24. Pirzad, A.; Shakiba, M.R.; Zehtab-Salmasi, S.; Mohammadi, S.A.; Darvishzadeh, R.; Samadi, A. Effect of water stress on leaf relative water content, chlorophyll, proline and soluble carbohydrates in *Matricaria chamomilla* L. *J. Med. Plants Res.* **2011**, *5*, 2483–2488.
25. Bradford, K.J.; Hsiao, T.C. *Physiological Responses to Moderate Water Stress. Physiological Plant Ecology II*; Volume 12/B of the Series Encyclopedia of Plant Physiology; Springer-Verlag: Berlin, Germany, 1982; pp. 263–324.
26. Chartzoulakis, K.; Noitsakis, B.; Therios, I. Photosynthesis, plant growth and dry matter distribution in kiwifruit as influenced by water deficits. *Irrig. Sci.* **1993**, *14*, 1–5. [CrossRef]
27. Hussein, M.M.; Zaki, S.S. Influence of water stress onphotosynthesis pigments of some Fenugreek varieties. *J. Appl. Sci. Res.* **2013**, *9*, 5238–5245.
28. Köksal, E.S.; Üstün, H.; Özcan, H.; Güntürk, A. Estimating water stressed dwarf green bean pigment concentration through hyperspectral indices. *Pak. J. Bot.* **2010**, *42*, 1895–1901.
29. Redillas, M.C.; Park, S.H.; Lee, J.W.; Kim, Y.S.; Jeong, J.S.; Jung, H.; Bang, S.W.; Hahn, T.R.; Kim, J.K. Accumulation of trehalose increases soluble sugar contents in rice plants conferring tolerance to drought and salt tress. *Plant. Biotechnol. Rep.* **2012**, *6*, 89–96. [CrossRef]

4

Studying Crop Yield Response to Supplemental Irrigation and the Spatial Heterogeneity of Soil Physical Attributes in a Humid Region

Amir Haghverdi [1,*], Brian Leib [2], Robert Washington-Allen [3], Wesley C. Wright [2], Somayeh Ghodsi [1], Timothy Grant [2], Muzi Zheng [2] and Phue Vanchiasong [2]

[1] Department of Environmental Sciences, University of California, Riverside, 900 University Avenue, Riverside, CA 92521, USA; somayehg@ucr.edu

[2] Department of Biosystems Engineering & Soil Science, University of Tennessee, 2506 E.J. Chapman Drive, Knoxville, TN 37996-4531, USA; bleib@utk.edu (B.L.); wright1@utk.edu (W.C.W.); tgrant7@vols.utk.edu (T.G.); mzheng3@vols.utk.edu (M.Z.); vanchias@gmail.com (P.V.)

[3] Department of Agriculture, Nutrition, and Veterinary Science (ANVS), University of Nevada, Reno, Mail Stop 202, Reno, NV 89557, USA; rwashingtonallen@unr.edu

* Correspondence: amirh@ucr.edu

Abstract: West Tennessee's supplemental irrigation management at a field level is profoundly affected by the spatial heterogeneity of soil moisture and the temporal variability of weather. The introduction of precision farming techniques has enabled farmers to collect site-specific data that provide valuable quantitative information for effective irrigation management. Consequently, a two-year on-farm irrigation experiment in a 73 ha cotton field in west Tennessee was conducted and a variety of farming data were collected to understand the relationship between crop yields, the spatial heterogeneity of soil water content, and supplemental irrigation management. The soil water content showed higher correlations with soil textural information including sand ($r = -0.9$), silt ($r = 0.85$), and clay ($r = 0.83$) than with soil bulk density ($r = -0.27$). Spatial statistical analysis of the collected soil samples (i.e., 400 samples: 100 locations at four depths from 0–1 m) showed that soil texture and soil water content had clustered patterns within different depths, but BD mostly had random patterns. ECa maps tended to follow the same general spatial patterns as those for soil texture and water content. Overall, supplemental irrigation improved the cotton lint yield in comparison to rainfed throughout the two-year irrigation study, while the yield response to supplemental irrigation differed across the soil types. The yield increase due to irrigation was more pronounced for coarse-textured soils, while a yield reduction was observed when higher irrigation water was applied to fine-textured soils. In addition, in-season rainfall patterns had a profound impact on yield and crop response to supplemental irrigation regimes. The spatial analysis of the multiyear yield data revealed a substantial similarity between yield and plant-available water patterns. Consequently, variable rate irrigation guided with farming data seems to be the ideal management strategy to address field level spatial variability in plant-available water, as well as temporal variability in in-season rainfall patterns.

Keywords: farming data; precision agriculture; site-specific irrigation

---

1. Introduction

1.1. Supplemental Irrigation Management in Humid Regions

Irrigated agriculture has been playing a globally significant role in providing roughly one-third of the total food and fiber supply [1]. While irrigated acreage is shrinking in some arid regions in the

US due to increasing competition for water, supplemental irrigation is expanding in humid regions as a means to avoid unpredicted periods of water stress and maintain high yields [2]. For example, in west Tennessee, row crop irrigation has expanded rapidly from twenty-five center pivot irrigation systems installed in 2007 to 270 systems installed in 2012. This represents an expansion of 16,000 ha of cropland per year under supplemental irrigation [3], which necessitates an essential demand to study supplemental irrigation management of different crops in this region.

Precipitation is the main source of moisture in west Tennessee. However, severe in-season drought conditions for short periods are likely to occur, which could substantially reduce yields under rainfed agricultural practices. Supplemental irrigation is an irrigation strategy that attempts to maintain maximum yield production by irrigating during periods of insufficient rainfall to fulfill the crop water requirements. The application of supplemental irrigation management is a complex problem in west Tennessee, where precipitation patterns are temporally variable within and across cropping seasons and interact with the spatial mosaic of the physical and hydrological attributes of alluvial and windblown loess deposited soils. Soil properties, such as texture and bulk density, greatly affect soil water retention and movement and govern readily available soil water for crop irrigation management. Excess water content within the root zone could occur if irrigation adds to unpredicted rainfall events. This may cause insufficient aeration and consequent yield reduction. Moreover, runoff and deep percolation may lead to accelerated nutrient loss and soil erosion that in turn, increases the risk of contamination of nearby surface and/or groundwater.

Crop yield has been proven to be strongly related to soil physical properties. For example, Ref. [4] considered plant-available water (PAW: volumetric water content between the field capacity and the permanent wilting point within the root zone) as an input predictor of the wheat yield. They reported PAW as one of the dominant factors governing the spatiotemporal variation of yields. Soil texture was discussed by [5] as one of the greatest factors affecting the cotton yield. They found a relatively stable spatial pattern of yield over time, although yield and soil properties had stronger relationships during dry seasons than wet seasons. Graveel et al. [6] studied the response of corn to variations in soil erosion and sandy and silt textured profiles in west Tennessee and found a substantial difference in yield.

Cotton is a major crop in west Tennessee that is grown in more than 15 states and is vital to the US economy because it is a critical export-oriented product [7]. Currently, some 40% of US cotton is under irrigation, with the area expanding throughout the mid to southern US. Given the limited water resources in many cotton-growing areas, a considerable amount of research has recently been performed on cotton irrigation to improve the water use efficiency [8]. However, inconsistent cotton yields have been observed in response to irrigation in the humid portion of the US [9]. Suleiman et al. [10] studied the use of cotton deficit irrigation in a humid climate using FAO's 56-crop coefficient method in Georgia and suggested establishing a 90 % irrigation threshold for the full irrigation of cotton in humid climates. Bajwa and Vories [11] evaluated the cotton canopy response to irrigation in a moderately humid area in Arkansas and found that under wet conditions, excessive irrigation decreased the yield of cotton lint. A similar result was reported by [12], who also found that excessive rainfall limited the yields from irrigation. Gwathmey et al. [13] conducted a four-year supplemental irrigation study in Jackson, Tennessee, and found a 38% improvement in lint yields at a 2.54 cm wk^{-1} supplemental irrigation rate compared to three of four years of the rainfed irrigation scenario. Grant et al. [14] used a surface drip irrigation system to investigate the response of the cotton yield to irrigation across different soil types with different PAW. This study illustrated that uniform irrigation is not the optimum management decision for the cotton wherever field-level soil heterogeneity affects the spatial distribution of PAW.

1.2. Farming Data and Precision Agriculture

Traditionally, irrigation studies were limited to small plots at research stations, mostly due to economic and computational limitations. Additionally, contemporary constraints to irrigation studies include the personnel time and expense for data collection, as well as the limitations of conventional computing infrastructure and statistical methods to analyze the increasingly larger spatiotemporal datasets that have inherent noise and uncertainty. In west Tennessee, the inherent heterogeneity and the spatiotemporal changes in soil and weather-related attributes of the region make it hard to extrapolate the results of design-based experiments on small plots to real field conditions. Supplemental irrigation scheduling is a site-specific irrigation management question where each field has its own irrigation management challenge that requires unique solutions. On-farm experimentation is an alternative for design-based experiments, since collecting site-specific information is becoming more and more common and affordable in US agriculture.

In contemporary agriculture, precision farming enables farmers to locally collect various site-specific information, such as the yield and soil apparent electrical conductivity (ECa). Crop yield maps provide valuable quantitative information on crop production, change in production, and the response of crop production to different agricultural inputs, including irrigation and fertilizer. Soil survey maps; soil sampling; on-the-go sensors; and remote sensing from field, airborne, and satellite sensors are the most widely used methods to obtain information on the spatial distribution of different soil attributes [15]. Soil sampling at the field-level provides valuable information on the spatial variation of soil attributes, but collecting this data has become laborious and expensive. ECa is a proxy for less accessible soil attributes, including soil texture and soil available water [16], and thus has created substantial interest in its use for soil mapping and management zone delineation in precision agriculture. ECa is measured in a simple and inexpensive way, where an electrical current is induced into the soil while the field is traversed. However, there are some inconsistencies in the literature concerning factors that affect the variability of ECa in non-saline fields [17]. This suggests the need to investigate the practical utility of using ECa for site-specific management in different regions, particularly because most of the supporting ancillary datasets including topographic edaphic features (e.g., elevation, slope, and aspect) are freely available. If not, these site-specific attributes can be measured and mapped without spending a considerable amount of time and money. Recently, new wireless technologies have enabled progressive farmers to remotely and continuously monitor soil properties over time, including soil temperature, soil water content, and soil matric potential.

Consequently, this study was carried out to understand the relationship between the spatial heterogeneity of soil and crop yields to better inform the management of site-specific supplemental irrigation in west Tennessee. The objectives of this study were to conduct an on-farm experiment and analyze yield maps to:

1. Assess the impact of the spatial heterogeneity of soil water content on the pattern of yield using on-farm data that was collected by the farmer's soil moisture sensors and yield monitor systems;
2. Compare the cotton lint yield under different supplemental irrigation regimes across different soil types;
3. Assess the temporal stability of low/high yield zones by combining the measured historical yield data of different crops with available cotton yield data.

2. Materials and Methods

2.1. Study Area

The study area was a 73 ha irrigated field that is located in southwestern Dyer County in west Tennessee along the Mississippi river (Figure 1). The field was equipped with two center pivot irrigation systems that were used for the irrigation of no-till cotton during each cropping season. The field is on Mississippi river terrace alluvial deposits from which Robinsonville loam and fine sandy loam, Commerce silty clay loam, and Crevasse sandy loam soils have been produced (Figure 1). Figure 2 illustrates the long-term variability in regional climate. The mean monthly growing season precipitation and temperature is 97-mm month^{-1} and 21 °C from May to November, respectively (Figure 2). Rainfall is relatively high, even in dry years. Temperature changes are less pronounced and to some extent, inversely proportional to rainfall. The supplemental irrigation strategy has been growing in this region since rainfall events are not usually temporally well-scattered to fulfill the crop water requirement over the entire growing season.

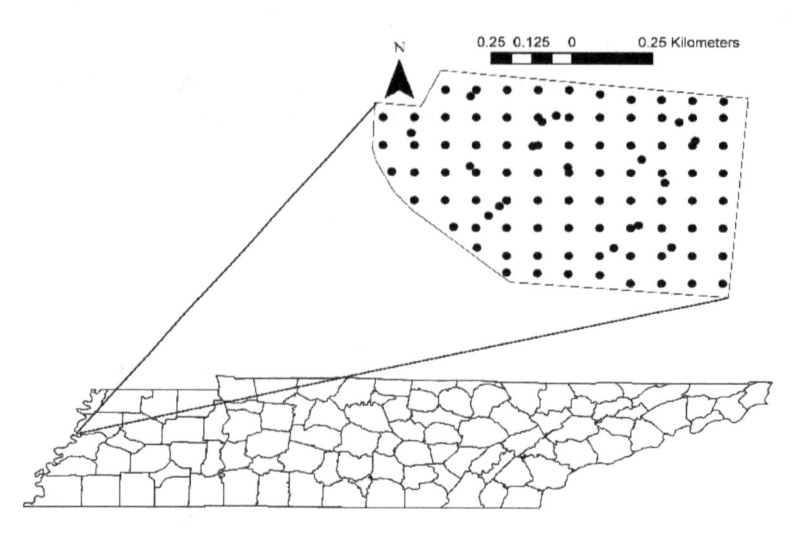

Figure 1. The 73-ha supplemental irrigation study field is located in southwestern Dyer County in west Tennessee along the Mississippi river. Soil samples were collected at four depths from 0–1 meter at 100 locations.

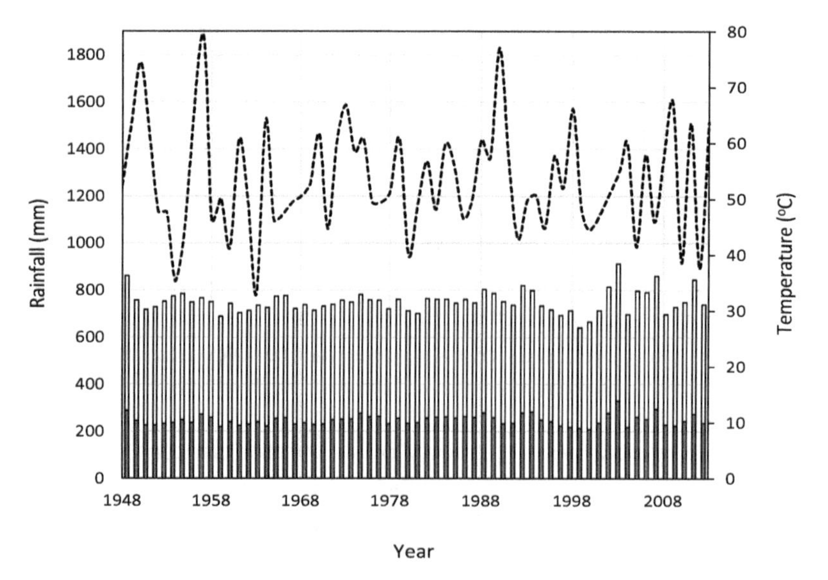

Figure 2. The long-term climatic variation in rainfall (dashed line) and temperature (column) in west Tennessee. Temperature columns show the mean monthly minimum (in black) and the mean monthly maximum (in white).

2.2. Soil Data Collection and Lab Analysis

Haghverdi et al. [18] described the soil data collection where one hundred undisturbed samples (100 cm deep) were collected by a truck-mounted soil sampler between 21 and 22 March 2014 (Figure 1). Some 86 of these samples were collected using a grid sampling scheme where samples were about 100-m apart (i.e., half the mean semivariogram range of proxies). The rest of the samples (=14) were randomly collected from underneath the center pivot circles. The field sampling occurred after rainfall events, when the soil water status was assumed to be close to the field capacity.

Each 67-mm diameter core was sub-sampled at four depths between 0–100 cm in 25-cm increments, i.e., 0–25 cm, 25–50 cm, 50–75 cm, and 75–100 cm, with adjustments in respect to the available horizons. The mean depth across samples approximated 25 cm for all the layers. Hereafter, the word "layer" is used to describe subsamples rather than real soil horizons. The soil texture of each depth was estimated in the laboratory using a hydrometer [19]. The soil water content was estimated by subtracting oven-dried weights from wet weights. Bulk density (BD) was estimated as the oven dry weight to volume of each subsample. ECa was collected using a Veris 3100 (Veris Technologies, Salina, KS, USA) instrument on March 20, 2014 with 10 m and 20 m spacing between points in the same row and adjacent rows, respectively. The Veris 3100 has six rolling coulters for electrodes and collects two simultaneous ECa measurements from shallow (~0–30 cm) and deep depths (~0–90 cm).

2.3. *Descriptive and Spatial Analysis of Soil Properties*

The correlation between the volumetric water content at the time of sampling and soil texture, i.e., sand, silt, and clay percentages, and bulk density was investigated. A soil texture triangle was plotted for each of the four depths, with each depth layer being approximately 25-cm thick. The relationship between ECa data and soil physical information, obtained from soil samples, was studied. To match ECa and soil basic data, the ECa data were interpolated to each sample using an ordinary kriging approach [20].

The spatial analysis was done using ARCGIS 10.2.2 [21]. To examine the spatial autocorrelation of the attributes, the semivariogram (Equation (1)) and Global Moran's I statistic (Equation (2), [22]) were obtained as follows:

$$\gamma(\mathbf{h}) = \frac{1}{2N(\mathbf{h})}\left\{\sum_{i=1}^{N(\mathbf{h})}[Z(\mathbf{x}_i+\mathbf{h})-Z(\mathbf{x}_i)]^2\right\} \quad (1)$$

where $\gamma(\mathbf{h})$ is the semivariance; $\mathbf{h}$ is the interval class; $N(\mathbf{h})$ is the number of pairs separated by the lag distance; and $Z(\mathbf{x}_i)$ and $Z(\mathbf{x}_i+\mathbf{h})$ are measured attributes at spatial location i and $i + \mathbf{h}$, respectively. The nugget effect, sill, and range are the basic parameters of a semivariogram to describe the spatial structure. The nugget effect mostly represents sampling/measurement errors and variation at scales smaller than the sampling interval. The total variance is called the sill and the range is the maximum distance at which variables are spatially dependent.

The Global Moran's I statistic is calculated as:

$$I = \frac{n}{\sum_{i=1}^{n}\sum_{j=1}^{n} w_{i,j}} \times \frac{\sum_{i=1}^{n}\sum_{j=1}^{n} w_{i,j} z_i z_j}{\sum_{i=1}^{n} z_i^2} \quad (2)$$

where z is the deviation of an attribute from its mean, $w_{i,j}$ is the spatial weight between the ith and jth point, and n is equal to the number of points. Moran's I is used to measure the degree of spatial autocorrelation or trend based on both feature locations and feature values simultaneously. Given a set of features and an associated attribute, it evaluates whether the pattern expressed is clustered, dispersed, or random [22]. The null hypothesis of this analysis states that the attribute being analyzed is randomly distributed among the features in the study area. Ordinary kriging was applied to samples of the ECa to generate maps that were compared and assessed against each other. A higher positive Moran's Index for an attribute indicates a stronger spatial structure. The z-score changes in line with the Moran's Index. A z-score from -1.65 to 1.65 shows that the spatial pattern is not significantly different than a random one. A z-score less than -1.65 is an indicator of a dispersed process, while a z-score greater than 1.65 displays a spatially clustered attribute.

2.4. *On-Farm Irrigation Experiment*

There were two center pivot systems available for irrigation within the 73-ha field. The on-farm experiment was conducted for two years and designed to study the supplemental irrigation-cotton lint yield relationship across different soil types. The farmer used a no-tillage method to plant

'PHY375' cotton variety on 30 May 2013 and 'Stoneville 4946' on 5 May 2014. The farmer used soil test recommendations for applications of variable rate potassium (K) and phosphorus (P). However, nitrogen was applied uniformly. Crop pest management was implemented following state extension recommendations and the field was harvested on 2 and 3 December 2013 and in the second year on 18–20 October 2014.

Throughout the experiment, we used the Management of Irrigation Systems in Tennessee (MOIST) program (http://www.utcrops.com/irrigation/irr_mgmt_moist_intro.htm) to discuss the efficiency of irrigation management with the farmer. MOIST is an irrigation decision support tool that delivers irrigation recommendations by simultaneously measuring and monitoring soil water status and calculating water balance through a deployed wireless soil sensor network. An on-farm weather and soil monitoring station contained a number of METER Devices (METER Group, Inc., Pullman, WA, USA), including an EM50G remote data logger, a VP-3 temperature and relative humidity sensor, an ECRN-100 high-resolution rain gauge, and a pyranometer: a solar radiation sensor, was installed in 2013 and run through 2014 using the MOIST program. Three additional stations with rain gauges and soil moisture sensors were added in 2014. Each station also had two MPS-2 soil matric potential and temperature sensors (METER Group, Inc., Pullman, WA, USA) installed at approximately 10 and 46 cm depths to monitor the soil water status. MOIST calculates the daily reference evapotranspiration (ET_{ref}) using Turc's 1961 equation (developed for regions with relative humidity > 50%, [23]) as follows [24]:

$$ET_{ref} = 0.013 \times \left(\frac{T}{T+15}\right) \times (R_s + 50) \quad (3)$$

where ET_{ref} is the daily reference evapotranspiration (mm d^{-1}), R_s is the daily solar radiation (Cal cm^{-2} d^{-1}), and T is the daily mean air temperature (°C). The data for each station were recorded once per hour, stored in the logger, and then transmitted to a web-based interface. The farmer managed irrigation applications. At the same time, we wanted to make sure that he was provided with sufficient information to irrigate appropriately, while maintaining statistical variability of the supplemental irrigation water applied (IW) across the field to fulfill our research purpose. In 2014, we started sending out weekly MOIST reports to the farmer. The report contained information on the soil water status and irrigation scheduling based on soil sensors and water balance calculations.

Two different methods were used to create irrigation application zones across the field: programming the two pivots (pie shape zones) and partially swapping the sprinkler nozzles (arc shape zones). Table 1 summarizes the information on the irrigation programs at each pivot. The farmer's routine irrigation schedule was 15.50 mm and 9.91 mm per revolution for the east and west pivots, respectively. The east (west) pivot panel was programmed to apply ±5.08 (±1.78) mm variation in irrigation per revolution on some pie shape zones. The control panels of pivots were Valley Select2 (Valmont Industries, Inc.) that were programmable for up to nine different pie shape zones. The program changes the irrigation rate by adjusting the pivot's travel speed, where speeding up the pivot causes less irrigation and slowing it down applies additional irrigation. Based on the pivots' characteristics and soil spatial variation, multiple banks of sprinklers were also selected and re-nozzled to form arc-shape irrigation zones. The center pivots can be operated both clockwise and counterclockwise, but were programmed only in the clockwise direction (Table 1).

Table 1. Detailed information on the supplemental irrigation programs for the two center pivots within the 73-ha supplemental irrigation field that is located in southwestern Dyer County in west Tennessee for one revolution.

	East Pivot			West Pivot		
Program Sector	**Start Angle [1] (degree)**	**Stop Angle (degree)**	**Depth of Water (mm)**	**Start Angle (degree)**	**Stop Angle (degree)**	**Depth of water (mm)**
1	90	110	10.41	275	315	9.91
2	110	0	15.49	315	335	11.68
3	0 [1]	20	20.57	335	355	8.38
4	20	40	10.41	355	235	9.91
5	40	70	15.49	235	255	11.68
6	70	90	20.57	255	275	8.38

[1] The zero degree was at north and pivots traveled clockwise.

We installed three Agspy (AquaSpy Inc., San Diego, CA, USA) soil moisture probes at three randomly selected points each year to monitor the soil water status, across pie-shape zones throughout the irrigation seasons. The AgSpy soil moisture capacitance probes were 1-m in length and obtained measurements at 10 depths at 0 to 100 cm, with 10 cm increments. The sensor output is a dimensionless number in the range 0 to 100, called the scaled frequency (*SF*), which is defined as:

$$SF = \frac{(F_a - F_s)}{(F_a - F_w)} \times 100 \quad (4)$$

where F_a is the frequency of oscillation in air (air count), F_s is the frequency of oscillation in soil (soil count), and F_w is the frequency of oscillation in water (water count). The F_a and F_w are calculated during the manufacturing of each sensor. The frequency of oscillation is related to the capacitance between sensor plates that is in turn influenced by the relative permittivity of the soil media. The relative permittivity of water is significantly greater than that of air and soil, thereby changes in soil water content will be detected by the sensor [25].

Table 2 summarizes irrigation and weather data for the 2013 and 2014 cropping seasons. The sensors were installed a couple of weeks after planting and were removed prior to the harvest period. Consequently, in situ data were not available for the whole cropping seasons. However, temperature and precipitation data from the closest weather station were obtained from the National Climate Data Center [26] to fill these gaps.

Table 2. Growing season summary of weather and supplemental irrigation data in the 73-ha study area for the 2013 and 2014 growing seasons, in comparison to the 30-year mean for these variables. The study area is located in southwestern Dyer County in west Tennessee.

Year	Variable	Month						
		May	**June**	**July**	**August**	**September**	**October**	**November**
2013	Rain, mm	23	150	190	95	79	112	63
	IW-East, mm			40	31	62		
	IW-West, mm			15	20	30		
	ET_{ref} [1], mm day^{-1}			4.33	4.43	3.92	2.49	1.28
2014	Rain, mm	143	172	56	124	120	18	
	IW-East, mm			62	31			
	IW-West, mm			20	30			
	ET_{ref} [1], mm day^{-1}	4.15	4.42	4.86	4.51	3.47	2.94	
30 year	Rain, mm	120	101	102	74	82	82	117
	Tmean, °C	21	25	27	26	22	16	10

[1] ET_{ref}: Reference evapotranspiration data that were calculated using the Turc equation (Equation (3)) from 19 July 2013 (7 May 2014) to 30 November 2013 (5 October 2014), IW: irrigation water applied by the farmer. The 30-year mean data collected from the closest weather station [26].

2.5. Multiyear Yield Data Analysis

To better understand the spatiotemporal dynamics of changes in yield, several years with different crops should also be considered [27]. Except for 2011, yield data from 2007 to 2012 (i.e., corn 2007, corn 2008, soybean 2009, cotton 2010, cotton 2012) had been collected by the producer using appropriate yield-monitor-equipped harvesters (Table 3). We combined these data with the 2013 and 2014 yield data to analyze the relative difference and temporal variance of yield on the study site under both rainfed and supplemental irrigation.

Table 3. Descriptive statistics on yield data (Mg ha^{-1}) at the field of study located in southwestern Dyer County in west Tennessee.

Year	Crop	Mean	SD
2007	Corn	7.137	4.158
2008	Corn	3.420	0.903
2009	Soybean	3.221	0.860
2010	Cotton	0.947	0.306
2012	Cotton	0.913	0.494
2013	Cotton	0.871	0.329
2014	Cotton	1.244	0.493

A multistep filtering process was designed and implemented in Microsoft Excel and ArcGIS 10.2.2 [21] to process the yield data and produce final yield maps. First, the yield maps were visually assessed using the farmer's knowledge of field conditions to identify potential unexpected patterns. Second, the data were color-coded based on harvest time to investigate the GPS tracks and movement of the harvester. Then, multiple filters were designed (e.g., using swath width, distance, speed of the harvester, change in speed) to remove outliers and erroneous data points. Last, yield data that were $>\pm 3$ standard deviations of the mean were assumed to be outliers and removed from the analysis. Then, the field was divided into 100 m^2 cells, and relative yield difference (Equation (5)) and yield temporal variance (Equation (6)) across years were calculated as follows [28]:

$$\overline{y}_i = \frac{1}{n}\sum_{k=1}^{n}\left[\frac{y_{i,k}-\overline{y}_k}{\overline{y}_k}\right] \times 100 \quad (5)$$

where n is the number of years with yield data available, $\overline{y}_i$ is the average percentage yield difference at cell i, $\overline{y}_k$ is the average yield (Mg ha^{-1}) across cells at year k, and $y_{i,k}$ is the yield value (Mg ha^{-1}) at cell i at year k.

$$\overline{\sigma_i^2} = \frac{1}{n}\sum_{k=1}^{n}\left(y_{i,k}-\overline{y}_{i,n}\right)^2 \quad (6)$$

where $\overline{\sigma_i^2}$ is the temporal variance at cell i, $\overline{y}_{i,n}$ is the average yield across the n years, and other variables are as previously defined.

3. Results and Discussion

3.1. Field-Level Soil Heterogeneity and Application of Soil ECa

Table 4 contains descriptive statistics for the measured soil properties. The BD had its highest mean value at the deepest layer, while the mean value was almost identical among other layers. The mean water content decreased with depth, while its standard deviation slightly increased. The higher water content in the surface layer is likely attributed to textural differences among layers and also rainfall events prior to the sampling, which built the moisture level up within the top layers, but perhaps did not fully penetrate to the deeper layers. The mean sand percentage increased with depth, which was inversely proportional to a decline in silt and clay. The mean and standard deviation of the deep ECa readings (27.52 ± 18.73) were greater than those of shallow readings (24.64 ± 10.66). The standard deviation among deep ECa reading was almost twice that of shallow readings. The same result was reported by [29] on differences between the standard deviation and distribution of shallow versus deep ECa readings.

Table 4. Descriptive statistics for selected soil properties from different soil sampling layers. Soil samples were collected at four depths from 0–1 meter at 100 locations.

Variable [1]	Layer	Min.	Max.	Mean	SD
BD, g cm^{-3}	1th	1.12	1.66	1.36	0.10
	2nd	1.11	1.70	1.35	0.12
	3rd	1.06	1.86	1.34	0.12
	4th	1.17	1.78	1.40	0.13
	total	1.06	1.86	1.36	0.12
WC, %	1th	10.75	59.74	28.35	7.43
	2nd	7.27	43.12	26.02	10.78
	3rd	5.98	42.38	21.64	11.08
	4th	5.67	45.32	20.18	11.15
	total	3.94	47.61	17.94	8.49
Sand, %	1th	8.77	88.25	38.07	20.11
	2nd	0.00	94.98	46.39	31.57
	3rd	2.50	95.70	61.38	31.10
	4th	5.46	96.86	69.90	26.09
Clay, %	1th	7.37	47.56	27.55	9.04
	2nd	2.50	56.60	22.18	14.17
	3rd	1.26	47.72	14.27	11.44
	4th	0.34	37.10	11.00	7.80
Silt, %	1th	4.38	54.06	34.38	12.75
	2nd	0.00	66.51	31.43	19.85
	3rd	0.00	72.81	24.35	21.76
	4th	0.00	69.23	19.10	19.83
ECa, mS m^{-1}	shallow	1.60	48.70	24.64	10.66
ECa, mS m^{-1}	deep	1.70	162.20	27.52	18.73

[1] BD: soil bulk density, WC: soil volumetric water content at the time of sampling, ECa: apparent soil electrical conductivity, SD: standard deviation.

The soil texture drastically varied across the field such that almost the entire soil texture triangle was covered by the collected samples, except for the silt and clay textures (Figure 3). There was a shift from fine to coarse textures by depth, with sand showing the greatest particle increase. The sand had the highest absolute correlation with the soil moisture of the samples, while there was a weak negative correlation between BD and the water content (Figure 4), showing that soil texture was the dominant attribute governing water content. There was a clear pattern in clay and silt percentage plots versus water content; the majority of the samples with lower clay and silt contents belonged to the deeper layers (a cluster of black dots in the soil texture triangle), while samples from the shallower layers were more likely to have higher clay and silt contents. The opposite was seen in the sand versus water content plot.

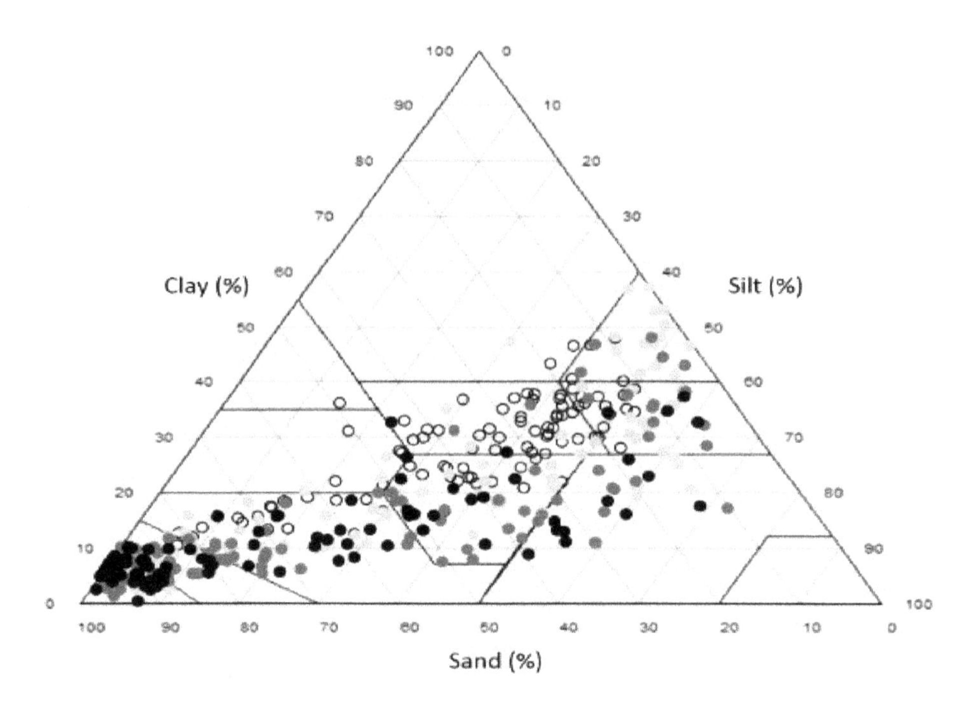

Figure 3. The textural distribution of soil samples from four different depths between 0 to 1 meter, where the darker colors correspond to the deepest depths. The samples were collected from a 73-ha two-pivot irrigation field that is located in southwestern Dyer County, Tennessee.

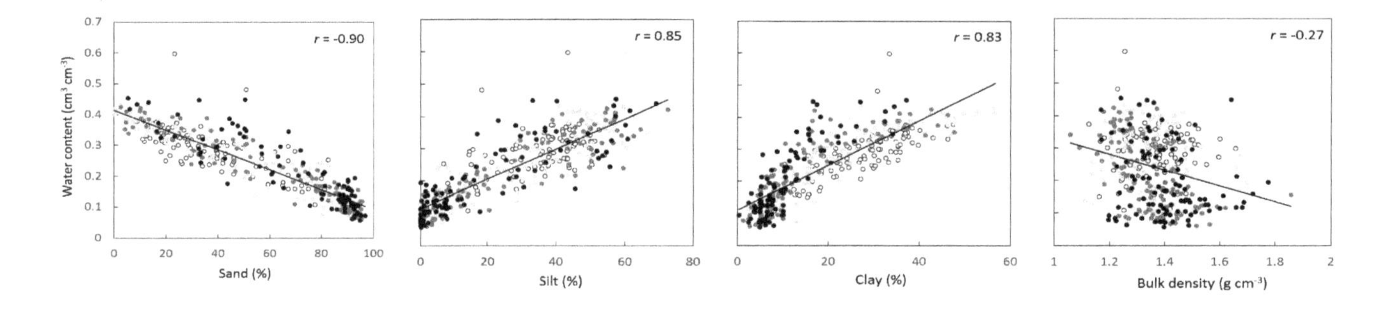

Figure 4. The relationship of 400 samples at four depths from 0–1 meter of soil texture (% Clay, Silt, & Sand) and bulk density (BD) to volumetric water content from a 73-ha two-pivot field in west Tennessee. The light to darker colors of the data markers correspond to 0–1 meter depths.

Table 5 presents the semivariogram and Global Moran's Index parameters for the selected attributes for each soil layer. The highest range did not belong to the same layer across soil properties. The average range varied from 200 m to 300 m among attributes, which was two to three times greater than the sampling intervals. The percent of nugget ranged from 18 to 50% among soil properties in the study of [30], who investigated the spatial variability of soil physical properties of alluvial soils in a 162 ha cotton field in Mississippi. This was somewhat similar to what was found for all the attributes except BD, which reached a nugget percent as high as 73 percent. The z-scores revealed all the attributes except BD within different layers had clustered patterns. BD only showed a clustered pattern at the third layer and had a random pattern at other layers.

Table 5. Semivariogram and Moran's I parameters of soil properties for different soil layers. Soil samples were collected at four depths from 0–1 meter at 100 locations.

Variable	Layer	Nugget	Sill	Range (m)	Moran's I	z-Score
* BD, g cm^{-3}	1th	0.008	0.011	526	0.087	1.181
	2nd	0.01	0.015	95	−0.086	−0.929
	3rd	0.011	0.016	280	0.137	1.802
	4th	0	0.017	100	0.091	1.221
	total	0	0.007	95	−0.007	0.038
WC, %	1th	0	44	100	0.175	2.266
	2nd	12	129	332	0.327	4.063
	3rd	0	131	206	0.284	3.545
	4th	56	125	212	0.284	3.556
	total	0	88	316	0.326	4.049
Sand, %	1th	115	446	360	0.421	5.213
	2nd	440	1119	300	0.365	4.510
	3rd	401	1037	219	0.320	3.978
	4th	413	717	260	0.300	3.747
Clay, %	1th	19	92	389	0.392	4.861
	2nd	123	215	428	0.239	3.016
	3rd	68	138	177	0.321	4.034
	4th	35	63	216	0.335	4.227
Silt, %	1th	39	174	334	0.382	4.740
	2nd	165	453	279	0.396	4.887
	3rd	211	484	200	0.270	3.366
	4th	6	10	341	0.266	3.332
ECa, mS m^{-1}	shallow	38	133	253	0.816	65.436
ECa, mS m^{-1}	deep	126	388	223	0.846	67.899

* BD: soil bulk density, WC: soil volumetric water content at the time of sampling, ECa: apparent soil electrical conductivity.

Figure 5 shows maps interpolated by ordinary kriging. The white strip expanding from the northwest to southeast of the field is a surface drainage pathway. There are three major sandy regions within the field of study at the surface layer located at: (i) surrounding pivot points at the eastern part of the field; (ii) south of the field, mostly outside of the irrigated zones; and (iii) northwest part of the field. The sequence of sand maps from the first to fourth layers illustrate how these coarse soil regions expanded across the field by depth such that sand covered the majority of the field in deeper layers. The sandy regions could be either river flood-induced sand boils or earthquake-induced sand blows. The vertical arrangement of soil textural components was not consistent across the field. The clay had its highest influence from 0–50 cm, yet sand was the dominant particle from 50–100 cm. The observed depth to sand during sampling ranged from 15–75 cm across the field, with an average depth of 40 cm for almost 40% of the sampling spots. For the rest of the samples (60%), either there was no clear immediate change from fine texture to coarse texture or sand appeared at the surface soil. The silt contributed highly in subsurface layers (25–75 cm), where it reached its highest quantity and SD (Table 4). The majority of the samples from subsurface layers (50–75 cm) with a high silt content were compacted to some extent. This compaction was also projected in relatively higher BD values from the same layers (Table 4). The BD map of the third layer corresponded well to the textural patterns, where higher BD matched coarse samples. However, it was difficult to identify a trend from the rest of the BD maps, as was expected from the results of the spatial analysis (Table 5).

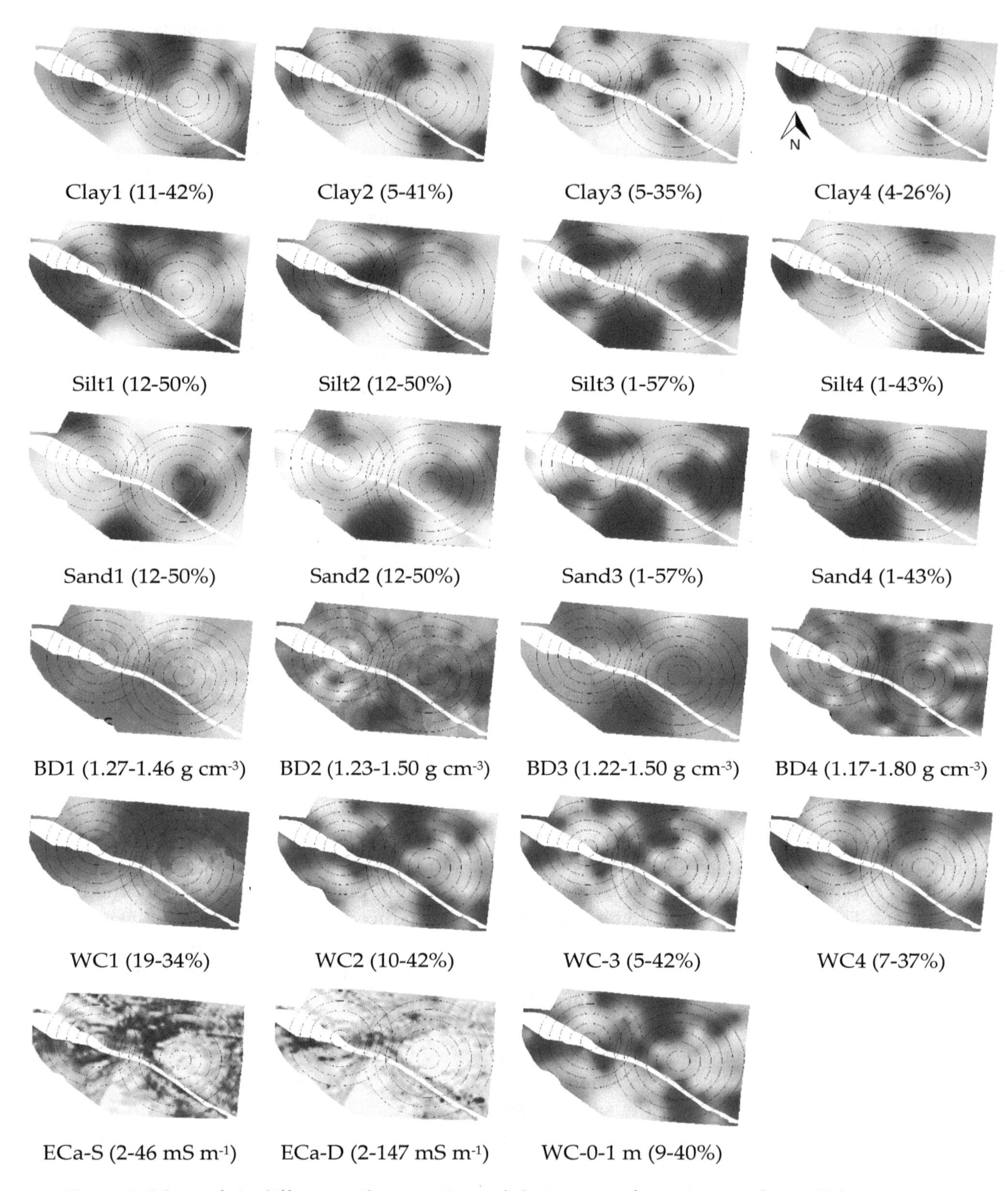

Figure 5. Maps of six different soil properties and their range of variation within a 73-ha two-pivot field in west Tennessee. These attributes include percent silt, sand, and clay, bulk density (BD, g cm^{-3}), volumetric water content (WC, %), and apparent electrical conductivity shallow (ECa-S, mS m^{-1}) and deep (ECa-D, mS m^{-1}). Numbers 1-4 denote layers 1–4 (layer 1: 0–25 cm, layer 2: 25–50 cm, layer 3: 50–75 cm, and layer 4: 75–100 cm). The maps were generated using ordinary kriging. The darker colors correspond to greater values for each attribute.

The soil water content is a dynamic property of soil with time. However, it is expected that a one-time measurement of water content values across a field provides a useful insight into the relative spatial pattern of soil hydraulic properties [31]. The water content map of the surface soil clearly showed the sandy textured areas as regions with a lower water content (Figure 5). At deeper depths, the water content maps almost exactly matched the spatial pattern of the sand maps. This occurred because coarse soils tend to dry out faster and hold less water than finer textured soils. This suggests that a one-time measurement of water content during the sampling process may be mathematically transformable to a PAW map. Overall, the water content map for the entire profile (0–100 cm) was similar to maps of individual layers.

The ECa maps tended to follow the same general spatial patterns as those for soil basic properties and water content (Figure 5). Table 6 illustrates the correlation coefficient between ECa values and soil basic data. The ECa data were moderately correlated to soil texture and water content information. The lowest correlation was between BD and ECa data. The correlation between shallow ECa and other attributes declined from layer 1 to 4, as expected, while the opposite was true for ECa deep readings. Sudduth et al. [29] showed that 90% of the shallow and deep reading responses in Veris machines were approximately obtained from the soil above the 30 cm and 100 cm depths, respectively. The sand increased with depth, hence regions with high conductivity became less pronounced in the deep ECa map as opposed to the shallow ECa map.

Table 6. Correlation coefficient between soil apparent soil electrical conductivity (ECa, mS m^{-1}) data and soil basic information at the four layers (L1–L4).

	Clay (%)				Sand (%)				Silt (%)			
	L1	L2	L3	L4	L1	L2	L3	L4	L1	L2	L3	L4
ECa-S	0.75	0.55	0.35	0.40	−0.75	−0.63	−0.45	−0.39	0.65	0.60	0.46	0.36
ECa-D	0.59	0.61	0.52	0.57	−0.62	−0.73	−0.63	−0.63	0.56	0.72	0.63	0.60

	* BD (g cm^{-3})				WC (%)			
	L1	L2	L3	L4	L1	L2	L3	L4
ECa-S	−0.01	−0.15	−0.31	−0.02	0.66	0.61	0.47	0.47
ECa-D	0.06	−0.20	−0.45	−0.13	0.63	0.71	0.64	0.65

* BD: soil bulk density, WC: soil volumetric water content at the time of sampling.

This study demonstrated that ECa is a useful surrogate for both soil texture and water content. Sudduth et al. [29] studied ECa readings on 12 fields in six states of the north-central US. They found a good relationship between ECa data and soil cation exchange capacity (CEC), as well as clay content at different times and locations, thus suggesting a general calibration equation of CEC and clay content to ECa readings. They found the most variation in ECa values in Iowa fields that had the widest range in soil texture, from loam to clay loam. In contrast, [17] reported water content as the main factor influencing ECa readings in a rainfed field, but they did not find soil texture to be a significant predictor of ECa. They found a weak correlation between water content and clay content and found that other factors including elevation and organic matter, may govern the amount of soil water content [17]. In theory, multiple factors, including the relative fractions occupied by soil, water and air, geometry and distribution of particles, and soil solution attributes, affect ECa [32]. The current introduced to measure the ECa of soil, in fact, travels through liquid, soil-liquid, and solid pathways [33]. We believe an accurate understanding of field-level soil texture variability and soil water status during the ECa measurement process is crucial to efficiently interpret ECa maps. Brevik et al. [34] studied the temporal stability of ECa data with respect to soil water content. They observed a strong influence of water content on ECa readings and found that ECa's power to differentiate soils was proportional to soil moisture. They mentioned that soil water content should be reported as an essential part of ECa studies. In this study site, the spatial heterogeneity of soil texture was the main factor that governed the spatial distribution of water content. Further studies, however, are needed to examine the efficacy of ECa for other typical field-level heterogeneity in the region, where the infiltration and redistribution of water within the root zone is governed by topography rather than soil textural variability.

3.2. Cotton Supplemental Irrigation

Table 7 summarizes the correlation coefficients between yield data and some soil properties across the field of study. To evaluate the effect of variable rate fertilizer application by the farmer on yield spatial variation, we also obtained correlation information for *P* and *K*. In general, correlation coefficients were higher for 2014 data than for 2013 data. The correlations between *P* and *K* with yield data were negligible. Given the high correlation between ECa data and PAW [18], the ECa data were used to group cotton lint yield information into four soil-based zones (Figure 6). The results illustrated in Figure 6 include irrigated areas, as well as corners of the field that were rainfed. In general, there was an increase in yield from soils with lower ECa to soils with higher ECa in 2014, but not in 2013. This is in line with the relatively higher correlation between water content/soil texture and the yield data in 2014 compared to 2013.

Table 7. Correlation coefficient values between cotton lint yield data (2013 and 2014 cropping seasons), soil properties at four depths from 0–1 m, and fertilizer application.

	2013					2014				
Layer	**1**	**2**	**3**	**4**	**Total**	**1**	**2**	**3**	**4**	**Total**
* BD, g cm^{-3}	−0.16	−0.04	−0.09	−0.07		0.00	−0.23	−0.49	−0.18	
WC, %	0.22	0.08	0.03	0.12		0.47	0.51	0.46	0.51	
Sand, %	−0.12	−0.03	−0.03	−0.08		−0.44	−0.52	−0.50	−0.53	
Clay, %	0.16	0.03	−0.03	0.01		0.40	0.44	0.42	0.47	
Silt, %	0.07	0.03	0.06	0.10		0.42	0.53	0.51	0.53	
WC33	0.18	0.06	0.05	0.12		0.40	0.50	0.52	0.51	
WC1500	0.19	0.05	0.01	0.11		0.40	0.48	0.48	0.50	
ECa-S, mS m^{-1}					0.12					0.49
ECa-D, mS m^{-1}					0.08					0.58
P, Mg ha^{-1}					−0.02					0.07
K, Mg ha^{-1}					−0.11					−0.23

* WC33 and WC1500: predicted volumetric water content at soil matric potentials −33 and −1500 kPa, respectively [18]; WC: volumetric water content at the time of sampling. Layers 1, 2, 3, and 4 were from 0–25 cm, 25–50 cm, 50–75 cm, and 75–100 cm, respectively. ECa shallow and deep readings represented approximately 0–30 cm and 0–90 cm of the soil profile, respectively.

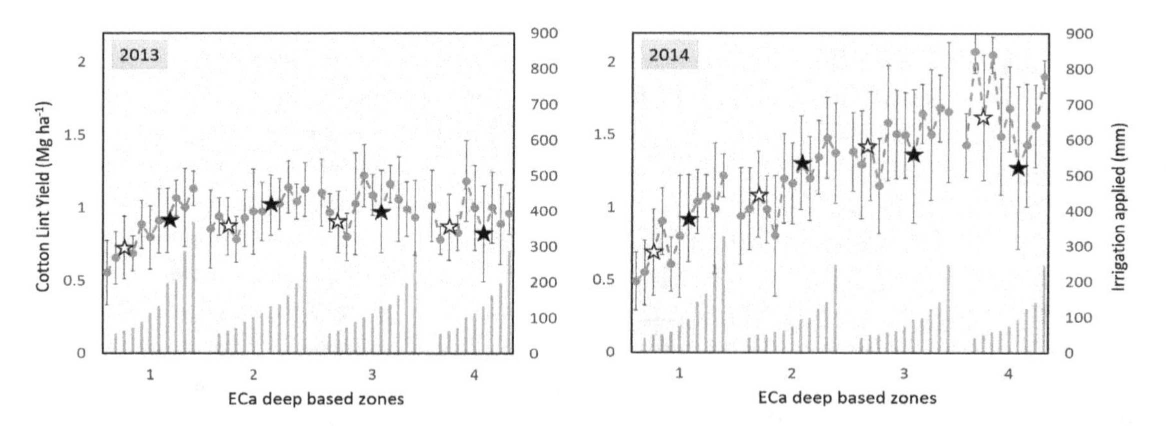

Figure 6. The effects of supplemental irrigation (blue bars, right-hand *y*-axis) on cotton lint yield (orange lines, left-hand *y*-axis) in 2013 and 2014 within a 73-ha two-pivot field in west Tennessee. The white and black five-point start symbols denote actual irrigation applications by the farmer for the west and east pivots, respectively. The deep ECa data were used to group cotton lint yield information into four zones, where ECa increases from zone 1 to zone 4, zone 1: 2–8 mS m^{-1}, zone 2: 8–18 mS m^{-1}, zone 3: 18–35 mS m^{-1}, and zone 4: 35–162 mS m^{-1}.

In 2013, only for coarse-textured soils (soils with lower ECa readings; zones 1 and 2), there was an overall positive response to supplemental irrigation. We observed no consistent positive response to higher irrigation levels for soils with higher ECa readings (i.e., zone 3 and 4 repressing fine-textured soil with higher PAW). In 2014, overall, cotton responded favorably to higher supplemental irrigation levels for the first three ECa zones, with coarse-textured soil showing the highest yield increase. The highest yield for the fine-textured zone (zone 4) did not belong to the greatest irrigation applications.

Figures 7 and 8 depict the dynamic of soil moisture (Agspy sensors) at different depths and locations over the 2013 and 2014 growing seasons, respectively. There were some missing data and bad readings mostly in 2014. In 2013, soil moisture probes *a*, *b*, and *c* were located underneath the east pivot, representing the pie shape zone with extra irrigation, pie shape zone with lower irrigation, and farmer's irrigation application, respectively. The soil moisture probes *a*, *b*, and *c* received 163 mm, 90 mm, and 133 mm seasonal IW, respectively. The average predicted PAW throughout the effective root zone (i.e., 1 m) was 0.28 $m^3\ m^{-3}$, 0.22 $m^3\ m^{-3}$, and 0.30 $m^3\ m^{-3}$ for probes *a*, *b*, and *c*, respectively [18]. The pattern of soil moisture was dynamic and varied among sensors at different locations and depths. At probe *a*, soil moisture depletion and replenishment occurred for sensors up to 40 cm deep during the monitoring period (i.e., days after planting (DAP): 42-133). Soil water status remained almost unchanged for deeper sensors for about 100 DAP, and then gradually exhibited a reduction, indicating that roots started to pull out water from deeper layers as the ET demand increased. At probe *b*, rainfall plus irrigation kept the soil moisture at a fairly constant level up to about 80 DAP for all sensors, while fluctuations decreased by depth, as expected. After that, there was a substantial depletion in soil moisture for sensors up to 50 cm, which even expanded to deeper sensors at about 95 DAP. At probe *c*, the overall trend was similar to that of probe *a*. Toward the end of the growing season, much rainfall at 112 DAP occurred that refilled the shallow layers for all soil moisture probes and also penetrated to deeper layers such that there was an increase in readings by the soil moisture sensors at 30 and 40 cm and no decrease for deeper sensors up to the end of the monitoring period.

In 2014, soil moisture probes *a*, *b*, and *c* were mainly located underneath the west pivot, representing the central area irrigated by both pivots, pie shape zone with lower irrigation, and farmer's irrigation application, respectively. Soil moisture probes *a*, *b*, and *c* (Figure 8) received 142, 42, and 50 mm of IW, respectively, during the 2014 cropping season. Within the effective root zone (i.e., 1m), the average PAW values predicted for probes *a*, *b*, and *c* were 0.19 $m^3\ m^{-3}$, 0.33 $m^3\ m^{-3}$, and 0.23 $m^3\ m^{-3}$, respectively [18]. Unlike 2013, most of the deeper sensors in 2014 showed some fluctuations starting at about 70 DAP, meaning that the crop started to use water from deeper layers as the crop water requirement increased. We attribute this to (i) bigger plants with larger canopies, and hence a higher ET demand; (ii) lower irrigation in 2014 compared to 2013; and (iii) lower irrigation under the west pivot (where we had sensors installed in 2014) compared to the east pivot (where we had sensors installed in 2013). Trends in probes *b* and *c* were similar. There were more fluctuations in the shallow sensors in probe *a*, since this sensor was irrigated by both pivots and was located on a coarse-textured soil with a low PAW. In both years, heavy rainfall events were responsible for big changes in the soil water status within the soil profile and considering sensor fluctuations, they usually penetrated deep down to 50 cm. Irrigation events, however, mostly refilled shallow layers up to 20 cm and barley influenced sensors deeper than 30 cm.

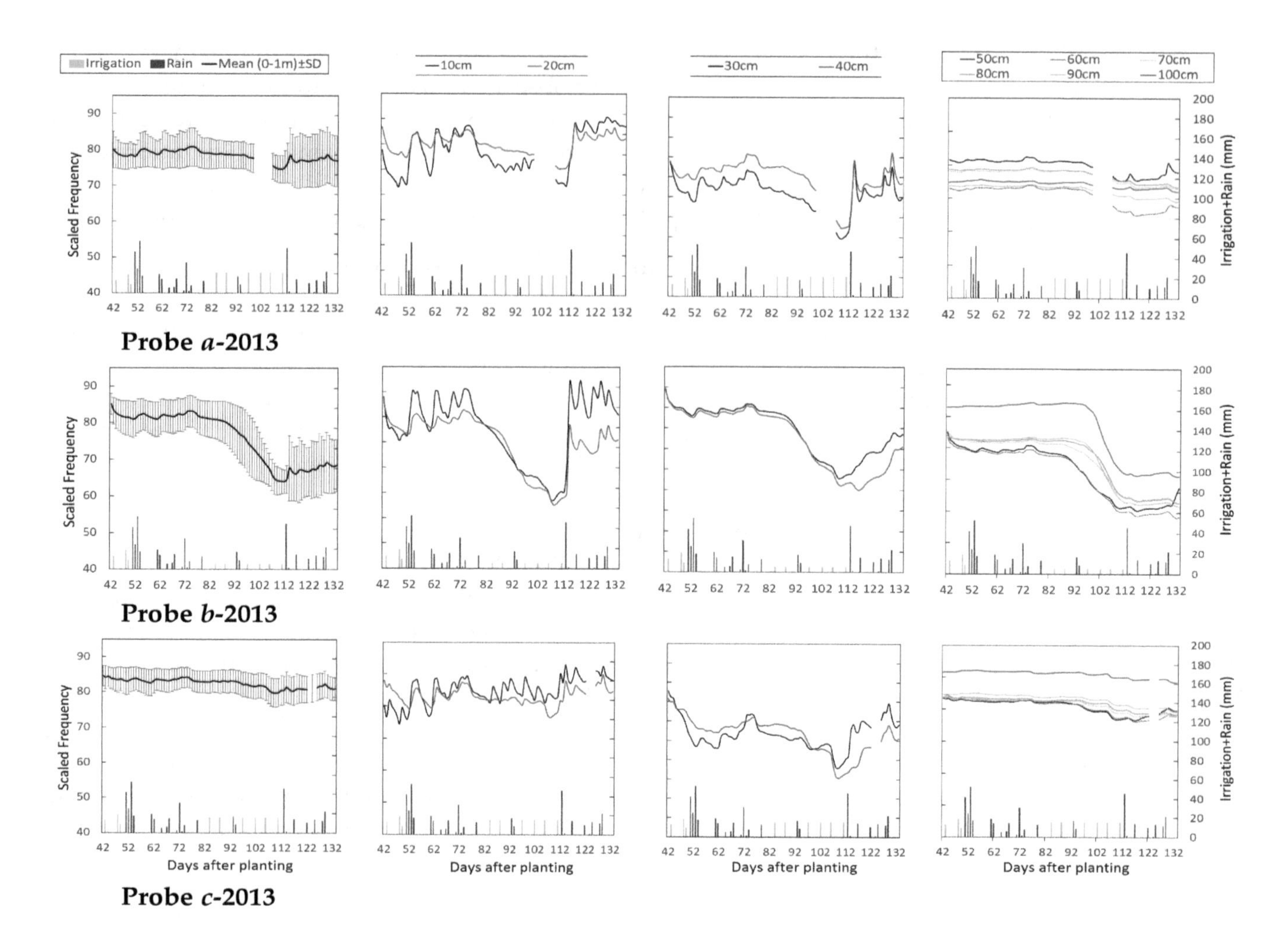

Figure 7. The change in soil moisture, measured as scale frequency (SF, Equation (4)), from 42 days after planting throughout the 2013 growing season in a 73-ha two-pivot supplemental irrigation field in west Tennessee. Light and dark blue bars show irrigation and rainfall, respectively. SF was measured using Agspy soil sensor probes at different locations and 10 depths from 0–1 m. In 2013, soil moisture probes *a*, *b*, and *c* were located underneath the east pivot, representing the pie shape zone with extra irrigation, pie shape zone with lower irrigation, and farmer's irrigation application, respectively.

The cotton lint response to supplemental irrigation differed across soil types. For soils with a lower PAW, there was a positive response to irrigation in comparison to rainfed, where soil moisture deficit is expected to reduce the boll number and yield [9]. The cotton response to irrigation was not consistent for soils with a higher PAW, except that a yield reduction occurred underneath both pivots for high irrigation rates in both cropping seasons. This is in line with the reported results in the literature, indicating that under wet conditions, excessive irrigation decreased the yield of cotton lint [11,12]. In 2013, the cotton lint yield was only 12% higher where we placed probe *a* (IW = 163 mm, predicted PAW = 0.28 m^3 m^{-3}) in comparison to the yield at probe *b* (IW = 90 mm, predicted PAW = 0.22 m^3 m^{-3}), even though there was a remarkable difference in soil water status throughout the growing season between the two soil moisture probes (Figure 7). On the other hand, in 2014, the yield difference between the exact same spots with a similar relative difference in IW increased by 44%. We attribute this to the wet season and delayed planting in 2013, which significantly affected the cotton response to irrigation. Delayed planting influences heat unit accumulation and distribution, which was underscored as an important factor for the short season cotton response to supplemental irrigation by [13]. Moreover, irrigation is expected to increase the number of bolls, but delays cutout (i.e., cessation of flowering) since irrigation continues the vegetative growth for a longer period. We believe that rapid canopy expansion occurred on soil with a higher PAW in 2013 due to excessive water within the crop effective root zone.

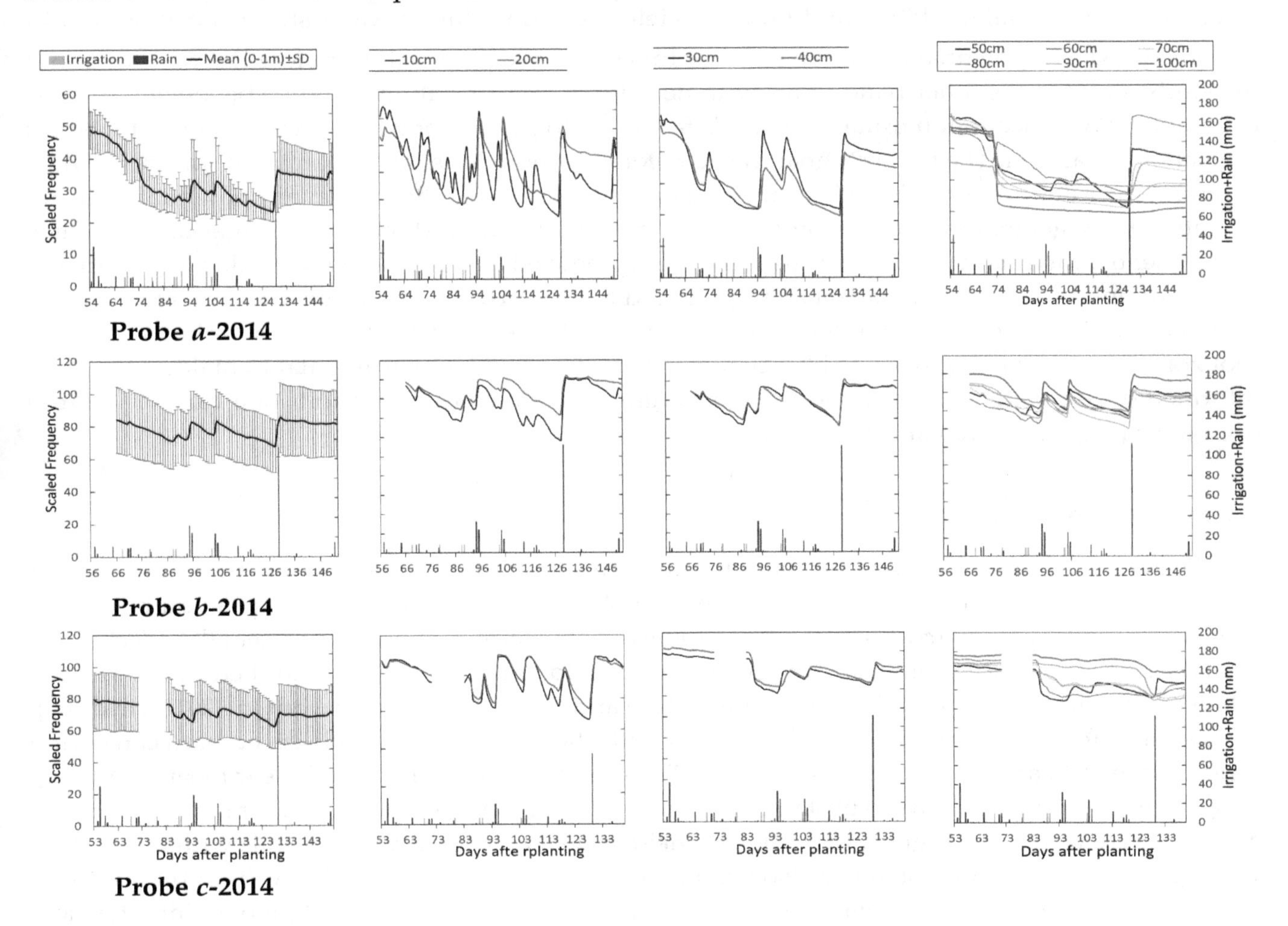

Figure 8. The change in soil moisture capacitance, measured as scale frequency (SF, Equation (4)), from 53 days after planting throughout the 2014 growing season in a 73-ha two-pivot supplemental irrigation field in west Tennessee. Light and dark blue bars show irrigation and rainfall, respectively. SF was measured using Agspy soil sensor probes at different locations and 10 depths from 0–1 m. In 2014, soil moisture probes *a*, *b*, and *c* were mainly located underneath the west pivot, representing the central area irrigated by both pivots, pie shape zone with lower irrigation, and farmer's irrigation application, respectively.

Both 2013 and 2014 were relatively wet years. In fact, the rainfall was always above the long term average, except for July 2014. There were some heavy rainfall events during the growing seasons, which caused a significant increase in the soil water content. This is a typical situation in west Tennessee, with unexpected rainfall events, where temporal changes in rainfall patterns significantly affect the yield response to supplemental irrigation across years. Bajwa and Vories [11] reported the same complexity on cotton irrigation scheduling in a moderately humid area in Arkansas when rainfall was plentiful and caused a yield reduction for high irrigated crops. Monitoring the soil water status revealed that rainfall events refilled the top soil and penetrated into deeper layers, while supplemental irrigation mostly influenced the shallow layer up to 20 cm. Therefore, any sustainable irrigation management in this region should take effective rainfall into account for site-specific irrigation scheduling. Sensors indicated fast depletion for soils with lower PAWs. This caused the crop to start using water from deeper layers as the cropping season advanced and ET demand increased. The sensor located on the overlap region of the two pivots showed that more frequent irrigation could prevent the shallow soil layer from substantial depletion and possible yield reduction due to water stress and thus should be considered as a potential irrigation strategy for coarse-textured soils with low PAWs throughout the study site. On the other hand, days with no rainfall and irrigation could be beneficial for soils with high PAWs since cotton usually responds favorably to periods of water stress adequate to reduce vegetation expansion. The farmer's goal was to tailor irrigation decisions to dominant soil areas, while avoiding over and under irrigation as much as possible. The results, however, show that uniform irrigation management can be detrimental to the field's overall production and water use efficiency. Consequently, variable rate irrigation strategies were predicted to enhance cotton production compared to current uniform irrigation management practiced by the grower [35]. There may be a requirement to practice a dynamic zoning strategy considering available water for the crop within its effective root zone during growing seasons [36]. Soil moisture data will be instrumental to making the most informed decisions for each soil and to determining how much different soils of a variable texture need to be irrigated to maximize production.

3.3. Multiyear Yield Analysis

Figure 9 illustrates the cotton yield maps for 2012, 2013, and 2014; mean yield map (from Equation (5)); and standard deviation yield map (from Equation (6)). We included the 2012 yield map, a year before this experiment when uniform irrigation had been applied by the farmer, to better represent the effect of soil spatial variation on the cotton yield. Thematic cotton yield maps for 2012 and 2014 cropping seasons followed the same spatial distribution as soil texture and water content maps, but the 2013 yield map showed a different pattern. This finding agrees with the low correlation observed between cotton yield data in 2013 and soil properties (Table 7). The spatial analysis of the multiyear mean yield map (ranged from −79 to 98) showed substantial similarity to the PAW map developed for the field of study by [18]. There were three regions with a lower yield and all on coarse-textured soils with lower PAW. The highest yield temporal stability also belonged to the regions with low PAW, (i) in the southern part of the field outside pivots coverage and (ii) in an area surrounding the east pivot point. The yield temporal stability was lower for other parts of the field, but it was hard to identify any cluster of cells with a similar temporal variance. We mainly attribute this to different rainfall patterns and irrigation regimes across years, and their effects on the yield across soil types, which caused substantial mean yield variability across years (Table 3). For instance, for cotton and corn, there was as much as 43% and 110% temporal differences between mean yields across years, respectively. However, it is known that other attributes related to management, water, crop, and climate affect the crop yield in a complex manner and the crop yield maps per se only provide limited information about the influence of each attribute [31]. In 2012 and 2014, the mean yield and standard deviation were higher than those for 2013, indicating a decline in yield on soils with higher available water in 2013 due to the delayed planting and excessive rainfall throughout the growing season.

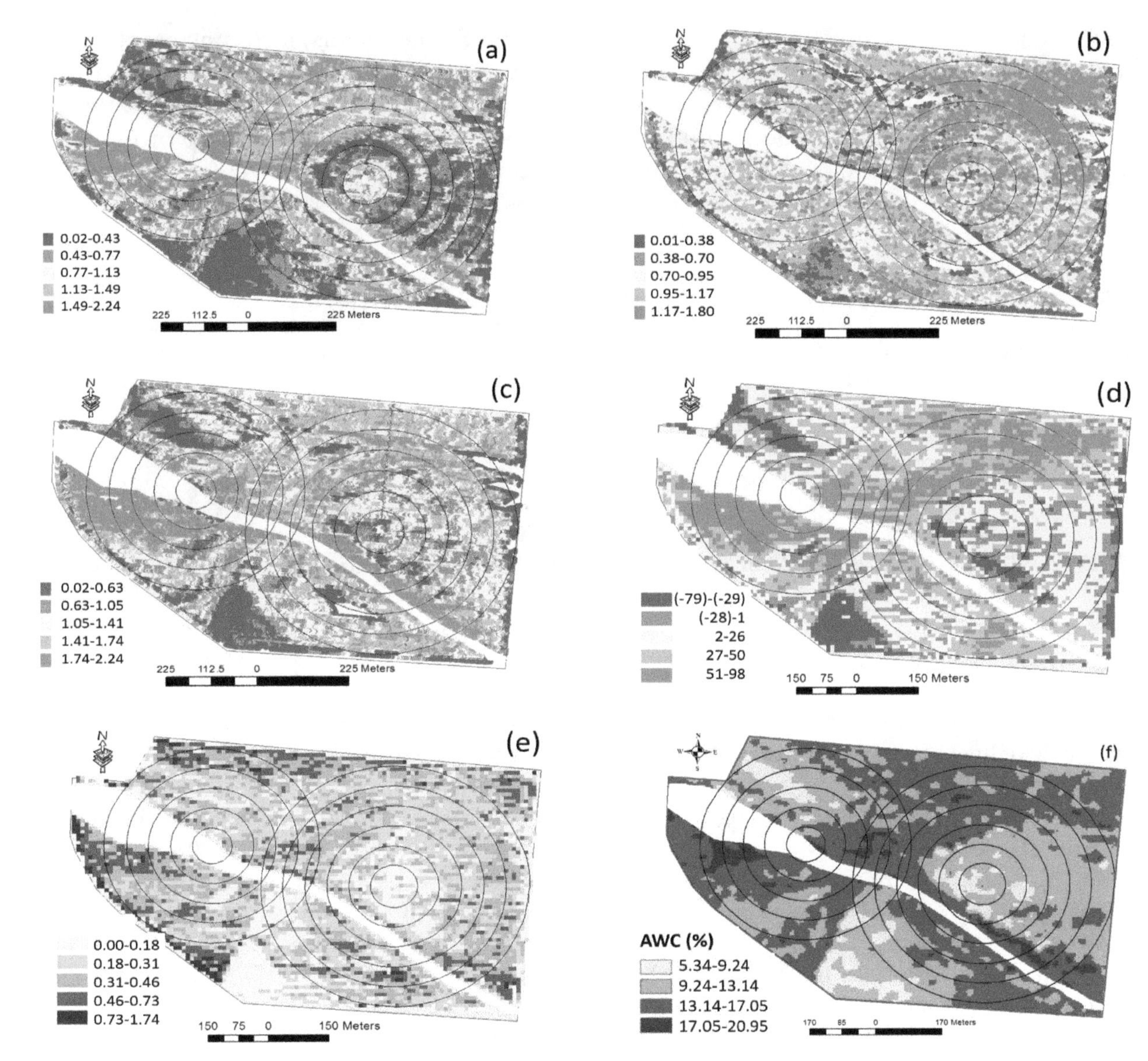

Figure 9. The cotton lint yield map (Mg ha^{-1}) time series from 2012 to 2014 (**a–c**), plus mean ((**d**), using Equation (5)) and standard deviation cotton yield maps ((**e**), using Equation (6)) that were derived from the producer's harvester data (Table 3, seven years of data for cotton, corn and soybean) for a 73-ha field in west Tennessee. The map of plant available water content (**f**) within the crop effective root zone (i.e., 100 cm) is adapted from the study by [18].

4. Conclusions

Irrigation investment has been expanding across the humid areas of the US Cotton Belt for the last 20 years because of the stabilization of yields and high commodity prices. Recent advances in modern instrumentation and measurement techniques, such as on-the-go sensing, remote sensing, and wireless networks of sensors, make site-specific on-farm experimentation possible for farmers. This is essential for field-level cotton irrigation management in humid areas as a complex problem due to substantial spatiotemporal heterogeneity in soil and weather-related parameters. In this study, we used a variety of information that is relatively easily collected by farmers, to investigate the impact of the spatial and temporal heterogeneity of soil attributes, including field-collected soil texture, moisture, ECa, and bulk density, on the cotton response to supplemental irrigation. We found that ECa was a useful proximal attribute to understand field-level spatial variability of alluvial soils in the region. Analyzing crop yield against maps of soil characteristics revealed that soil texture and soil water content remarkably influenced yield patterns, suggesting that variable rate irrigation is the appropriate irrigation scenario for this mixture of soils. We also found that other factors, including cropping season length and in-season rainfall pattern, may change or even reverse the expected lint yield from an irrigation treatment for a specific soil type. While soil variation is inherent and not controlled by farmers, irrigation, if well-scheduled, could be the key factor to optimizing the crop production and water use efficiency. The use of soil moisture sensors can help monitor soil water status and adjust irrigation application based on in-season rainfall patterns and within-field soil variability.

Author Contributions: Conceptualization, A.H. and B.L.; methodology, A.H. and B.L.; software, A.H. and B.L.; validation, A.H. and B.L.; formal analysis, A.H.; investigation, A.H.; resources, B.L..; Data curation, A.H., B.L., W.C.W., S.G., T.G., M.Z., and P.V.; writing—original draft preparation, A.H.; writing—review and editing, A.H., B.L., R.W-A., and T.G.; visualization, A.H.; supervision, B.L.; funding acquisition, B.L.

Acknowledgments: We appreciate Pugh Brothers Farms for allowing us to implement this project on their property.

References

1. FAO (Food and Agriculture Organization of the United Nations). *Statistical Yearbook 2013: World Food and Agriculture*; FAO: Rome, Italy, 2013.
2. NASS (National Agricultural Statistics Service). 2007 Census of Agriculture: Farm and Ranch Irrigation Survey, National Agricultural Statistics Service, USDA. 2010. Available online: http://www.agcensus.usda.gov/index.php (accessed on 22 March 2019).
3. Tennessee Farm Bureau Federation. Irrigation: Solving Potential Challenges: Policy Development 2013. 2013. Available online: https://www.tnfarmbureau.org/wp-content/uploads/2010/10/Irrigation.pdf (accessed on 22 March 2019).
4. Wong, M.T.F.; Asseng, S. Determining the Causes of Spatial and Temporal Variability of Wheat Yields at Sub-field Scale Using a New Method of Upscaling a Crop Model. *Plant Soil* **2006**, *283*, 203–215. [CrossRef]
5. Guo, W.; Maas, S.J.; Bronson, K.F. Relationship between cotton yield and soil electrical conductivity, topography, and Landsat imagery. *Precis. Agric.* **2012**, *13*, 678–692. [CrossRef]
6. Graveel, J.G.; Fribourg, H.A.; Overton, J.R.; Bell, F.F.; Sanders, W.L. Response of Corn to Soil Variation in West Tennessee, 1957–1980. *JPA* **1989**, *2*, 300–305. [CrossRef]
7. Perry, C.; Barns, E. *Cotton Irrigation Management for Humid Regions*; Cotton Inc.: Cary, NC, USA, 2012.
8. Vellidis, G.; Liakos, V.; Perry, C.; Tucker, M.; Collins, G.; Snider, J.; Andreis, J.; Migliaccio, K.; Fraisse, C.; Morgan, K.; et al. A smartphone app for scheduling irrigation on cotton. In Proceedings of the 2014 Beltwide Cotton Conference, New Orleans, LA, USA, 6–8 January 2014; Boyd, S., Huffman, M., Robertson, B., Eds.; National Cotton Council: Memphis, TN, USA, 2014.
9. Pettigrew, W.T. Moisture Deficit Effects on Cotton Lint Yield, Yield Components, and Boll Distribution. *Agron. J.* **2004**, *96*, 377. [CrossRef]
10. Suleiman, A.A.; Soler, C.M.T.; Hoogenboom, G. Evaluation of FAO-56 crop coefficient procedures for deficit irrigation management of cotton in a humid climate. *Agric. Water Manag.* **2007**, *91*, 33–42. [CrossRef]
11. Bajwa, S.G.; Vories, E.D. Spatial analysis of cotton (*Gossypium hirsutum L.*) canopy responses to irrigation in a moderately humid area. *Irrig. Sci.* **2007**, *25*, 429–441. [CrossRef]
12. Bronson, K.F.; Booker, J.D.; Bordovsky, J.P.; Keeling, J.W.; Wheeler, T.A.; Boman, R.K.; Parajulee, M.N.; Segarra, E.; Nichols, R.L. Site-Specific Irrigation and Nitrogen Management for Cotton Production in the Southern High Plains. *Agron. J.* **2006**, *98*, 212. [CrossRef]
13. Gwathmey, C.O.; Leib, B.G.; Main, C.L. Lint yield and crop maturity responses to irrigation in a short-season environment. *J. Cotton Sci.* **2011**, *15*, 1–10.
14. Grant, T.J.; Leib, B.G.; Duncan, H.A.; Main, C.L.; Verbree, D.A. A deficit irrigation trial in differing soils used to evaluate cotton irrigation scheduling for the Mid-South. *J. Cotton Sci.* **2017**, *21*, 265–274.
15. Khosla, R.; Westfall, D.G.; Reich, R.M.; Mahal, J.S.; Gangloff, W.J. Spatial Variation and Site-Specific Management Zones. In *Geostatistical Applications for Precision Agriculture*; Springer Nature: Berlin, Germany, 2010; pp. 195–219.
16. Gooley, L.; Huang, J.; Page, D.; Triantafilis, J. Digital soil mapping of available water content using proximal and remotely sensed data. *Soil Use Manag.* **2013**, *30*, 139–151. [CrossRef]

17. McCutcheon, M.C.; Farahani, H.J.; Stednick, J.D.; Buchleiter, G.W.; Green, T.R. Effect of Soil Water on Apparent Soil Electrical Conductivity and Texture Relationships in a Dryland Field. *Biosyst. Eng.* **2006**, *94*, 19–32. [CrossRef]
18. Haghverdi, A.; Leib, B.G.; Washington-Allen, R.A.; Ayers, P.D.; Buschermohle, M.J. High-resolution prediction of soil available water content within the crop root zone. *J. Hydrol.* **2015**, *530*, 167–179. [CrossRef]
19. Blake, G.R.; Hartge, K.H. Bulk density. In *Methods of Soil Analysis. Part 1*, 2nd ed.; Agron. Monogr. 9; Klute, A., Ed.; ASA and SSSA: Madison, WI, USA, 1986; pp. 363–375.
20. Cressie, N. Spatial prediction and ordinary kriging. *Math. Geol.* **1988**, *20*, 405–421. [CrossRef]
21. ESRI. *ArcMap (Version 10.2.2)*; Environmental Systems Resource Institute: Redlands, CA, USA, 2014.
22. Getis, A.; Ord, J.K. The Analysis of Spatial Association by Use of Distance Statistics. *Geogr. Anal.* **2010**, *24*, 189–206. [CrossRef]
23. Turc, L. Estimation of irrigation water requirements, potential evapotranspiration: A simple climatic formula evolved up to date. *Ann. Agron.* **1961**, *12*, 13–14.
24. Lu, J.; Sun, G.; McNulty, S.G.; Amatya, D.M. A comparison of six potential evapotranspiration methods for regional use in the southeastern United States. *J. Am. Water Resour. Assoc.* **2005**, *41*, 621–633. [CrossRef]
25. Paterson, N.D.; Richard, J.C.; Neil, M.W. Soil Moisture Sensor with Data Transmitter. U.S. Patent 12/310,946, 10 December 2009.
26. National Climate Data Center. National Climate Data Center Home Page. 2015. Available online: http://www.ncdc.noaa.gov/data-access/land-based-station-data (accessed on 13 June 2015).
27. Joernsgaard, B.; Halmoe, S. Intra-field yield variation over crops and years. *Eur. J. Agron.* **2003**, *19*, 23–33. [CrossRef]
28. Basso, B.; Bertocco, M.; Sartori, L.; Martin, E.C. Analyzing the effects of climate variability on spatial pattern of yield in a maize–wheat–soybean rotation. *Eur. J. Agron.* **2007**, *26*, 82–91. [CrossRef]
29. Sudduth, K.; Kitchen, N.; Wiebold, W.; Batchelor, W.; Bollero, G.; Bullock, D.; Clay, D.; Palm, H.; Pierce, F.; Schuler, R.; et al. Relating apparent electrical conductivity to soil properties across the north-central USA. *Comput. Electron. Agric.* **2005**, *46*, 263–283. [CrossRef]
30. Iqbal, J.; Thomasson, J.A.; Jenkins, J.N.; Owens, P.R.; Whisler, F.D. Spatial Variability Analysis of Soil Physical Properties of Alluvial Soils. *Soil Sci. Soc. Am. J.* **2005**, *69*, 1338. [CrossRef]
31. Corwin, D.L.; Lesch, S.M.; Shouse, P.J.; Ayars, J.E.; Soppe, R.; Ayars, J.E. Identifying Soil Properties that Influence Cotton Yield Using Soil Sampling Directed by Apparent Soil Electrical Conductivity. *Agron. J.* **2003**, *95*, 352–364. [CrossRef]
32. Friedman, S.P. Soil properties influencing apparent electrical conductivity: A review. *Comput. Electron. Agric.* **2005**, *46*, 45–70. [CrossRef]
33. Rhoades, J.D.; Corwin, D.L.; Lesch, S.M. Geospatial measurements of soil electrical conductivity to assess soil salinity and diffuse salt loading from irrigation. In *Solar Eruptions and Energetic Particles*; American Geophysical Union (AGU): Washington, DC, USA, 1999; Volume 108, pp. 197–215.
34. Brevik, E.C.; Fenton, T.E.; Lazari, A. Soil electrical conductivity as a function of soil water content and implications for soil mapping. *Precis. Agric.* **2006**, *7*, 393–404. [CrossRef]
35. Haghverdi, A.; Leib, B.G.; Washington-Allen, R.A.; Buschermohle, M.J.; Ayers, P.D. Studying uniform and variable rate center pivot irrigation strategies with the aid of site-specific water production functions. *Comput. Electron. Agric.* **2016**, *123*, 327–340. [CrossRef]
36. Haghverdi, A.; Leib, B.G.; Washington-Allen, R.A.; Ayers, P.D.; Buschermohle, M.J. Perspectives on delineating management zones for variable rate irrigation. *Comput. Electron. Agric.* **2015**, *117*, 154–167. [CrossRef]

5

Feasibility of Moderate Deficit Irrigation as a Water Conservation Tool in California's Low Desert Alfalfa

Ali Montazar [1,*], Oli Bachie [1], Dennis Corwin [2] and Daniel Putnam [3]

[1] University of California Division of Agriculture and Natural Resources, UC Cooperative Extension Imperial County, 1050 East Holton Road, Holtville, CA 92250, USA; obachie@ucanr.edu

[2] USDA-ARS, United States Salinity Laboratory, 450 West Big Springs Road, Riverside, CA 92507, USA; Dennis.Corwin@usda.gov

[3] Department of Plant Sciences, University of California Davis, One Shields Ave., Davis, CA 95616, USA; dhputnam@ucdavis.edu

* Correspondence: amontazar@ucanr.edu.

Abstract: Irrigation management practices that reduce water use with acceptable impacts on yield are important strategies to cope with diminished water supplies and generate new sources of water to transfer for other agricultural uses, and urban and environmental demands. This study was intended to assess the effects of moderate water deficits, with the goal of maintaining robust alfalfa (*Medicago sativa* L.) yields, while conserving on-farm water. Data collection and analysis were conducted at four commercial fields over an 18-month period in the Palo Verde Valley, California, from 2018–2020. A range of deficit irrigation strategies, applying 12.5–33% less irrigation water than farmers' normal irrigation practices was evaluated, by eliminating one to three irrigation events during selected summer periods. The cumulative actual evapotranspiration measured using the residual of energy balance method across the experimental sites, ranged between 2,031 mm and 2.202 mm, over a 517-day period. An average of 1.7 and 1.0 Mg ha^{-1} dry matter yield reduction was observed under 33% and 22% less applied water, respectively, when compared to the farmers' normal irrigation practice in silty loam soils. The mean dry matter yield decline varied from 0.4 to 0.9 Mg ha^{-1} in a clay soil and from 0.3 to 1.0 Mg ha^{-1} in a sandy loam soil, when irrigation water supply was reduced to 12.5% and 25% of normal irrigation levels, respectively. A wide range of conserved water (83 to 314 mm) was achieved following the deficit irrigation strategies. Salinity assessment indicated that salt buildup could be managed with subsequent normal irrigation practices, following deficit irrigations. Continuous soil moisture sensing verified that soil moisture was moderately depleted under deficit irrigation regimes, suggesting that farmers might confidently refill the soil profile following normal practices. Stand density was not affected by these moderate water deficits. The proposed deficit irrigation strategies could provide a reliable amount of water and sustain the economic viability of alfalfa production. However, data from multiple seasons are required to fully understand the effectiveness as a water conservation tool and the long-term impacts on the resilience of agricultural systems.

Keywords: Colorado River Basin; drought; irrigation management strategy; water deficit; water productivity

---

1. Introduction

Due to recurring droughts and altered weather patterns, the Colorado River Basin is facing increasing uncertainty concerning water supplies. Hence, implementing impactful agricultural water conservation tools and strategies might have a significant value to the resiliency and profitability of agricultural systems in the low desert region of California and Arizona. It is likely that water deficits will be the reality for agriculture in the future, mostly affecting agronomic crops such as alfalfa (*Medicago sativa* L.), which accounts for about 28% of the crops grown in the area [1] and is the dominant water user in the region.

Currently, efficient use of irrigation water and improved irrigation management strategies are the most cost-effective tools to address water conservation issues. While more than 95% of California's low desert alfalfa (nearly 80,000 hectares) is currently irrigated by surface irrigation systems [2], one strategy to enhance water-use efficiency and on-farm water conservation in alfalfa fields is through improved technology of water delivery. Improved systems, such as subsurface drip irrigation, overhead linear move sprinkler irrigation, automated surface irrigation, and tailwater recovery systems might enable more precise control of irrigation water. Many of these technologies were already adopted by local farmers, although the process of adoption is continuing.

Another strategy is deficit irrigation of alfalfa, applying less water than the full crop water requirements for a season. Deficit irrigation was investigated as a valuable and sustainable crop production strategy over a wide variety of crops, including alfalfa, to maximize water productivity and to stabilize—rather than maximize—yields while conserving irrigation water [3–7].

The overall effect of deficit irrigation highly depends on the type of crop and adopted irrigation strategy. Although alfalfa is frequently criticized for its high seasonal water requirements, it has positive biological features, environmental benefits, and greater yield potential than many other crops under water stressed conditions [8], such as deep-rootedness, high yield and harvest index, contribution to wildlife habitat, and ability to survive a drought. If water allotments to alfalfa are significantly curtailed, it will result in reduced evapotranspiration (ET), CO_2 exchange, symbiotic N_2 fixation, and dry matter yield [9]. There is a positive relationship between alfalfa yield and its ET [10,11], but alfalfa yields are not always reduced in direct proportion to the reduction in applied water during droughts [12,13]. Non-stressed total season ET values are greater than most crops because of long periods of effective ground cover. Alfalfa, being a herbaceous crop, exhibits rapid growth characteristics and its yield is linearly related to ET [3,14], under optimum growing conditions. Dry matter per unit of water used in alfalfa is compared favorably with other C_3 plants.

Several studies investigated the effect of mid-summer irrigation cut-off (no irrigation after June until the following spring) on alfalfa yield. Ottman et al. [15] found that alfalfa yield under mid-summer deficit irrigation in Arizona was very low and did not recover in sandy soil, but summer irrigation termination had less effect on sandy-loam soils. At a site in the San Joaquin Valley of California, yields of a mid-summer irrigation treatment were 65% to 71% of that of a fully irrigated alfalfa, over a 2-year period [16]. Mid-summer deficit irrigation in the Imperial and Palo Verde Valleys of California reduced yields to 53–64% [17] and 46% [18], relative to a fully irrigated alfalfa. Studies in multiple environments of California showed reductions in alfalfa yield, significant conservation of water, and the consistent ability of the crop to recover after drought [3]. In long-term studies conducted on the western slope of the Rocky Mountains of Colorado, researchers determined that late-season deficits impacted yields in some cases, but not all, and that full recovery in the year following re-watering occurred in most cases [19]. In a study conducted in Nevada, significant yield reductions occurred from deficit irrigation over a three-year period, but yields were recovered under adequate irrigation during the fourth year [20]. Another study conducted in Kansas [21] showed no yield reductions in alfalfa hay yields, under 20% and 30% sustained deficit subsurface drip irrigation over the season. The researchers suggested that rainfall had a major contribution to the crop water use, with precipitation contributing to about 25% of seasonal crop water use between June and October, for the Kansas experiment.

The main purpose of this study was to identify and optimize moderate summer deficit irrigation strategies with profitable and sustainable alfalfa forage production, while conserving water under limited water conditions. The study intended to develop a dataset that could serve as a reference for further studies and an opportunity to better understand this underutilized water conservation strategy in the desert environment.

2. Materials and Methods

2.1. Field Experiments

Studies were conducted at four commercial alfalfa fields (designated "sites A1" through "A4") in the desert environment of the Palo Verde Valley, CA, USA. The study area had a true desert climate with an annual average air temperature, total annual precipitation, and ET_o of 21.4 °C, 78.2 mm, and 1782 mm, respectively (Table 1). Soil characteristics for all experimental sites pertaining to four generic horizons are provided in Table 2. Dominant soil type ranged from loams at sites A1 and A2, to clay at site A3, and sandy loam at site A4. Soil cation exchange capacity (CEC) ranged between 7.9 and 12.7 meq/100 g at site A4 to between 9.2 and 22.4 meq/100 g at site A3. The Colorado River was the source of irrigation water with an average pH of 8.1 and an average electrical conductivity (EC_w) of 1.1 dS m^{-1} for all fields.

Table 1. Monthly mean long-term (10-year, 2009–2018) climate data for Blythe, CA, USA, data from CIMIS (California Irrigation Management Information System) station NE #135.

Month	Rain (mm)	Solar Radiation (W m^{-2})	Average Air Temp. (°C)	Average Dew Point Temp. (°C)	Average Wind Speed (m s^{-1})	Total ET_o (mm)
Jan	13.4	126.4	11.1	2.8	2.1	65.2
Feb	10.2	171.9	13.5	4.2	2.2	86.2
Mar	6.4	228.8	17.5	5.4	2.4	144.0
Apr	1.3	282.3	21.0	5.8	2.8	188.7
May	2.3	322.3	24.3	8.8	2.7	227.5
Jun	0.0	331.2	29.4	13.6	2.3	232.4
Jul	8.3	295.0	32.3	19.5	2.3	221.5
Aug	9.0	275.7	32.1	20.0	2.2	202.4
Sep	5.6	232.0	28.4	17.1	2.0	157.8
Oct	6.0	186.9	22.0	11.2	1.9	122.9
Nov	3.2	140.9	14.9	5.4	1.9	76.3
Dec	12.5	111.7	10.5	2.7	2.0	58.0

Table 2. Physical and chemical properties of the soil of the four experimental sites. CEC represents the cation exchange capacity (meq/100 g). A1, A2, A3, and A4 represent alfalfa experimental sites.

Experimental Site	Generic Horizon (m)	Soil Texture			Organic Matter (%)	CEC (meq/100 g)	pH
		Sand (%)	Clay (%)	Silt (%)			
A1	0–0.3	44.2	11.1	44.7	1.7	20.6	8.0
	0.3–0.6	46.9	8.1	45.0	0.8	17.4	8.1
	0.6–0.9	41.7	7.7	50.5	0.9	18.6	8.2
	0.9–1.2	47.8	5.9	46.3	0.7	16.3	8.2
A2	0–0.3	39.7	20.6	39.7	2.3	22.4	8.1
	0.3–0.6	50.1	11.9	37.9	0.7	14.1	8.0
	0.6–0.9	75.0	5.1	19.9	0.9	8.0	8.2
	0.9–1.2	83.7	4.1	12.2	0.8	7.2	8.2
A3	0–0.3	31.5	13.2	55.3	1.3	18.1	8.0
	0.3–0.6	26.1	18.7	55.2	0.8	22.4	8.1
	0.6–0.9	88.6	2.8	8.6	0.5	10.1	8.2
	0.9–1.2	94.8	1.4	3.8	0.5	9.2	8.4
A4	0–0.3	69.3	13.1	17.6	1.7	12.7	7.9
	0.3–0.6	91.9	2.8	5.3	0.8	8.1	8.2
	0.6–0.9	82.0	8.3	9.7	1.2	8.7	8.1
	0.9–1.2	85.9	5.7	8.4	0.9	6.4	8.3

All four fields were planted in October 2018. Low desert alfalfa fields are typically harvested in a 28-day to 33-day cycle during spring and summer, while a total of 8–10 harvests per year is common in the region. Eleven harvest cycles were investigated in this study.

The experimental fields represent soil types (a wide range from sandy loam to clay) and irrigation management practices in the low desert region of California. The surface irrigation practices consisted of straight border irrigation (sites A3 and A4) and graded furrow irrigation (sites A1 and A2). An average of 75% may be assumed as the irrigation efficiency of the irrigation systems [22]. Sites A2 and A4 were divided into three individual sections (large plots) and sites A1 and A3 were divided into six individual plots. Five irrigation strategies were studied within the experimental plots, including normal farmer irrigation practice (NI, as control strategy) and four summer deficit irrigations (DI1, DI2, DI3, DI4; Table 3). The deficit irrigation strategies were implemented by eliminating irrigation events during summer harvest cycles. For the harvest cycles of July–September, three irrigation events per cycle is common irrigation practice in the Palo Verde Valley. Three plots were accommodated for each of the deficit irrigation strategies.

Table 3. Description of irrigation management strategies imposed at the experimental sites over the study period (18-month).

Site	Number of Irrigation Events					Description of Irrigation Management Strategies
	NI	DI1	DI2	DI3	DI4	
A1	30	27	28	-	-	DI1: irrigation events were eliminated 20 July 2019, 23 August 2019, and 26 September 2019 DI2: irrigation events were eliminated 23 August 2019 and 26 September 2019
A2	31	28	29	-	-	DI1: irrigation events were eliminated 21 July 2019, 24 August 2019, and 26 September 2019 DI2: irrigation events were eliminated 23 August 2019 and 26 September 2019
A3	24	-	-	22	23	DI3: irrigation events were eliminated 31 July 2019 and 2 September 2019 DI4: irrigation event was eliminated 2 September 2019
A4	25	-	-	23	24	DI3: irrigation events were eliminated 19 July 2019 and 24 August 2019 DI4: irrigation event was eliminated 24 August 2019

NI: following normal farmer irrigation practice over the study period
DI1: 33% less applied water than corresponding NI strategy during selected summer period
DI2: 22% less applied water than corresponding NI strategy during selected summer period
DI3: 25% less applied water than corresponding NI strategy during selected summer period
DI4: 12.5% less applied water than corresponding NI strategy during selected summer period

2.2. ET Monitoring and Data Processing

The actual evapotranspiration (ET_a) was measured using the residual of energy balance (REB) method with a combination of surface renewal (SR) and eddy covariance (ECov) techniques. The SR and ECov are well-recognized methods to estimate the sensible heat flux density (H) and to calculate the latent heat flux density (LE), using the REB approach [23–28].

A full flux density tower was set up in the plots under normal farmers' irrigation practice at each of the experimental site, totaling four towers (Figure 1). In each tower, several sensors were set up. An NR LITE 2 net radiometer (Kipp & Zonen, Ltd., Delft, The Netherlands) was used to measure net radiation (Rn). Two 76.2 μm diameter, type-E, chromel-constantan thermocouples model FW3 (Campbell Scientific, Inc., Logan, UT, USA) were used to measure high frequency temperature data for computing uncalibrated sensible heat flux (H0), using the SR technique. An RM Young Model 81000RE sonic anemometer (RM Young Inc., Traverse City, MI, USA) was used to collect high frequency wind velocities in three orthogonal directions at 10 Hz, to estimate H for the latent heat flux density calculations using the ECov technique.

Each tower also consisted of three HFT3 heat flux plates (REBS Inc., Bellevue, WA, USA) inserted at a 0.05 m depth below the soil surface, to measure soil heat storage at three different locations; three 107 thermistor probes (Campbell Scientific, Inc., Logan, UT, USA) to measure soil temperature at three depths in the soil layer above the heat flux plates; three EC5 soil moisture sensors (METER Groups Inc., Pullman, WA, USA) to measure soil volumetric water content at soil depths and locations near the heat flux plate and the thermistor probes; EE181 temperature and RH sensor (Campbell Scientific, Inc., Logan, UT, USA) to measure air temperature and relative humidity; an SP LITE 2 Pyranometer (Kipp & Zonen, Ltd., Delft, The Netherlands) to measure solar radiation; and a TE525MM tipping-bucket rain gauge with magnetic reed switch to measure precipitation.

Figure 1. A fully automated surface renewal and eddy covariance evapotranspiration (ET) tower in the plot under normal farmer irrigation practice (NI) at site A3.

Except for the soil sensors, all other sensors were set up at 1.8 m above the ground surface. The data were recorded using a combination of a Campbell Scientific CR1000X data logger and a CDM-A116 analog input module. Direct two-way communication with each monitoring flux tower was possible using a cellular phone modem model CELL210 (Campbell Scientific, Inc., Logan, UT, USA). The data of the sonic anemometer and fine wire thermocouples were collected at a 10 Hz sampling rate and the data of the other sensors were sampled once per minute. Half-hourly data were archived for later analysis.

Both the EC and SR techniques were individually employed to determine sensible heat flux density. The available energy components, R_n and G (ground heat flux density) were also measured throughout the study period. After acquiring the half-hourly H_0 data, a calibration factor (α) was established by determining the slope through the origin H values from the ECov technique versus H_0 from the SR technique, separately for the positive and negative values of H_0. The calibrated SR H value was finally estimated as $H = \alpha \cdot H_0$. The SR H and the ECov H were used to determine the LE values. The advantage from using both the ECov and SR methods was that they are independent and similar results provide a high level of confidence in the data used [26,28,29]. Latent heat flux density was calculated using the Residual of Energy Balance equation, as follows:

$$LE = Rn - G - H \tag{1}$$

where LE, G, H are positive away from the surface, and Rn is positive towards the surface. G is the ground heat flux density at the soil surface. It is assumed that Rn, G, and H are measured accurately. While use of the full eddy covariance method often does not demonstrate closure, Twine et al. [30] recommended that ECov results could be forced to have closure by holding the measured Bowen ratio (H/LE) constant, and increasing the H and LE values until $Rn - G = H + LE$. Twine et al. [30] also reported that using the REB method provides nearly the same accuracy as using the Bowen ratio correction. After determining LE, ET_a in mm d^{-1} was calculated by dividing the LE in MJ m^{-2} d^{-1} by 2.45 MJ kg^{-1}, to obtain the ET values in kg m^{-2} d^{-1}, which was equivalent to mm d^{-1}.

A Tule sensor (Tule technologies, Inc., Oakland, CA, USA) was used in each of the deficit irrigation plots at the experimental sites, to estimate ET_a using a surface renewal technique. The estimated daily ET_a from the Tule sensors at each site was verified by comparing with the ET_a measured from the full flux density towers.

Using the daily ET_a determined in each experimental site and the daily reference ET (ET_o) retrieved from the spatial CIMIS (California Irrigation Management Information System) data [31] for the coordinates of the monitoring station, the daily actual crop coefficient K_a (=$K_s \times K_c$) was calculated using Equation (2):

$$K_a = ET_a/ET_o \quad (2)$$

The daily stress coefficient (K_s) represents water and salt stresses, management, and environmental multipliers. To obtain the actual ET, K_s is needed to adjust crop coefficient (K_c). Spatial CIMIS combines remotely sensed satellite data with traditional CIMIS stations data, to produce site-specific ET_o on a 2-km grid, which provides a better estimate of ET_o for the individual sites.

2.3. Canopy Temperature and Soil Moisture Monitoring

Two SI-411fixed view-angle infrared thermometers (IRTs, Apogee Instruments, Logan, UT, USA) were used to measure canopy temperature in each experimental plot. The IRTs were installed on a pole with a 47.5° angle below horizon, in opposite direction, viewing north and south to match for consistency. The IRTs were installed 1.8 m from the ground surface. The average temperatures of IRTs viewing north and south were considered to be the canopy temperatures. Canopy temperature was scanned with the IRTs units every minute and readings were averaged over a 30-min interval, using ZL6 cellular data logger (METER Groups Inc., Pullman, WA, USA).

Crop Water Stress Index (*CWSI*) was estimated using the difference between measured canopy and air temperatures (dT_m), using Equation (3):

$$CWSI = \frac{(dT_m - dT_{LL})}{(dT_{UL} - dT_{LL})} \quad (3)$$

The dT_m was compared against lower (dT_{LL}) and upper (dT_{UL}) limits of the canopy–air temperature differential, which could be reached under non-water-stressed and non-transpiring crop conditions. The Idso et al. approach [32] was used for estimating dT_{LL} and dT_{UL}.

Watermark Granular Matrix Sensor (Irrometer company, Inc., Riverside, CA, USA) was used to measure soil water tension at multiple depths of 15, 30, 45, 60, 90, and 120 cm, on a continuous basis. The data of Watermark sensors were recorded by a 900M Monitor data logger (Irrometer company, Inc., Riverside, CA, USA), on a 30-min basis.

2.4. Soil Salinity Assessment

Soil properties were surveyed and characterized within an approximate footprint area of 200 m × 200 m, around the ET monitoring stations in each plot, to assess soil salinity following deficit irrigation regimes. Surveys of apparent soil electrical conductivity (EC_a) were conducted in October 2019 (right after the alfalfa harvest), using mobile electromagnetic induction (EMI) equipment, following the guidelines developed by the U.S. Salinity Laboratory of the United States Department of Agriculture, for field-scale salinity assessment [33–36]. EC_a measurements were taken with a dual-dipole EM38 sensor (Geonics Ltd., Mississauga, ON, Canada), in horizontal (EM_h) and vertical (EM_v) dipole modes, to provide shallow (0 to 0.75 m) and deep (0 to 1.5 m) measurements of EC_a, respectively. At each of the plots, soil cores at four distinct depth ranges (0–0.3, 0.3–0.6, 0.6–0.9, and 0.9–1.2 m) were taken from 6 sampling locations, which were selected using the ESAP software to reflect the spatial variability of root zone soil salinity. A comprehensive laboratory analysis was conducted on all soil samples.

2.5. Yield Measurements and Plant Stand Evaluation

Yield sampling from the sub-plots was conducted on the same day or the day before, when the participating growers scheduled to harvest the entire experimental fields. In each irrigation plot, yield samples were taken from 12 sub-plots with a dimension of 1.5 m wide and 2.0 m long (Figure 1). The sub-plots were harvested using a hand cutter. A portable PVC quadrate was used to accurately sample uniform sub-plot sizes. Plant cutting height was 6–8 cm. Fresh weights of plants harvested within the quadrate was recorded, after which samples were dried for three days in conventional oven at 60 °C and recorded for alfalfa dry matter (DM). The significance of deficit irrigation strategies on mean dry matter yields was evaluated using a *t*-test.

Forage quality test for each collected sample was conducted using the Near Infrared Reflectance Spectroscopy (NIRS) method [37], to determine Crude Protein (CP), Acid Detergent Fiber (ADF), and Lignin percentage.

All irrigation plots were evaluated for plant stand density in February 2020, four months after switching the deficit irrigated plots back to normal irrigation practices. A portable PVC quadrate of 0.6 m wide and 0.6 m long was used to count the number of plants from the center of the 12 sub-plots that were used for yield measurements. Mean plant numbers per hectare were compared to the plots under normal farmers' irrigation practices.

2.6. Water Productivity

Two water productivity indices were calculated to compare the efficiency of irrigation water use and actual ET, using Equations (4) and (5):

$$\text{Irrigation water productivity (IWP)} = \frac{Alfalfa\ dry\ matter}{Irrigation\ water\ applied} \quad (4)$$

$$\text{Evapotranspiration water productivity (ETWP)} = \frac{Alfalfa\ dry\ matter}{ETa} \quad (5)$$

where the unit of alfalfa dry matter is kg ha^{-1}, and the units of irrigation water applied and ET_a are mm.

2.7. Statistical Analysis

The statistical significances of mean dry matter alfalfa yield and mean forage quality indices were performed using *t*-tests.

3. Results

3.1. Weather Parameters

There was higher than normal rainfall between November 2019 and March 2020 (Figure 2 and Table 1). Total precipitation was 105 mm over these five months. This precipitation amount was more than a long-term annual rainfall of the study area. The average daily air temperature, dew point temperature, solar radiation, and wind speed for the 2019 season (January 2019–December 2019) were approximately 21.0 °C, 234.8 W m^{-2}, 7.5 °C, and 2.3 m s^{-1}, respectively. However, peak daily temperatures of 30 °C to 42 °C during the late summer months when deficit treatments were imposed were common for this desert environment (Figure 2). More windy days were observed during the first half of the 2019 season, when compared with the 2020 season. The average daily wind speed was nearly 9% higher from January–June 2019 than the same season in 2020, although maximum average wind speed for the 2020 season, as recorded on 4th of February 2020 was 6.7 m s^{-1}. During the 2019 season, the month of June had the highest average daily solar radiation of 334.1 W m^{-2}, followed by May (324 W m^{-2}) and July (310 W m^{-2}).

3.2. Applied Water

The cumulative applied water (irrigation water + rainfall) for the different irrigation strategies at each site is shown in Figure 3. The total applied water was 3006, 2933, 2783, and 3125 mm in irrigation strategy NI for sites A1, A2, A3, and A4, respectively, over the 2019 season. These amounts were 788, 897, 716, and 1007 for sites A1, A2, A3, and A4, respectively, during the study period of the 2020 season. Although, most water amounts were provided by the irrigation events, the number of irrigation events were slightly different among the study sites. A total of 23 irrigation events occurred at sites A1 and A2 during the 2019 season, in irrigation strategy NI. The number of irrigation events was 22 at site A3 and 21 in at site A4.

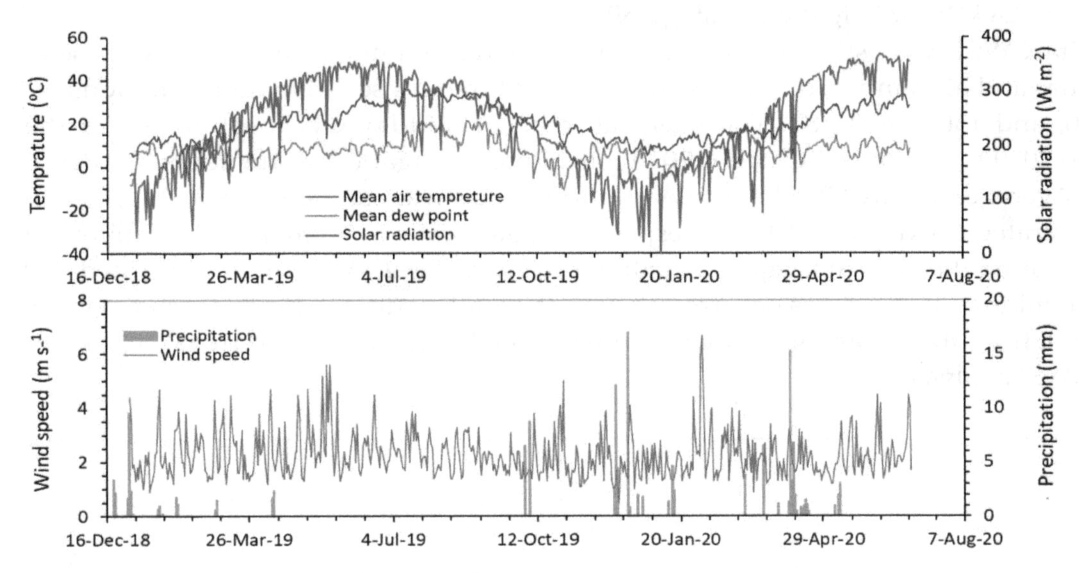

Figure 2. Daily weather data of the study area over the 18-month period of this experiment (January 2019 to June 2020).

The total reduction in the water used in the deficit irrigation strategies (DI1 to DI4) when compared with strategy NI was different at each of the experimental fields. For example, it was 314 mm in strategy DI1 and was 203 mm in strategy DI2 at site A1. The reduction in applied water was 310 mm in strategy DI1 at site A2, 185 mm in strategy DI3 at site A3, and 239 mm in strategy DI3 at site A4. The results demonstrated that the greatest amount of water conservation with the proposed deficit irrigation strategies was achieved under irrigation strategy DI1 at site A1, and the least amount was conserved in strategy DI4 at site A3.

3.3. Soil Moisture Status and Crop Water Stress

The soil moisture sensors placed within the effective root zone (15–120 cm) provided a representative condition of the soil water status. Figure 3 depicts the half-hourly soil water tension for irrigation strategy NI at sites A2 and A4, from March 2019 to June 2020. Due to the necessity of drying soils during and after alfalfa harvest, and to allow equipment to be used on alfalfa fields and for drying of hay, irrigation events typically stopped approximately 4–5 days before cuttings and resumed 4–5 days after cuttings. Therefore, changes in soil water tension could be observed before and after these dry-down periods (Figure 4). Soil water tension responded most within the top 60 cm depth, while some responses were also observed at deeper soil depths over time. For example, water tension values at site A4 sharply increased to more than 200 kPa at the topsoil (15–45 cm), before the first irrigation events, right after the alfalfa harvests. Soil water tension readings below 60 cm indicated that soil moisture was effectively maintained at a relatively uniform and desirable level during the study period, even at site A4, with a sandy loam soil where soil water tension values were less than 25 kPa.

Soil moisture data indicated that the soil at site A2, which had a silty loam soil, was generally not within the stressed range. However, alfalfa at site A4 might occasionally experience moderate water stress around cuttings. The recommended average soil water tension levels within the effective alfalfa root zone at which irrigation was triggered on loamy and sandy loam soil was at 60–90 kPa and 40–50 kPa ranges, respectively [12]. The insufficient soil moisture levels at site A4 during summer harvest cycles, from July through September (Figure 4) might have caused some mild alfalfa water stress. Soil moisture was clearly impacted by irrigation strategy DI3 at this site (Figure 5). However, additional potential mild water stress could have occurred in the middle of the harvest cycles due to halted irrigation water. For instance, soil water tension values increased to 134 and 93 kPa on 24 July 2019, at 30 and 60 cm of soil depth, as a result of eliminating the irrigation event on 19 July 2019 (Figure 5). Similarly, soil moisture readings for the same soil depths and dates at this site was less than 28 kPa for irrigation strategy NI.

Alfalfa CWSI was estimated for different irrigation strategies at each site, based on canopy temperature and air temperature measurements for three consecutive hourly periods of 1100–1200, 1200–1300, and 1300–1400 PST. The seasonal trend of midday CWSI estimated for the period of early-June 2019 through mid-October 2019 for different irrigation strategies at site A4, is shown in Figure 6. Average midday CWSI at site A4 for the period was estimated to be 0.13, 0.15, and 0.14 for irrigation strategies NI, DI3, and DI4, respectively, suggesting a similar, but relatively lower CWSI responses within the normal farmers irrigation practices. The CWSI values illustrated that moderate short-term midday water stress occurred around the alfalfa cuttings (and mid-harvest cycles) of July and August, thus, there was a good match between the findings obtained from the soil moisture data and the CWSI analysis.

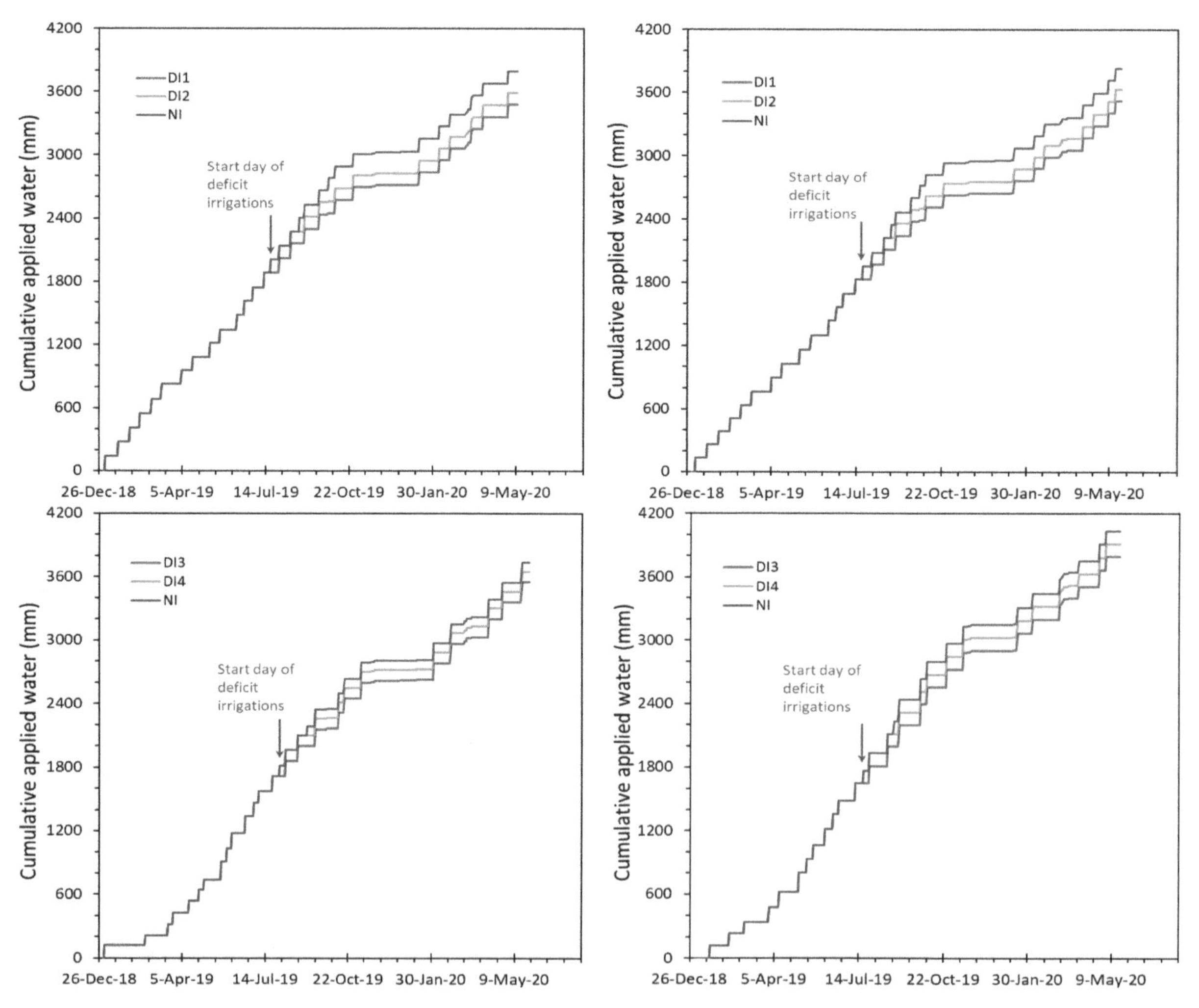

Figure 3. Cumulative applied water in each of the experimental irrigation strategies (NI, DI1–DI4) and sites (A1–A4) over the study period.

3.4. *Actual Evapotranspiration and Crop Coefficients*

ET_a was determined by calculating half-hourly latent heat flux density, using the REB approach with the SR and EC techniques. The daily ET_o, ET_a from the SR calculations, and K_a values for the irrigation strategy NI, at sites A1 and A4 are shown in Figure 7. The ET_a varied widely for each crop harvest cycle and throughout the experimentation seasons at both sites. For example, the ET_a

at site A4 ranged between 3.4 mm d^{-1} after alfalfa cutting and 9.3 mm d^{-1} at midseason full crop canopy, from June through August. The maximum and minimum ET_a at site A1 were 2.6 mm d^{-1} and 0.3 mm d^{-1} during three-months of the study period, November 2019 to January 2020.

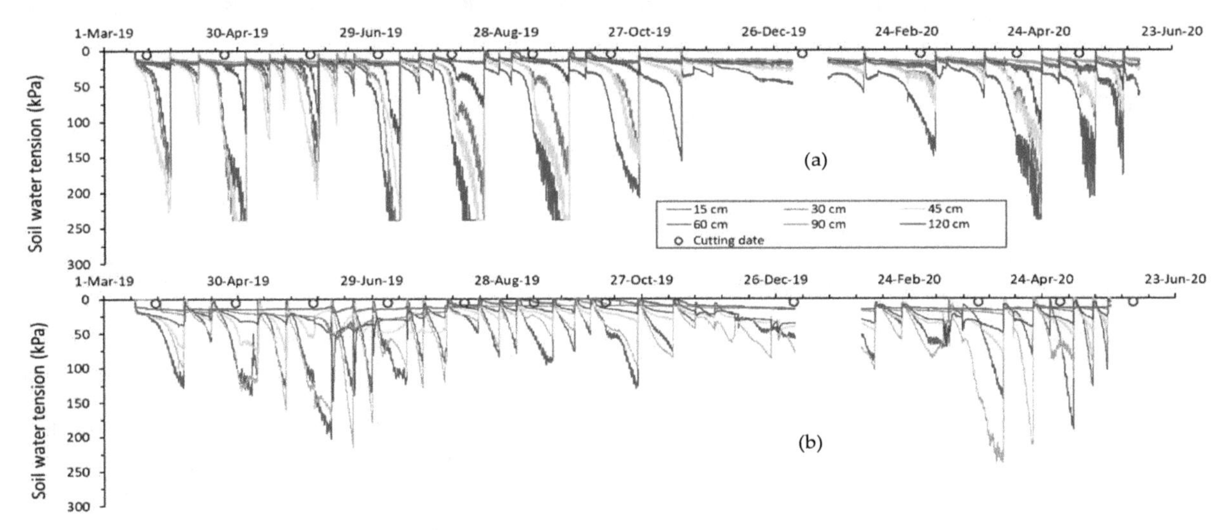

Figure 4. Half-hourly soil water tension (kPa) measured at multiple depths of 15 cm, 30 cm, 45 cm, 60 cm, 90 cm, and 120 cm in plots under normal farmer irrigation practice (NI) at—(**a**) site A4 and (**b**) site A2, from March 2019 to June 2020. Cutting dates are demonstrated with circles on the x-axes.

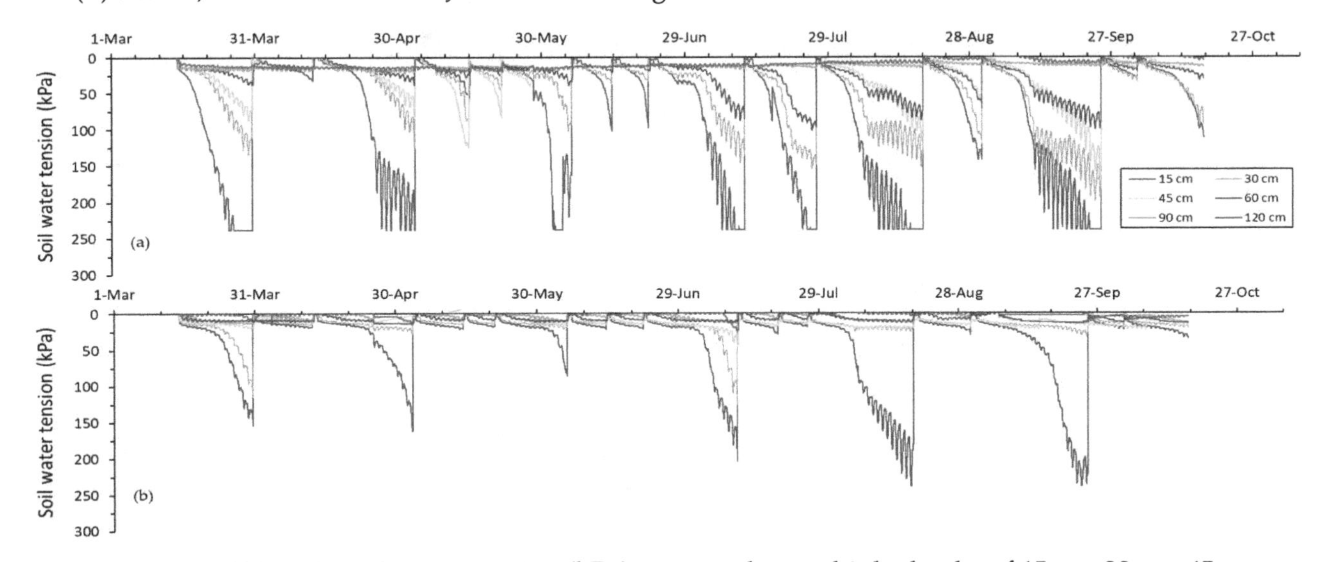

Figure 5. Half-hourly soil water tension (kPa) measured at multiple depths of 15 cm, 30 cm, 45 cm, 60 cm, 90 cm, 120 cm in plots under (**a**) irrigation strategy DI3 at site A4 and (**b**) irrigation strategy DI4 at site A4, from March 2019 to mid-October 2019.

The cumulative ET_a (CET_a) in irrigation strategy NI at sites A1–A4, for a 517-day period (1 January 2019 to 31 May 2020) was 2202, 2187, 2031, and 2175 mm, respectively (Figure 8). For a one-season 12-month irrigation period (2019) for the same irrigation strategy, the cumulative ETa was 1596 mm at site A1, 1582 mm at site A2, 1423 mm at site A3, and 1558 mm at site A4. Comparing the total applied water and the cumulative ET_a under normal farmer irrigation strategies indicated that the plots under NI irrigation strategy remained over-irrigated during the whole study period. However, moderate water stress was occasionally observed for the NI irrigation strategy, particularly at sites A3 and A4, because of the dry-down around alfalfa harvests or the unprecedented delays in irrigation schedules.

The daily K_a values for the sites A1 and A4 (irrigation strategy NI) over the study period are shown in Figure 7. The K_a value depended on alfalfa growth stages, ranging from smallest during initial growth stage, just after each harvest, and reaching the maximum when the crop height was at mid and full canopy development stages, attained prior to each harvest cycle. Large K_a values were attained at both sites during March and April (ranging from 0.47 after harvest to 1.24 at full canopy). Growth of alfalfa in early season (January) and late season (November to December) was slower due to cooler weather and lower solar radiation, in which lower K_a values were observed. The average K_a values of alfalfa sites over the study period varied from 0.8 at site A3 to 0.87 at site A1 (Figure 8 and Table 4).

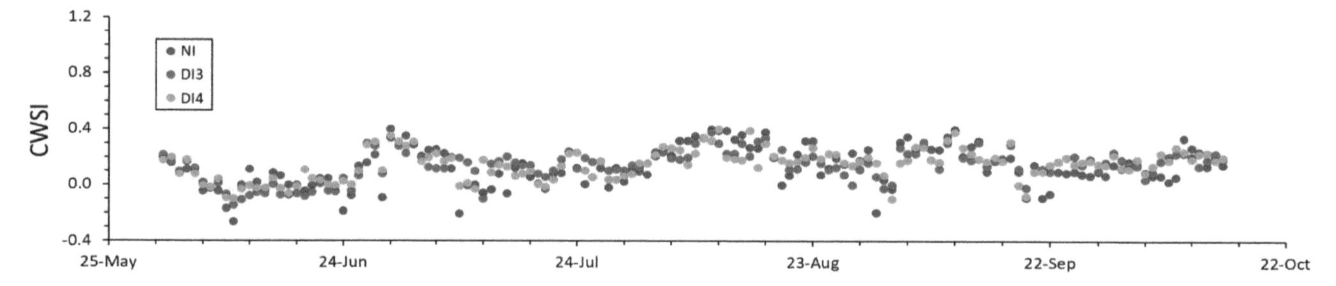

Figure 6. Daily crop water stress index (CWSI) values for the plots under different irrigation strategies (NI, DI3, and DI4) at site A4. NI, DI3, and DI4 represent normal farmer irrigation practice, applying 25% less water than NI, and applying 12.5% less water than NI, respectively.

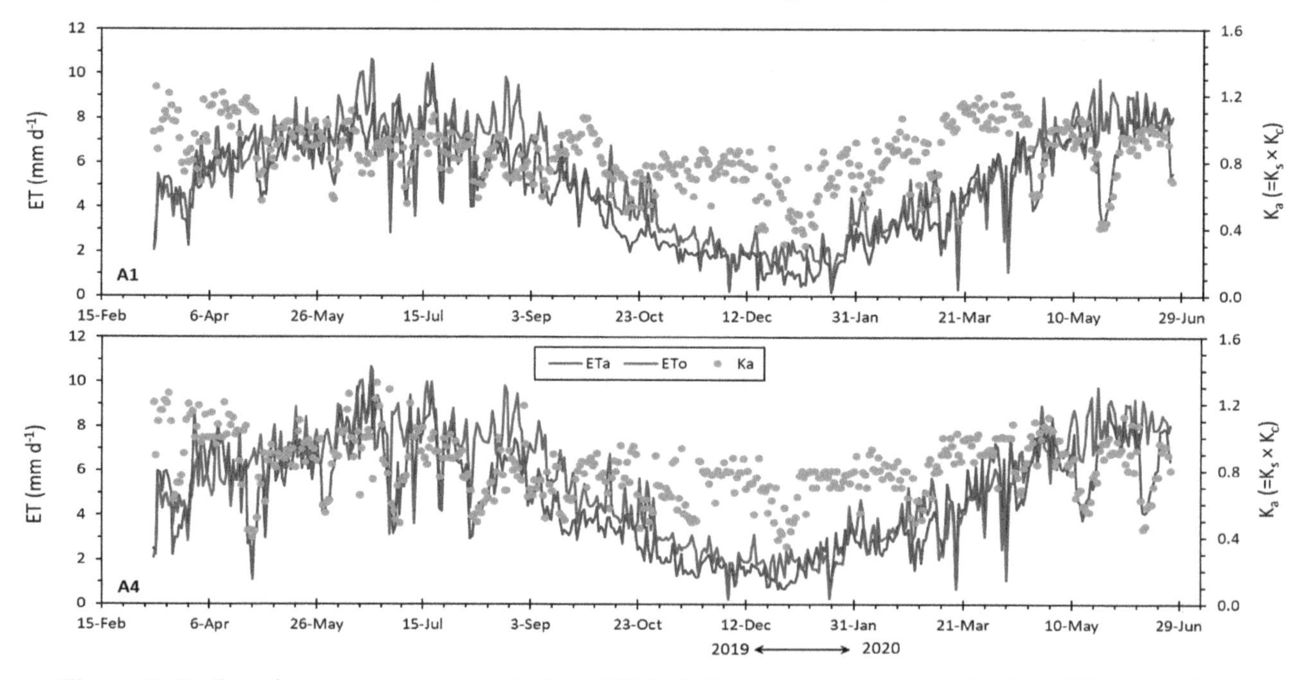

Figure 7. Daily reference evapotranspiration (ET_o), daily actual evapotranspiration (ET_a), and daily actual crop coefficient (K_a) at sites (**A1**) and (**A4**), from March 2019 to June 2020.

Average K_a values at harvest cycles (eight cuttings in 2019 and three cuttings in 2020) for each experimental site are provided in Table 4. The results demonstrated seasonal variabilities in the harvest cycle K_a values. With an average seasonal K_a value of 0.87 for the 2019 season at the site A1, the average cutting cycle K_a values varied from 0.72 (cutting cycle 8) to 1.0 (cutting cycle 2). Average seasonal K_a values for the 2019 season at sites A2, A3, and A4 were 0.86, 0.79, and 0.84, respectively. There was a considerable increase in average K_a value (12.5%) at site A3 from the 2020 cropping season compared to 2019. Lower K_a values in the first three harvest cycles of the 2019 season might have been due to poor irrigation management. Site A3 received 424 mm water during the first three months of 2019, which was the lowest amount of water applied amongst the experimental sites (Figure 3). While trivial differences (an average of 5%) were found in the average K_a values at sites A1 and A2 during harvest cycles of June–August (cutting cycles 4 to 6) at site A1 and A2, substantial differences (average of 20%) were obtained in the mean K_a values of cutting cycles at sites A3 and A4.

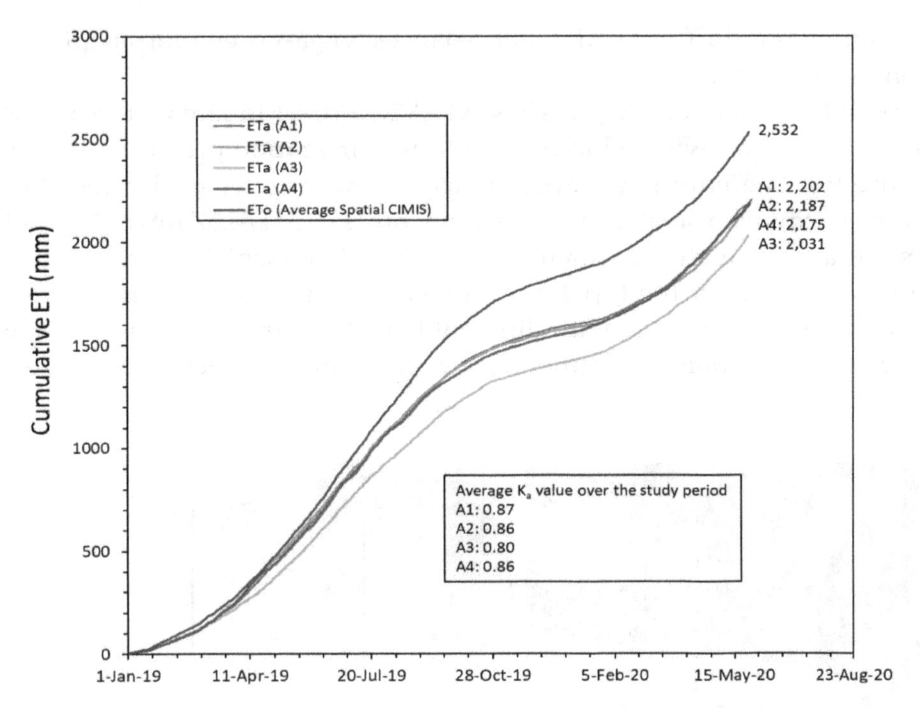

Figure 8. Cumulative reference evapotranspiration (Spatial CIMIS ET_o) and cumulative actual evapotranspiration (ET_a) at each of the experimental sites (A1–A4). The cumulative ET_a are provided for the plots under normal farmer irrigation practices at each site. The average spatial CIMIS ET_o of the four sites is provided as ET_o.

Table 4. Mean (±standard deviation) K_a values of harvest cycles for each experimental site (A1–A4) determined from surface renewal measurements. The values are reported for the normal farmer irrigation practices over eight cuts in the season 2019 (Year 1) and three cuts in the season 2020 (Year 2).

Cut—Year Number	Harvest Time	K_a			
		A1	A2	A3	A4
Cut 1—Year 1	23 Mar–4 Apr	0.81 (±0.13)	0.81 (±0.14)	0.79± (0.16)	0.80 (±0.14)
Cut 2—Year 1	24 Apr–8 May	1.00 (±0.14)	1.03 (±0.15)	0.83 (±0.16)	1.01 (±0.11)
Cut 3—Year 1	1 Jun–12 Jun	0.94 (±0.13)	0.92 (±0.13)	0.78 (±0.14)	0.83 (±0.14)
Cut 4—Year 1	3 Jul–12 Jul	0.89 (±0.12)	0.87 (±0.13)	0.85 (±0.13)	0.98 (±0.12)
Cut 5—Year 1	5 Aug–16 Aug	0.88 (±0.14)	0.86 (±0.11)	0.83 (±0.10)	0.86 (±0.15)
Cut 6—Year 1	5 Sep–19 Sep	0.85 (±0.10)	0.82 (±0.11)	0.75 (±0.12)	0.77 (±0.11)
Cut 7—Year 1	9 Oct–29 Oct	0.78 (±0.12)	0.74 (±0.13)	0.74 (±0.11)	0.73 (±0.12)
Cut 8—Year 1	31 Dec–15 Jan	0.72 (±0.14)	0.73 (±0.14)	0.71 (±0.13)	0.70 (±0.15)
Cut 1—Year 2	4 Mar–30 Mar	0.78 (±0.16)	0.79 (±0.15)	0.83 (±0.17)	0.87 (±0.14)
Cut 2—Year 2	20 Apr–5 May	1.02 (±0.13)	0.93 (±0.15)	0.97 (±0.15)	0.94 (±0.15)
Cut 3—Year 2	22 May–14 Jun	0.90 (±0.14)	0.89 (±0.13)	0.90 (±0.12)	0.90 (±0.12)
Average	-	0.87 (±0.10)	0.86 (±0.09)	0.82 (±0.07)	0.86 (±0.10)

The observed average K_a value from this study was lower than the value (0.95) suggested by Doorenbos and Kassam [38] for dry climate and 0.99 reported by Hanson et al. [3] for the Sacramento Valley. However, the average K_a value was about the same as what was found by Kuslu et al. [39].

3.5. Soil Salinity and Water Availability Features

It is well-known that salinity associated problems are a major challenge for global food production, with particularly critical impact in the low desert region. Applications of excess water to control soil root zone salinity is an important agricultural practice for these regions and considered a 'beneficial use' of water, since soils and crop production can only be sustained by controlling salinity. Buildup of salinity might be considered a serious concern and likely a key limitation for any reduced water demand strategies in the region. Therefore, it is important to understand the impact of deficit irrigation on potential soil salinity buildup and soil water balances, vis-à-vis evapotranspiration and devising optimal irrigation management.

Spatially interpolated map of EM_V at sites A1, A2, and A4 in late October of 2019, just before all deficit irrigated plots were switched to normal farmer irrigation practice after the 1.5-year study, is shown in Figure 9. The entire surveyed areas that were affected by the different irrigation strategies at these sites exhibited small EM_V measurements (11.13–174.57 mS m^{-1} or 0.11–1.17 dS m^{-1}). These measurements approximate the differences in EM values (which is affected by salinity, texture, and moisture) in approximately the top 1.2 m of soil. Inconsequential differences were observed among EM_V values of the plots treated with different irrigation strategies. However, we should point out that these were moderate deficit treatments, and more severe deficits might cause greater excess salinity buildup.

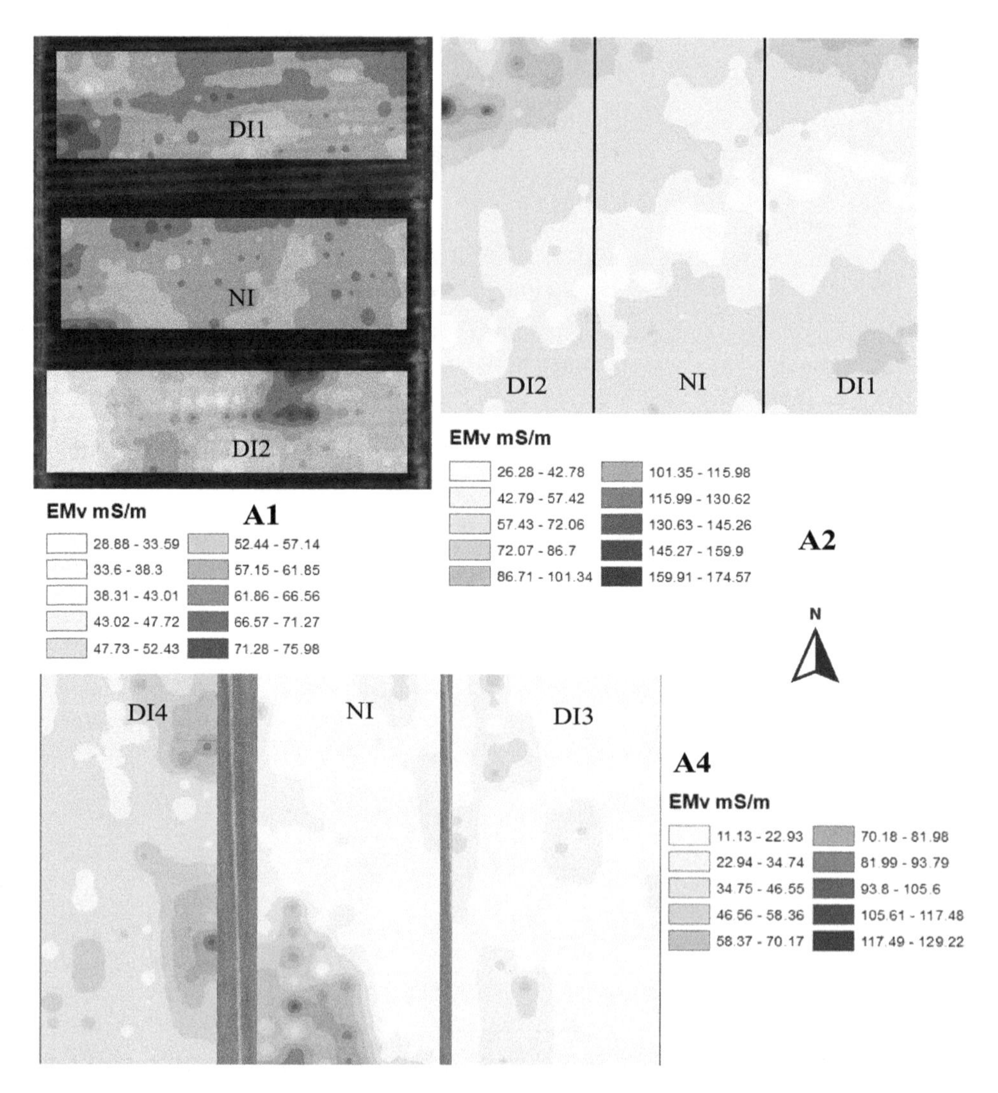

Figure 9. Spatial distribution of ancillary variable EM_V (electromagnetic induction measurement in the vertical coil orientation) at sites (**A1**), (**A2**), and (**A4**). 100 m S m^{-1} is equal 1 dS m^{-1}.

In this study, the mean EC_e at the effective crop root zone (30–120 cm), across the experimental sites in late October 2019, demonstrated that deficit irrigation strategies had some impacts on the soil salinity (Figure 10), however, these values were always in the 'acceptable' range for alfalfa. A higher level of ECe values were observed in plots under irrigation strategy DI1, in comparison to plots under irrigation strategies NI and DI2, at sites A1 and A2 (furrow irrigated alfalfa fields). For instance, the mean ECe at site A2 was 2.2 dS m^{-1} and 4.2 dS m^{-1} at 60 and 90 cm soil depths, respectively, in irrigation strategy DI1. However, the values were 1.47 and 1.76 dS m^{-1} in irrigation strategy NI, and were 1.45 and 2.2 dS m^{-1} in irrigation strategy DI2, at the corresponding depths, respectively.

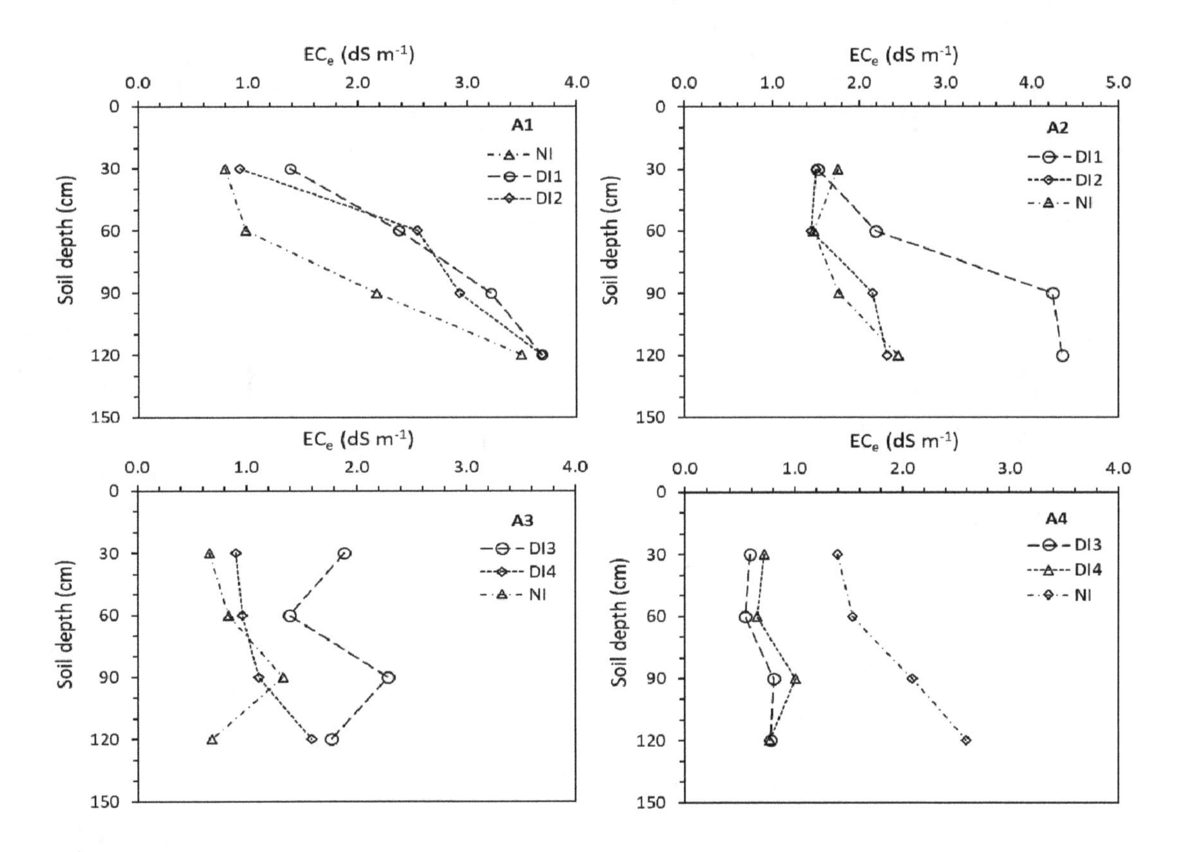

Figure 10. Whole-soil profile representations of mean EC_e (electrical conductivity of the saturation extract) distribution of observed values in different irrigation strategies at four experimental sites. The complete data (collected in late October 2019) from the six soil core sampling locations in plots under different irrigation strategies at each site were used to plot EC_e.

3.6. *Alfalfa Yield Responses*

Alfalfa dry matter values for eight seasonal cuttings 2019 (Y1) and the first three cuttings for 2020 (Y2) under each irrigation strategies of all experimental sites are illustrated in Figure 11. The results indicated that the first four cuttings were the most productive for all sites in 2019. For instance, mean DM yield at site A2 from irrigation strategy NI were 4.7, 4.0, 4.7, and 3.9 Mg ha^{-1} for first, second, third, and fourth cuttings, respectively. For sites A1, A2, and A4, mean DM values for the first three cuttings in 2020 were lower than the corresponding DM values in 2019. At site A3, alfalfa DM yields were relatively lower than the other sites over the first three cuttings of the 2019 season.

Alfalfa mean DM yields from the 5th to 7th cuttings in 2019 were much lower than the first four cuttings even under full irrigation practices. Yields in late summer were moderately affected by deficit irrigation strategies. For example, yield reduction at site A1 from cuttings 5, 6, and 7 in irrigation strategy DI1, compared to irrigation strategy NI were 0.3, 0.8, and 0.6 Mg ha^{-1}, respectively (Figure 11). The reduction in DM yield was 0.5 Mg ha^{-1} for cutting 6 and was 0.4 Mg ha^{-1} for cutting 7 in irrigation strategy DI2. Yields summed over the 11 cuttings, over 1.5 years, were reduced from 0.7% to 4.6% (0.3 to 1.7 Mg ha^{-1}) by the deficit irrigation practices at the four sites.

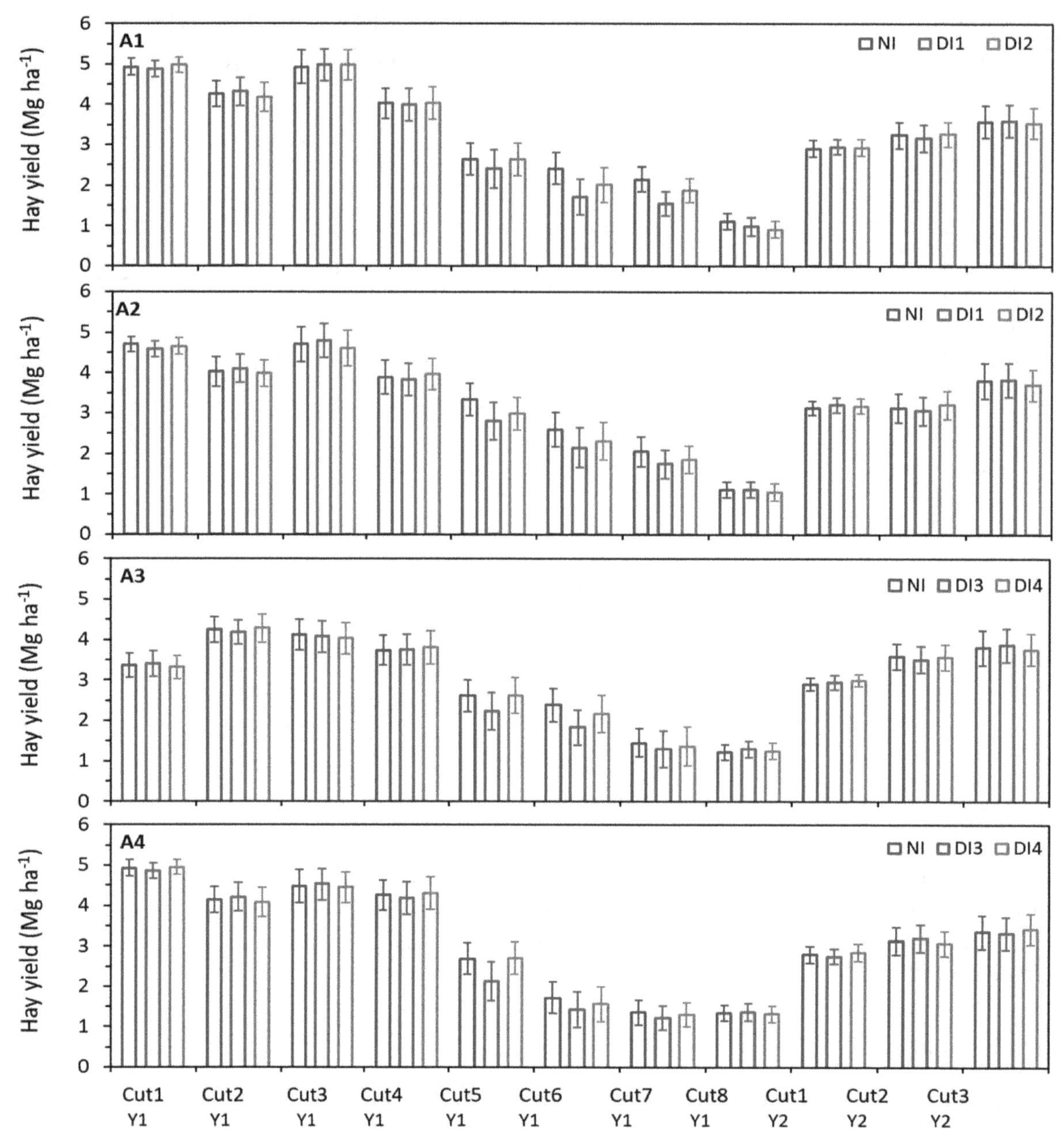

Figure 11. Mean dry matter (DM) yields of each irrigation strategy (NI, DI1–DI4) at the experimental sites (**A1–A4**) for eight cuttings in the season 2019 (Cu1-Y1 to Cut8-Y1) and three cuttings in the season 2020 (Cut1-Y2 to Cut3-Y2). The bars demonstrate the standard deviation of DM values.

3.7. *Forage Quality*

There was a trend for a small improvement in forage quality due to the deficit irrigation strategies, but not at all sites (Table 5). A significant reduction in acid detergent fiber was observed in deficit irrigation strategies DI1 at site A2 and DI2 at site A1, compared to normal irrigation practice (p values of 0.001 for DI1 at site A2, 0.02 for DI2 at site A1). Significant crude protein increase was also found from implementing deficit irrigation regimes DI1 at site A2 (p value of 0.04) and DI3 at site A4 (p value of 0.01), but not at other sites. No significant impact was observed on lignin percentage. The improved forage quality might be attributed to a reduction in stem growth (increase % leaf) under such irrigation practices. Small improvements in alfalfa forage quality under deficit irrigation regimes was also reported by other researchers [5,40,41].

Table 5. Mean forage quality indices (Acid Detergent Fiber (ADF), Crude Protein (CP), and Lignin) of normal farmer irrigation practices against deficit irrigation strategies. The forage quality data of June through September 2019 was used for this analysis (t-test).

Site	ADF (%)					CP (%)					Lignin (%)				
	NI	DI1	DI2	DI3	DI4	NI	DI1	DI2	DI3	DI4	NI	DI1	DI2	DI3	DI4
A1	29.1	28.9 ^{ns}	27.8 *	-	-	21.0	20.7 ^{ns}	20.8 ^{ns}	-	-	5.2	5.1 ^{ns}	5.2 ^{ns}	-	-
A2	31.2	26.4 *	28.2 ^{ns}	-	-	19.6	22.5 *	21.2 ^{ns}	-	-	5.2	5.1 ^{ns}	5.1 ^{ns}	-	-
A3	27.6	-	-	25.7 ^{ns}	27.5 ^{ns}	18.3	-	-	19.4 ^{ns}	18.4 ^{ns}	4.5	-	-	4.6 ^{ns}	4.7 ^{ns}
A4	31.2	-	-	26.5 ^{ns}	30.0 ^{ns}	17.6	-	-	20.6 *	18.3 ^{ns}	5.1	-	-	4.9 ^{ns}	5.0 ^{ns}

^{ns} Non-significant. * Significant at the 5% level of probability.

3.8. Water Conservation Versus Yield Reduction

Deficit strategies with alfalfa are primarily feasible due to the seasonal yield patterns of the crop, with heavy yields during early season, and very light yields in late summer. Approximately 73–74% of total alfalfa seasonal yield productivity at the experimental sites occurred by mid-July (20 July 2019), right before starting summer deficit irrigation strategies (Figure 12). This finding is similar to results from research reported for the Sacramento Valley of California [5,39]. The deficit irrigation strategies could affect the 5th through 7th cuttings, while only 21–22% of the annual DM yield was produced during this period.

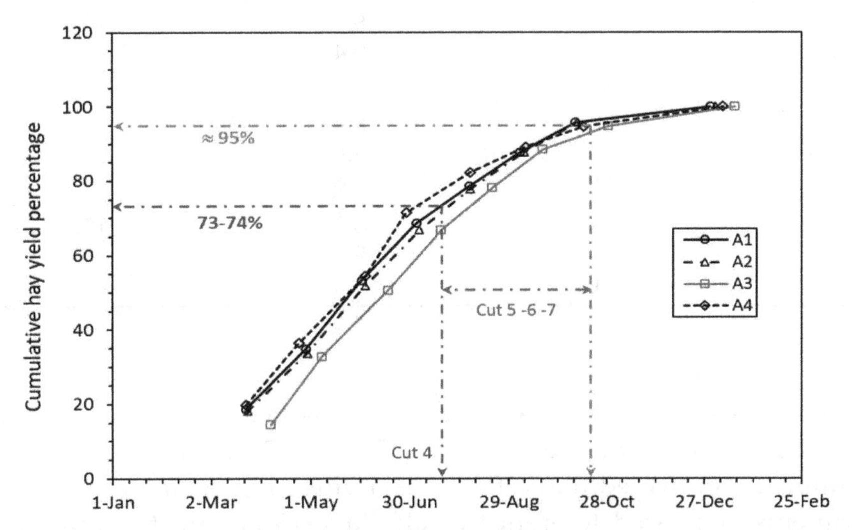

Figure 12. Cumulative alfalfa yield percentage over the growing season 2019 at the experimental sites (A1–A4). Results are provided for the normal irrigation practices at each site.

A significant DM reduction was observed in deficit irrigation strategies DI1 and DI2, compared to normal irrigation practice at site A2 (p value of 0.0004 for DI1 and 0.002 for DI2). There was also a significant yield reduction in deficit irrigation DI1 at site A1 (p value of 0.0005) (Table 6). No significant DM reduction was affected by deficit irrigation regimes DI3 and DI4 at sites A3 and A4; and deficit irrigation DI2 at site A2. The findings suggest an average of 1.7 Mg ha^{-1} and 1.0 Mg ha^{-1} dry matter yield reduction in deficit irrigation strategies DI1 and DI2 at sites A1 and A2, respectively (Table 6). The average DM yield reduction at sites A3 and A4 was nearly 1.0 Mg ha^{-1} in deficit irrigation strategy DI3 and 0.4 Mg ha^{-1} in deficit irrigation strategy DI4.

The total amount of conserved water across the experimental sites varied from 83 mm (3.0%) at site A3 to 314 mm (10.5%) at site A1, relative to what was used under seasonal water applied in normal irrigation practice. Summer deficit irrigation strategies enhanced the IWP values, but not the ETWP values (Table 7). For instance, irrigation strategies DI1 and DI2 at site A2 improved the IWP value by 0.5 and 0.4 kg ha^{-1} mm^{-1}, respectively, compared to the irrigation strategy NI (with an IWP of 8.9 kg ha^{-1} mm^{-1}).

Table 6. The total mean dry matter yield (Mg ha^{-1}) for different irrigation management strategies and total mean dry matter yield in normal irrigation practice at the experimental sites (18-month period). The significance of independent *t*-tests is provided.

Site	Normal Irrigation Practice (NI)	Deficit Irrigation Strategy			
		DI1	DI2	DI3	DI4
A1	36.2	34.6 *	35.4 ns	-	-
A2	36.6	35.0 *	35.5 *	-	-
A3	33.5	-	-	32.6 ns	33.1 ns
A4	34.3	-	-	33.2 ns	34.0 ns

ns Non-significant. * Significant at the 5% level of probability.

Table 7. The irrigation water productivity (IWP) and actual evapotranspiration (ET) water productivity (ETWP) values in different irrigation strategies (NI, DI1–DI4) at each of the experimental sites (A1–A4).

Site	Irrigation Strategy	*IWP* (kg ha^{-1} mm^{-1})	*ETWP* (kg ha^{-1} mm^{-1})
A1	NI	8.9	16.6
	DI1	9.3	16.1
	DI2	9.3	16.4
A2	NI	8.9	16.4
	DI1	9.4	16.1
	DI2	9.3	16.3
A3	NI	8.4	16.1
	DI3	8.6	15.8
	DI4	8.6	16.1
A4	NI	8.0	15.9
	DI3	8.3	15.6
	DI4	8.2	15.9

4. Discussion

Alfalfa was historically reported to be a moderately sensitive crop to salinity, with estimated yield declines above a saturated soil extract (EC_e) of 2.0 dS m^{-1} [42]. However, more recent reports and experiments in California confirmed that alfalfa has a much higher tolerance of salinity. Field and greenhouse experiments estimated tolerance of alfalfa varieties up to approximately 6.0 EC_e or higher [43]. The findings from this study suggest that the proposed deficit irrigation strategies might cause salt accumulation at the crop root zone (particularly for furrow irrigated alfalfa fields) and this practice might elevate soil salinity class from a non-saline soil (0–2 dS m^{-1}) to slightly saline (2–4 dS m^{-1}) condition. Soil salinity can be managed by switching to farmer irrigation practices in early fall (mid- to late-October), with no need for the excessive water to leach salts.

The deliberate re-filling of the soil profile with irrigation water after implementing summer deficits should be considered, both in terms of water availability, salinity management, and water-use policy. The continuous soil moisture readings in this study verified that there was insignificant soil moisture depletion from deficit irrigations to require some recharge (Figures 4 and 5). As can be seen from the soil water tension data plots, the average soil water tension was kept constant at about 88 kPa, at the top 30 cm, and maintained at about 15 kPa, for the 45–120 cm soil depth. Consequently, farmers might confidently refill the soil profile after implementing summer deficit irrigation strategies and switching to their normal irrigation practices with little excessive water needs in the fall.

At sites A1–A4, the total mean annual DM yield in irrigation strategy NI for 2019 was 26.2, 26.5, 23.2, and 25.0 Mg ha^{-1}, respectively (Figure 11). These alfalfa yield values from the long-seasoned low desert sites are generally higher than the average alfalfa yield in California, which ranges from 14.6 to 16.1 Mg ha^{-1}, and a thirty-year (1984–2013) statewide average yield of 15.3 Mg ha^{-1} [44]. Dry matter yields of 11.2 Mg ha^{-1} was reported in an earlier deficit study that imposed severe water deficits, restricting applications to 1,249 mm seasonal water use in the Imperial Valley [3].

Alfalfa plant stand evaluation conducted on 18 February 2020 showed no significant differences in the plant population between the deficit irrigation strategies and normal irrigation practices, suggesting that there is no evidence of losing the alfalfa plant density from the implemented deficit irrigation strategies. For instance, plant population per hectare at site A1 was estimated to be 103×10^6, 105×10^6, and 102×10^6 in the plots under irrigation strategies NI, DI1, and DI2, respectively. Additionally, no yield reduction was observed from the summer deficit irrigation strategies within the first three harvest cuttings of the 2020 season (Figure 10), indicating a full recovery of the crop upon re-watering.

None of the deficit irrigation strategies produced severe water or salinity stress at the experimental sites. Soil water availability in non-sandy soils (sites A1–A3) was retained at a desired level during and after summer irrigation strategies. At site A4 (sandy loam soil), the residual soil moisture below the depth of 45 cm was consistent at a non-stressed level, and supplied enough water for alfalfa, suggesting that this might be a primary reason why there was only a small reduction in total ET_a of deficit irrigated plots. The maximum ET_a reduction (54 mm) was observed in irrigation strategy DI1 at site A2, when compared with strategy NI. The findings revealed that ET_a was not directly reduced in relation to the level of reduction in applied water. This might be part of the reason why no improvement was gained in the ETWP values from moderate deficit irrigation strategies. Further measurement is necessary to provide a more solid conclusion on the impact of the deficit irrigation strategies on ET_a. Overall, the ETWP values computed for the normal farmer irrigation practices and deficit irrigation trials at the experimental sites (an average of 16.1 kg ha^{-1} mm^{-1}) were as high as the values predicted by Montazar et al. [45] for alfalfa fields, under subsurface drip irrigation in the low desert of California.

Imposition of summer water deficits is likely to result in yield reductions, a finding that was similar to other researches [3–7,17]. The findings suggest that conserving 0.083–0.314 (ha.m) ha^{-1} water through summer deficit irrigation strategies might result in 0.3–1.7 Mg ha^{-1} yield loss in California's low desert alfalfa production system. Therefore, while insignificant yield reduction is unpreventable, the proposed management strategies could serve as an effective water conservation tool. The hay yield reduction might be a consequence of reduced water distribution uniformity caused by the deficit irrigation regimes. Large-scale farming systems in the low desert, along with the use of surface irrigation methods, resulted in lower water distribution uniformity values (over time and space) with deficit irrigation strategies. Results expected from these deficit irrigation strategies would likely be more favorable in more efficient irrigation systems such as subsurface drip irrigation systems or advanced overhead sprinkler systems.

The practice of filling the soil profile so that it holds as much water as possible would be an effective early-season alfalfa irrigation strategy. Such practice might allow alfalfa to take full advantage of the available water and promote its rapid, early season growth, when the yield potential was highest, and when soil and water temperatures were not likely to be high enough to stress the crop and limit crop productivity. Consequently, combining full irrigation in winter-spring with moderate deficit irrigation during summer could be an efficient approach in conserving water than continuously irrigating (or over irrigating) for the entire season.

5. Conclusions

This study aimed at assessing the effectiveness of moderate deficit irrigation strategies during summer harvest cycles (less applied water than normal farmers' irrigation practices) on conserving water and maintaining a robust hay production.

The proposed deficit irrigation strategies conducted showed a promising and decent amount of water conservation and simultaneously generated desirable hay yields and quality. However, yield penalties of this practice must be considered. These moderate deficit irrigation practices resulted in an average of 1.47 Mg ha^{-1} and 0.31 Mg ha^{-1} hay yield reduction, but used

33% (≈0.31 (ha.m) ha^{-1}) and 12.5% less applied-water than normal farmer practice over the summer (≈0.10 (ha.m) ha^{-1}), respectively.

Several cautionary notes need to be considered with the data reported and the analysis provided in this study:

- The ET and crop coefficient values reported herein are referred to as observed or actual values, which are limited by water deficits and salinity, and the 'dry down' required for frequent harvests. The maximum crop ET (ET_c) is limited only by energy availability to vaporize water and not soil hydrology or salinity, or droughts imposed by harvest scheduling. Imposed stress, such as this, is common to almost all alfalfa growing regions.
- Although stand density under desert conditions decays more rapidly than other locations, there were negligible differences between different deficit irrigation strategies and normal farmers' irrigation practices in this study after one year of water deficits. However, it is uncertain whether multiple years of summer irrigation strategies might threaten the long-term viability of the crop stand and yields.
- Although, it might be unlikely to prevent salinity buildup due to summer water deficits, salinity issues are likely to be managed through irrigation practices that flush salts in the months after implementing deficit irrigations. Continuous monitoring of soil salinity is recommended to ensure flushing/leaching salts out of root zone over multiple deficit irrigation seasons.
- The importance of re-filling of the soil profile with water, in the year after implementing summer irrigation strategies, need to be considered both in terms of water availability, crop production, and policy. Such practices might enable shifting water demand to water-rich time periods in early spring. This practice would benefit both early season growth and salinity management in subsequent years. In this study, continuous soil moisture readings verified that soil moisture was insignificantly depleted in the deficit irrigation fields. However, data from multiple irrigation seasons are required to fully certify this conclusion.
- Implementation of the proposed summer deficit irrigation strategies on alfalfa could provide a reliable source of seasonally available water as well sustain the economic viability of agriculture in the region. These strategies might be sustainable as an effective water conservation tool if such measures provide adequate economic incentives to the participating farmers. Incentive programs to farmers must offset the risk of implementing the proposed practices (even trivial production loss), as a tool for adopting water conservation practices.

Author Contributions: Conceptualization, A.M. and D.P.; Data curation, A.M.; Formal analysis, A.M.; Funding acquisition, A.M.; Investigation, A.M., O.B., and D.C.; Methodology, A.M., D.C., and D.P.; Project administration, A.M.; Supervision, A.M.; Writing—original draft, A.M.; Writing—review & editing, A.M., O.B., D.C., and D.P. All authors have read and agreed to the published version of the manuscript.

Acknowledgments: The authors would like to thank the local NRCS office in Blythe, Palo Verde Resource Conservation District, and Palo Verde Irrigation District for providing their support and input. The authors gratefully acknowledge the Seiler Farms and Chaffin Farms for their sincere collaboration during this study, and for allowing us to implement the project in their agricultural operations. The authors wish to acknowledge the technical support of Tayebeh Hosseini and Tait Rounsaville, whose conscientious works in the field and in the laboratory were crucial to the success of this study.

References

1. Montazar, A. Update alfalfa crop water use information: An estimation for spring and summer harvest cycles in California low desert. *Agric. Briefs-Imp. Cty.* **2019**, 22, 202–205.
2. Montazar, A. Deficit Irrigation Program Study. In Proceedings of the Law of Colorado River Conference, Scottsdale, AZ, USA, 12–13 March 2020.

3. Hanson, B.; Putnam, D.; Snyder, R. Deficit irrigation of alfalfa as a strategy for providing water for water-short areas. *Agric. Water Manag.* **2007**, *93*, 73–80. [CrossRef]
4. Orloff, S.; Hanson, B. Conserving water through deficit irrigation of alfalfa in the intermountain area of California. *Forage Grazing Lands* **2008**. [CrossRef]
5. Orloff, S.; Putnam, D.; Bali, K. *Drought Strategies for Alfalfa*; ANR Publication no. 8522; University of California: Oakland, CA, USA, 2015.
6. Montazar, A.; Radawich, J.; Zaccaria, D.; Bali, K.; Putnam, D. Increasing water use efficiency in alfalfa production through deficit irrigation strategies under subsurface drip irrigation. Water shortage and drought: From challenges to solution. In Proceedings of the USCID Water Management Conference, San Diego, CA, USA, 17–20 May 2016.
7. Montazar, A. Partial—Season Irrigation as an Effective Water Management Strategy in Alfalfa. *Agric. Briefs* **2017**, *10*, 3–7.
8. Peterson, P.R.; Sheaffer, C.C.; Hall, M.H. Drought Effects on Perennial Forage Legume Yield and Quality. *Agron. J.* **1992**, *84*, 774–779. [CrossRef]
9. Kirkham, M.B.; Johnson, D.E.; Kanemasu, E.T.; Stone, L.R. Canopy temperature and growth of differentially irrigated alfalfa. *Agric. Meteorol.* **1983**, *19*, 235–246. [CrossRef]
10. Grimes, D.W.; Wiley, P.L.; Sheesley, W.R. Alfalfa yield and plant water relations with variable irrigation. *Crop Sci.* **1992**, *32*, 1381–1387. [CrossRef]
11. Asseng, S.; Hsiao, T. Canopy Co_2 assimilation, energy balance, and water use efficiency of an alfalfa crop before and after cutting. *Field Crops Res.* **2000**, *67*, 191–206. [CrossRef]
12. Orloff, S.; Putnam, D.; Bali, K.; Hanson, B.; Carlson, H. Implications of deficit irrigation management on alfalfa. In Proceedings of the California Alfalfa and Forage Symposium, Visalia, CA, USA, 12–14 December 2005.
13. Ottman, M. Irrigation cutoffs with alfalfa-what are the implications? In Proceedings of the 2011 Western Alfalfa and Forage Conference, Las Vegas, NV, USA, 11–13 December 2011.
14. Sheaffer, C.C.; Tanner, C.B.; Kirkham, M.B. Alfalfa water relations and irrigation. In *Alfalfa and Alfalfa Improvement*; Hanson, A.A., Barnes, D.K., Hill, R.R., Eds.; Agronomy Monographs, ASA-CSSA-SSSA; The American Society of Agronomy: Madison, WI, USA, 1988; Volume 29, pp. 373–409.
15. Ottman, M.J.; Tickes, B.R.; Roth, R.L. Alfalfa yield and stand response to irrigation termination in an arid environment. *Agron. J.* **1996**, *88*, 44–48. [CrossRef]
16. Frate, C.A.; Roberts, B.A.; Marble, V.L. Imposed drought stress has no long-term effect on established alfalfa. *Calif. Agric.* **1991**, *45*, 33–36. [CrossRef]
17. Robinson, F.E.; Teuber, L.R.; Gibbs, L.K. *Alfalfa Water Stress Management During Summer in Imperial Valley for Water Conservation*; Research Report; Desert Research and Extension Center: El Centro, CA, USA, 1994.
18. Putnam, D.; Takele, E.; Kallenback, R.; Graves, W. *Irrigating Alfalfa in the Low Desert: Can Summer Dry-Down be Effective for Saving Water in Alfalfa?* United States Bureau of Reclamation: Washington, DC, USA, 2000.
19. Cabot, P.; Brummer, J.; Hansen, N. Benefits and impacts of partial season irrigation on alfalfa production. In Proceedings of the 2017 Western Alfalfa and Forage Conference, Reno, NV, USA, 28–30 November 2017.
20. Guitjens, J.C. Alfalfa irrigation during drought. *J. Irrig. Drain. Eng.* **1993**, *119*, 1092–1098. [CrossRef]
21. Lamm, F.L.; Harmoney, K.R.; Aboukheira, A.A.; Johnson, S.K. Alfalfa production with subsurface drip irrigation in the Central Great Plains. *Trans. ASABE* **2012**, *55*, 1203–1212. [CrossRef]
22. Irmak, S.; Odhiambo, L.O.; Kranz, W.L.; Eisenhauer, D.E. *Irrigation Efficiency and Uniformity, and Crop Water Use Efficiency*; Extension Circular EC732; University of Nebraska-Lincoln: Lincoln, NE, USA, 2011.
23. Snyder, R.L.; Spano, D.; Pawu, K.T. Surface Renewal analysis for sensible and latent heat flux density. *Bound. Layer Meteorol.* **1996**, *77*, 249–266. [CrossRef]
24. Shaw, R.H.; Snyder, R.L. Evaporation and eddy covariance. In *Encyclopedia of Water Science*; Stewart, B.A., Howell, T., Eds.; Marcel Dekker: New York, NY, USA, 2003.
25. Paw, U.K.T.; Snyder, R.L.; Spano, D.; Su, H.-B. Surface renewal estimates of scalar exchange. In *Micrometeorology in Agricultural Systems*; Hatfield, J.L., Baker, J.M., Eds.; Agronomy Society of America: Madison, WI, USA, 2005; pp. 455–483.
26. Montazar, A.; Rejmanek, H.; Tindula, G.N.; Little, C.; Shapland, T.M.; Anderson, F.E.; Inglese, G.; Mutters, R.G.; Linquist, B.; Greer, C.A.; et al. Crop coefficient curve for paddy rice from residual of the energy balance calculations. *J. Irrig. Drain. Eng.* **2017**, *143*, 04016076. [CrossRef]

27. Shapland, T.M.; McElrone, A.J.; Paw, U.K.T.; Snyder, R.L. A turnkey data logger program for field-scale energy flux density measurements using eddy covariance and surface renewal. *Ital. J. Agrometeorol.* **2013**, *1*, 1–9.
28. Snyder, R.L.; Pedras, C.; Montazar, A.; Henry, J.M.; Ackley, D.A. Advances in ET-based landscape irrigation management. *Agric. Water Manag.* **2015**, *147*, 187–197. [CrossRef]
29. Montazar, A.; Krueger, R.; Corwin, D.; Pourreza, A.; Little, C.; Rios, S.; Snyder, R.L. Determination of Actual Evapotranspiration and Crop Coefficients of California Date Palms Using the Residual of Energy Balance Approach. *Water* **2020**, *12*, 2253. [CrossRef]
30. Twine, T.E.; Kustas, W.P.; Norman, J.M.; Cook, D.R.; Houser, P.R.; Meyers, T.P.; Prueger, J.H.; Starks, P.J.; Wesely, M.L. Correcting eddy-covariance flux underestimates over a grassland. *Agric. Forest Meteorol.* **2000**, *103*, 279–300. [CrossRef]
31. CIMIS. Available online: http://wwwcimis.water.ca.gov/SpatialData.aspx (accessed on 7 September 2020).
32. Idso, S.B.; Jackson, R.D.; Pinter, P.J., Jr.; Reginato, R.J.; Hatfield, J.L. Normalizing the stress-degree-day parameter for environmental variability. *Agric. Meteorol.* **1981**, *24*, 45–55. [CrossRef]
33. Corwin, D.L.; Lesch, S.M. Application of soil electrical conductivity to precision agriculture: Theory, principles, and guidelines. *Agron. J.* **2003**, *95*, 455–471. [CrossRef]
34. Corwin, D.L.; Lesch, S.M. Characterizing soil spatial variability with apparent soil electrical conductivity: I. Survey protocols. *Comput. Electron. Agric.* **2005**, *46*, 103–133. [CrossRef]
35. Corwin, D.L.; Lesch, S.M. Apparent soil electrical conductivity measurements in agriculture. *Comput. Electron. Agric.* **2005**, *46*, 11–43. [CrossRef]
36. Corwin, D.L.; Scudiero, E. Field-scale apparent soil electrical conductivity. In *Methods of Soil Analysis*; Logsdon, S., Ed.; Soil Science Society of America: Madison, WI, USA, 2016; Volume 1.
37. Norris, K.H.; Barnes, R.F.; Moore, J.E.; Shenk, J.S. Predicting forage quality by near infrared reflectance spectroscopy. *J. Anim. Sci.* **1976**, *43*, 889–897. [CrossRef]
38. Doorenbos, J.; Kassam, A.H. Yield response to water. In *FAO Irrigation and Drainage Paper 66*; FAO: Rome, Italy, 1979; p. 193.
39. Kuslu, Y.; Sahin, U.; Tunc, T.; Kiziloglu, F.M. Determining water-yield relationship, water use efficiency, seasonal crop and pan coefficient for alfalfa in as semiarid region with high altitude. *Bulg. J. Agric. Sci.* **2010**, *16*, 482–492.
40. Jones, L.P. Agronomic Responses of Grass and Alfalfa Hayfields to No and Partial Season Irrigation as Part of a Western Slope waTer Bank. Master's Thesis, Colorado State University, Fort Collins, CO, USA, 2015.
41. Putnam, D.H.; Ottman, M.J. *Your Alfalfa Crop Got the Blues? It May be 'Summer Slump'*; Cooperative Extension; University of California: Half Moon Bay, CA, USA, 2013.
42. Ayers, R.S.; Westcott, D.W. Water Quality for Agriculture. In *FAO Irrigation and Drainage Paper No. 29*; FAO: Rome, Italy, 1985; p. 174.
43. Putnam, D.; Benes, S.; Galdi, G.; Hutmacher, B.; Grattan, S. Alfalfa (*Medicago sativa* L.) is tolerant to higher levels of salinity than previous guidelines indicated: Implications of field and greenhouse studies. In Proceedings of the 19th EGU General Assembly, Vienna, Austria, 23–28 April 2017.
44. U.S. Department of Agriculture. Farm and Ranch Irrigation Survey, 2012. In *Census of Agriculture*; USDA: Washington, DC, USA, 2014; Volume 3.
45. Montazar, A.; Putnam, D.; Bali, K.; Zaccaria, D. A model to assess the economic viability of alfalfa production under subsurface Drip Irrigation in California. *Irrig. Drain.* **2017**. [CrossRef]

6

Research Advances in Adopting Drip Irrigation for California Organic Spinach

Ali Montazar [1,*], Michael Cahn [2] and Alexander Putman [3]

[1] Division of Agriculture and Natural Resources, University of California, UCCE Imperial County, 1050 East Holton Road, Holtville, CA 92250, USA
[2] Division of Agriculture and Natural Resources, University of California, UCCE Monterey County, 1432 Abbott Street, Salinas, CA 93901, USA
[3] Department of Microbiology and Plant Pathology, University of California, Riverside, 900 University Avenue, Riverside, CA 92521, USA
* Correspondence: amontazar@ucanr.edu

Abstract: The main objective of this study was to explore the viability of drip irrigation for organic spinach production and the management of spinach downy mildew disease in California. The experiment was conducted over two crop seasons at the University of California Desert Research and Extension Center located in the low desert of California. Various combinations of dripline spacings and installation depths were assessed and compared with sprinkler irrigation as control treatment. Comprehensive data collection was carried out to fully understand the differences between the irrigation treatments. Statistical analysis indicated very strong evidence for an overall effect of the irrigation system on spinach fresh yields, while the number of driplines in bed had a significant impact on the shoot biomass yield. The developed canopy crop curves revealed that the leaf density of drip irrigation treatments was slightly behind (1–4 days, depending on the irrigation treatment and crop season) that of the sprinkler irrigation treatment in time. The results also demonstrated an overall effect of irrigation treatment on downy mildew, in which downy mildew incidence was lower in plots irrigated by drips following emergence when compared to the sprinkler. The study concluded that drip irrigation has the potential to be used to produce organic spinach, conserve water, enhance the efficiency of water use, and manage downy mildew, but further work is required to optimize system design, irrigation, and nitrogen management practices, as well as strategies to maintain productivity and economic viability of utilizing drip irrigation for spinach.

Keywords: downy mildew; drip irrigation; irrigation management; organic production; spinach

1. Introduction

Spinach (*Spinacia oleracea* L.) is a fast-maturing, cool-season vegetable crop. In California, spinach is mainly produced in four areas of the southern desert valleys, the southern coast, the central coast, and the central San Joaquin Valley. The farm gate value of Californian non-processing spinach from a total planted area of 13,557 ha was nearly 242 million dollars in 2017 [1]. The area in California under spinach production increased more than 30% over the last 10 years; the planted area increased from 1478 ha in 2008 to 3893 ha in 2017 in the Imperial Valley [2,3].

Water and nitrogen are generally essential drivers for plant growth and survival, and for spinach, are two key factors that considerably affect yield and quality [4]. Spinach is highly responsive to nitrogen fertilization [5] and accumulates as much as 134 kg ha^{-1} of nitrogen in 30 days [6]. Spinach yield was shown to increase as a result of high nitrogen application rates but decreased when nitrogen application rate was excessive [6]. Researchers reported that with the increase of nitrogen fertilizer application, nitrogen use efficiency of spinach was significantly decreased; however, water use efficiency (as the ratio of yield to the seasonal crop water use) of spinach was increased in most cases [5,7]. Reducing nitrogen fertilizer use in spinach will be a challenge. The crop has a shallow root system, a high N demand, which occurs over a short period, and strict quality standards for a deep green color that tends to encourage N applications beyond the agronomic requirements to maximize yield [6]. Addressing these challenges requires that nitrogen fertilizer is applied at the optimal time and rate based on crop uptake and the soil–nitrogen test, and that irrigation water is efficiently applied to minimize leaching. The unpublished data from a survey conducted by the University of California Cooperative Extension Monterey county indicates that spinach-growers generally apply more water than that needed by crop water (100% to 150% more).

Downy mildew on spinach is a widespread and very destructive disease in California. It is a disease that is the most significant issue facing the spinach industry, and crop losses can be significant in all areas where spinach is produced [8]. *Peronospora effusa* is an obligate oomycete that causes downy mildew of spinach. Spinach growers rely on host resistance and fungicides to manage downy mildew. This dependence on resistance was relatively effective until recently, when malignant forms of *P. effusa* began to emerge in rapid succession [9]. In recent years, several new downy mildew races have appeared in the state of California, raising concerns about the ability to manage this threat and causing the industry to consider research strategies to address the problem. Organic spinach producers are especially vulnerable to these virulent strains because synthetic fungicide use is prohibited, and choice in regard to variety is determined at planting. This has led to significant yield losses in organic spinach production.

Like all downy mildew pathogens, *P. effusa* requires cool and wet conditions for infection and disease development [8,10]. The California industry is known for using very high planting densities and a large number of seed lines per bed [10]. For the baby leaf or clipped markets, the planting density is usually 8.6–9.8 million seeds per hectare [11]. The dense canopy of spinach retains much moisture, and creates ideal conditions for infection and disease development. Spores (called sporangia) are dispersed at short distances via wind or splashing water, or at medium distances via wind. In addition, most conventional and organic spinach fields are irrigated by sprinkler irrigation in California. Overhead irrigation deposits free moisture on leaf surfaces and increases relative humidity in the canopy, which contributes to the speed and severity of downy mildew epidemics within a field when other conditions, such as temperature, are favorable. Production practices that reduce the favorability of the spinach canopy for downy mildew development are needed to reduce losses to downy mildew in both organic and conventional production.

Because most conventional and organic spinach fields are irrigated by sprinkler irrigation in California, overhead irrigation could contribute to the speed and severity of downy mildew epidemics within a field when other conditions, such as temperature, are favorable. While preventing downy mildew on spinach is of high interest of organic vegetable production in California, integrated approaches are greatly required to reduce yield losses from this major disease. New irrigation management techniques and practices in spinach production may have a significant economic impact to the leafy greens industry through the control of downy mildew. In addition to losses from plant pathogens, new irrigation practices could reduce risks to food safety caused by overhead application of irrigation water.

Drip irrigation has revolutionized crop production systems in western states of the US by increasing yields and water-use efficiency in many crops [12]. Sub-surface drip irrigation was successfully implemented on vegetable crops, such as processing tomato, where yields have increased by 20–50% over recent years since this practice was adapted [13]. While sub-surface drip irrigation has been utilized primarily for high-value specialty crops, several studies have reported the benefits of its application for agronomic low-value crops [12–16]. Drip irrigation was evaluated versus micro-sprinkler irrigation in spinach [17]. The drip was found to be better than the micro-sprinkler because of greater yields and the lower installation cost. The effectiveness of drip fertigation for reducing nitrate in spinach was reported by Takebe et al. [18]. Drip fertigation was considered to reduce nitrate more stably. In addition to higher yield, sub-surface and surface drip irrigation can reduce risks of plant diseases in vegetables by minimizing leaf wetness and waterlogged soil conditions.

Adapting drip irrigation for high-density spinach plantings may be a possible solution to reduce losses from downy mildew, improve crop productivity and quality, and conserve water and fertilizer. Currently, drip irrigation is not used for producing spinach in California, and there is a lack of information on the viability of this technology and optimal practices for irrigating spinach with drips. In fact, to our knowledge, few studies have been conducted to assess the potential benefit of drip irrigation for this high-density planted spinach. As an initial test, the main objective of this project was to evaluate the viability of adapting drip irrigation for organic spinach production. The project was particularly aimed at understanding the system design to successfully produce spinach, and to conduct a preliminary assessment on the impact of drip irrigation on the management of spinach downy mildew.

2. Materials and Methods

The field experiments were carried out over two crop seasons at the organic field of the University of California Desert Research and Extension Center (UC DREC) (32°48′35″ N; 115° 26′39″ W; 22 m below mean sea level) located in the Imperial Valley, California. The field had a silty clay soil (the top 30 cm soil surface contains 14% sand, 42% silt, and 44% clay). Soil characteristics referring to three genetic horizons selected from the soil survey are presented in Table 1. Soil pH ranged between 7.94 and 8.04, and its electric conductivity was from 1.3 to 2.9 dS m^{-1}. The area has a desert climate with a mean annual rainfall of 76 mm and air temperature of 22 °C. Spinach can typically grow from October to March in this region. The average pH and electrical conductivity of water supply (surface water from the Colorado River) was 8.0 and 1.18 ds/m during the experiment, respectively.

Table 1. Soil characteristics of the study site.

Soil Parameters	Soil Depth		
Generic horizon (cm)	0–30	30–60	60–90
Texture	silty clay	clay loam	clay
Sand (%)	14	21	8
Silt (%)	42	41	34
Clay (%)	44	38	58
pH	7.94	8.04	8.02
EC (dS m^{-1})	1.3	1.4	2.9
Cation exchange capacity (meq/1000 g)	48.6	50.7	54.0
Bulk density (g cm^{-3})	1.54	1.48	1.45
Exchange sodium percentage (%)	3.5	4.1	6

2.1. Experimental Design and Treatments

Experiment 1 (fall experiment): Land preparation was conducted in late September 2018, and untreated Viroflay spinach seeds were planted at a rate of 37 kg ha^{-1} on 31st October. For the research trial (Figure 1a), five irrigation system treatments consisted of two drip depths (driplines on the soil surface and driplines at 3.8 cm depth), two dripline spacings (three driplines on a 203 cm bed and four driplines on an 203 cm bed), and sprinkler irrigation (203 cm bed). The experiment was arranged in a randomized complete block with four replications. Each drip replication had three beds, and each sprinkler replication had six beds. The beds were 61 m long. Figure 2 presents the individual experimental beds with four driplines at 3.8 cm and on the soil.

All treatments were germinated by sprinklers (two sets of five-hour irrigations). A flow control drip tape from the Toro company was used with a hose diameter of 15.88 mm, wall thickness of 0.154 mm, emitter spacing of 20.3 cm, and operating emitter flowrate of 0.49 L/h using the Nelson sprinklers R200WF Rotator (280 L/h @ 3.45 Bar) were used with spacings of 12.2 m × 9.1 m. Water distribution uniformity was measured for the sprinkler irrigation using the ASABE (American Society of Agricultural and Biological Engineers) standard method [19]. The average distribution uniformity for the sprinkler irrigation was 80%, and an average of 92% was assumed as water distribution uniformity of the drip irrigation.

True 6-6-2 (a homogeneous pelleted fertilizer from True Organic Products) was applied at a rate of 89 kg of N per hectare as pre-plant fertilizer, and True 4-1-3 (a liquid fertilizer from True Organic Products) was applied as complementary fertilizer through injection into the irrigation system. For the drip treatments, True 4-1-3 was applied three times after germination (by crop harvest) at a rate of 45, 33, and 44 kg of N per hectare. This liquid fertilizer was applied at a rate of 56, 42, and 50 kg of N per hectare for the sprinkler irrigation system. Soil tests conducted before planting were used to determine the status of available nitrogen/nutrients to develop fertilizer recommendations to achieve optimum crop production and prevent nitrogen deficiency. Initially, the crop was irrigated with more water than required, as determined by following crop evapotranspiration (ET, as the sum of soil evaporation and plant transpiration), and using soil moisture data, we tried to irrigate spinach trials more than crop water requirements to make sure there was no water stress for the entire crop season. However, according to our data, this led to over-irrigation at some points in the early and mid-crop season. Irrigation events were typically scheduled once a week for the sprinkler treatments and twice a week for the drip treatments at the required running times of each system/treatment.

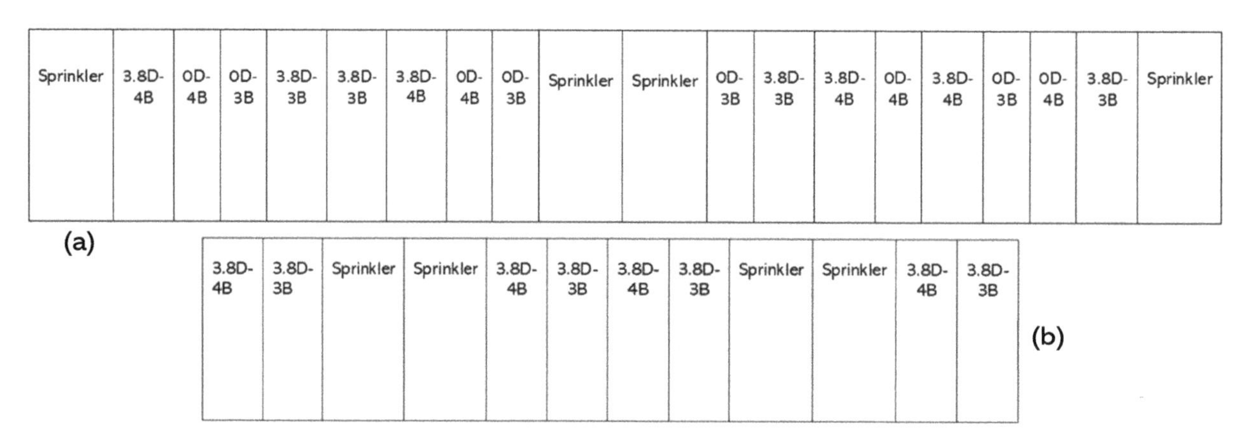

Figure 1. The fall (**a**) and winter (**b**) experiment layouts (not to scale). Sprinkler: treatment irrigated by sprinklers; 3.8D-3B: treatment with three driplines in each bed installed at 3.8 cm depth; 3.8D-4B: treatment with four driplines in each bed installed at 3.8 cm depth; 0D-4B: treatment with four driplines in each bed on the soil surface; 0D-3B: treatment with three driplines in each bed on the soil surface.

Figure 2. Beds with four driplines in the fall crop season. The pictures demonstrate individual beds with four driplines at 3.8 cm depth 30 days after planting (**a**) and on the soil surface 12 days after planting (**b**).

Experiment 2 (winter experiment)**:** Untreated Viroflay spinach seeds were planted at a rate of 38 kg ha^{-1} on 28th January. For the research trial (Figure 1b), three irrigation system treatments consisted of two dripline spacings (three driplines on a 203 cm bed and four driplines on a 203 cm bed), and sprinkler irrigation (203 cm bed). All driplines were installed at 3.8 cm depth. The experiment was arranged in a randomized complete block with four replications. Each drip and sprinkler treatment replication consisted of three beds × 61 m length. All treatments were germinated by sprinklers (two sets of five hours). The buffer beds located between sprinkler and drip treatments were not planted. The drip treatments were more frequently irrigated (three times a week with shorter irrigation events) than the drip treatments in the fall trial.

True 6-6-2 organic fertilizer was applied before planting at a rate of 78 kg of N per hectare, and the crop was supplemented with True 4-1-3 liquid fertilizer injection into irrigation system. For the drip system, True 4-1-3 was applied four times after germination (by crop harvest) at a rate of 34, 45, 33, and 44 kg of N per hectare. In the sprinkler system, the same fertilizer was applied at a rate of 45, 45, 50, and 45 kg of N per hectare.

2.2. Field Measurements and Analysis

The actual crop ET (evapotranspiration) was measured using Tule Technology sensor (www.tuletechnologis.com) which uses the residual of energy balance method as surface renewal equipment. Using the actual crop water-use data measured and spatial California Irrigation Management Information System (CIMIS) data (http://wwwcimis.water.ca.gov/SpatialData.aspx), the actual crop coefficient curve was developed for each crop season. The reference ET (ET_o) was retrieved from spatial CIMIS as well. Images were taken on weekly basis utilizing an infrared camera (NDVI digital camera, NDVI stands for the Normalized Difference Vegetation Index) to quantify the development of the crop canopy of each treatment over the crop seasons. The images were analyzed using PixelWrench2 software. Decagon 5TE sensors were installed at three depths (20, 30, and 45 cm) to monitor soil water content on a continuous basis. The numerous horizontal roots of spinach typically remain in the top 30 cm of soil with a spread of approximately 40 cm, while soil water storage at the top 45 cm (spinach crop root zone) was calculated using the soil water content data. The applied water for the irrigation treatments was measured throughout the crop seasons using magnetic flowmeters. The NDVI values and plant leaf wetness values were measured using Spectral Reflectance sensors (combination of SRS-Pi Hemispherical Sensor and SRS-Pr Field Stop Sensor) and dielectric leaf wetness sensors (PHYTOS 31), respectively (METER Group, Inc. USA) on a continuous basis. Leaf chlorophyll was measured using an *atLEAF CHL STD* sensor (FT Green LLC: Wilmington, DE, USA) on a weekly basis.

Shoot biomass (sum of the weight of leaves and stem represents shoot biomass) measurements at the final harvests were carried out in three sample areas of 0.56 m^2 (0.92 m × 0.61 m) per replicate and treatment. The bed located in the center of each replication in each of the treatments was selected as the sample bed (four sample beds per each treatment, a total of 20 sample beds for five irrigation treatments in the fall trial and 14 sample beds for three irrigation treatments in the winter trial). Fresh weight was measured in order to determine shoot biomass accumulation. Leaf chlorophyll content was measured on 20 individual leaves in each of the sample beds. Weekly plant samples (shoots) were analyzed for total nitrogen at the UC Davis Analytical Lab. The statistical significances were performed using generalized linear mixed model using the GLIMMIX (generalized linear mixed models) procedure in SAS 9.4. (SAS Institute Inc., Cary, NC, USA)

2.3. Downy Mildew

Both replicate runs of the experiment were visually scouted by walking down arbitrarily selected rows of all treatments. When disease symptoms were observed, the number of plants exhibiting downy mildew symptoms was counted for the entire length of two of the three beds in each plot. To calculate disease incidence, or the percentage of plants affected by downy mildew, the number of plants in each bed was divided by the estimated plant population. The estimated plant population was determined from the seeding rate and the germination rate averaged over each treatment at emergence. Downy mildew incidence was analyzed in a generalized linear mixed model using the GLIMMIX procedure in SAS 9.4. A block was treated as a random effect, and options in the model statement included use of the beta distribution and the logit link function. Due to evidence for a significant effect of treatment, means were separated using the *lsmeans* statement with the Tukey-Kramer adjustment for multiple comparisons.

3. Results and Discussion

3.1. Weather Conditions

The daily air temperature, relative humidity, and wind speed variations were different over the fall and the winter experiments (Figure 3). At the fall trial, we observed a mean daily air temperature of 18.8 °C at the early- and mid-seasons when the temperature stayed above 6.5 °C at nighttime. The mean daily temperature decreased to 12.1 °C during the late season, and relative humidity dramatically increased over the last 10 days before the final harvest. An average daily wind speed of 1.5 m s^{-1} was observed during the fall experiment.

While a more variable air temperature was observed at the winter experiment, the mean daily temperature was 12.6 °C at the early- and mid-seasons, and the temperature fell below 1.0 °C for a few nights. Higher daytime relative humidity was measured during the early- and mid-seasons compared to the fall. Although there was not a significant difference between the average wind speed of the crop seasons, several windy days occurred during the winter season (hours with a wind speed of more than 8 m s^{-1}). The cumulative ET_o for the fall and winter crops was 122 and 177 mm, respectively.

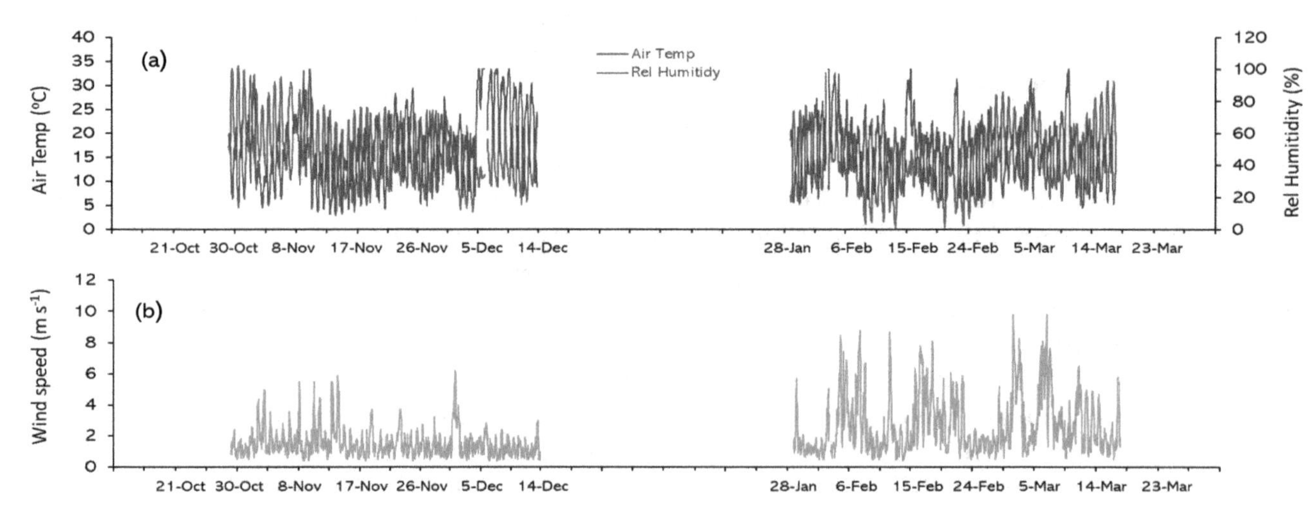

Figure 3. Daily air temperature (**a**), relative humidity (**a**), and wind speed (**b**) over the fall season (October through December, left side of the figures) and the winter season (January through March, right side of the figures).

3.2. Crop Water Use and Applied Water

Total crop water use (seasonal ET) of 92.3 and 154.9 mm was measured in the fall and the winter trials, respectively. Variable daily crop ET and crop coefficient values were measured during the entire crop season (Figure 4). The maximum and minimum crop ET observed was 3.8 and 0.7 mm d^{-1} in the fall experiment and 5.3 and 0.3 mm d^{-1} in the winter experiment, respectively. A similar range of crop coefficient value (0.2–1.2) was obtained for both crop seasons. Piccinni et al. reported a crop water use of 157.5 mm and crop coefficient range of 0.2–1.5 for winter spinach in Texas using lysimeter measurements [20].

The total applied water for the sprinkler treatment was 144.8 and 225.2 mm in the fall and the winter trials, respectively. The total applied water for the drip treatments was 130.5 mm over the fall season and 202.4 mm over the winter season. Overall, an average of 11% more water was applied through the sprinkler system compared with the drip system to compensate for the lower water application uniformity of sprinkler irrigation system. A well-designed and properly managed drip irrigation usually has high distribution uniformity, and unlike conventional sprinkler irrigation systems, has the potential to conserve water because of a lower potential for tailwater runoff and deep percolation losses.

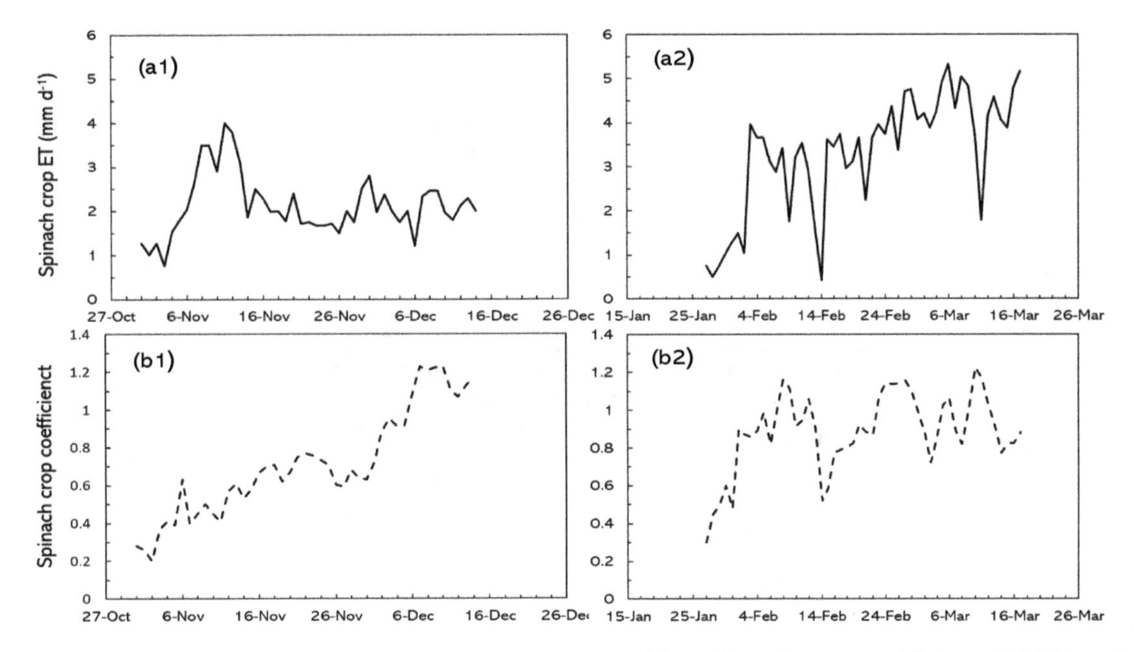

Figure 4. Daily spinach crop evapotranspiration or ET (**a1**,**a2**) and crop coefficient (**b1**,**b2**) values in the experimental seasons. The subfigures a1 and a2 show crop ET in the fall and spring trials, respectively, and the subfigures b1 and b2 show crop coefficient values in the fall and spring experimental seasons, respectively.

Soil water storage in the profile was calculated, assuming that the water content sensed at each depth was representative of the soil from that depth to the midpoint between the next upper and lower depths. A comparison of the soil water storage between sprinkler and drip (3.8–4B) treatments versus soil water storage in the average field capacity (FC) of the top 45 cm of the soil is shown in Figure 5. Even though there were differences between the daily available soil water in both treatments and crop seasons, the soil water storage amounts were uniformly maintained at around the average soil water storage of field capacity (19.35 cm at the top 45 cm of the soil). More uniform soil water availability in the drip treatment over the winter experiment could be as a result of more frequent irrigation events with shorter durations.

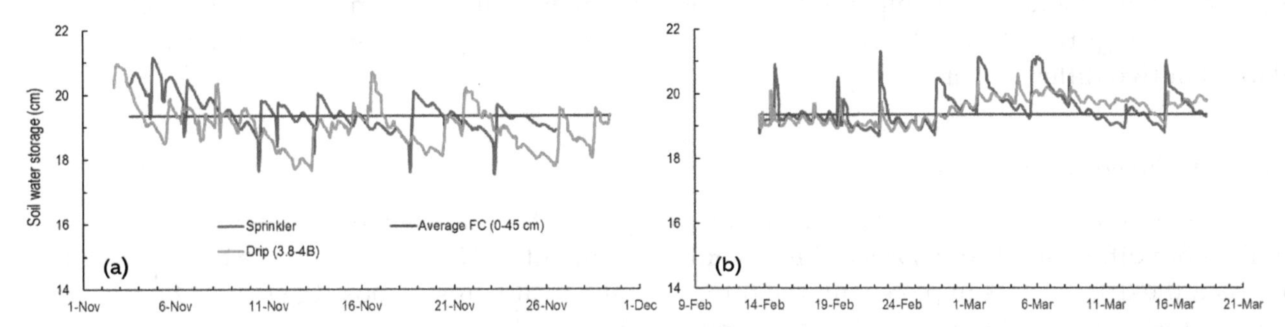

Figure 5. Soil water storage at the surface to a 45 cm-deep profile at the drip (3.8–4B) and sprinkler treatments in the fall (**a**) and winter (**b**) crop seasons. The blue line shows the soil water storage calculated using an average field capacity of the top 45 cm of the soil (0.43 m^3 m^{-3}).

3.3. Crop Canopy over the Season

Crop canopy cover is defined as the percentage of plant material which covers the soil surface, and can be a very useful index for estimating the crop coefficient. Here, the canopy cover percentage was developed for each of the irrigation treatments (Figure 6). Canopy cover percentages show that the leaf density of drip irrigation treatments was slightly behind (1–4 days depending upon the irrigation treatment and crop season) than that of the sprinkler irrigation treatments.

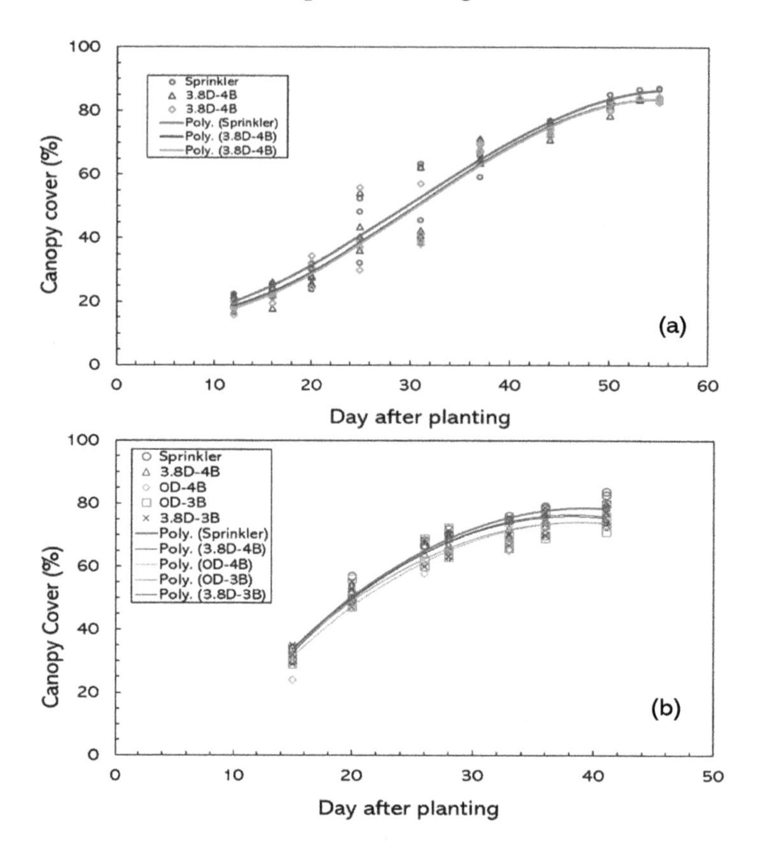

Figure 6. Canopy crop curve for the different irrigation treatments in the (**a**) winter crop season and (**b**) the fall crop season.

The individual canopy cover curves for each season demonstrate that spinach crop water requirements and irrigation scheduling could be different in fall and winter seasons. For instance, in this study, an average of 52% and 70% of canopy crop coverage was observed 30 days after planting in the winter and the fall crop season, respectively. We may expect a longer season for spinach planted in the winter than the fall in the Imperial Valley, though it may change depending on the specific weather conditions. Figure 6 shows the canopy cover for individual experimental beds with four driplines planted at two different dates.

3.4. Crop Growth and Greenness

Few differences between drip and sprinkler treatments were visible for the winter planted trial, while more differences were observed in the fall experiment in November. Leaves began to yellow in between the drip laterals in plots with the three-dripline treatments. A possible reason for this may be that the fertigation did not move the N in between the driplines. The total plant tissue N content of the drip treatments with three driplines in bed was less than the sprinkler treatment on the 30th of November (1.8% N for the drip vs. 3.2% N for the sprinkler), while the difference in N content of the tissue was less between the sprinkler treatment and the drip treatment with four driplines in the bed (Figure 7). Overall, a higher plant-tissue nitrogen content was observed for the sprinkler treatment compared to drip treatments.

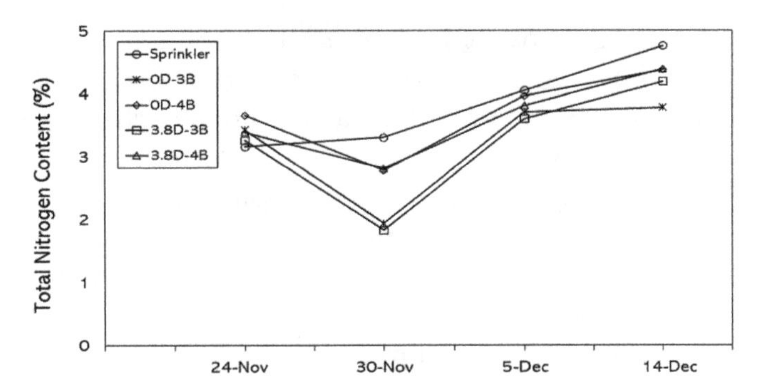

Figure 7. The trends of total plant nitrogen content over the fall crop season. The results are presented for the individual irrigation treatments.

The leaf chlorophyll content (Figure 8) was also higher in the sprinkler treatments compared to the drip treatments, though the drip treatment with four driplines at 3.8 cm depth was numerically more similar to sprinkler treatments and sometime had a greater leaf chlorophyll content during the crop seasons. A greater level of leaf chlorophyll content in the late season was observed at the winter experiment than the fall experiment for both sprinkler and drip treatments. For instance, the total leaf chlorophyll content of the sprinkler treatment two days before the final harvest at the winter trial was 4 μg cm^{-2} more than leaf chlorophyll content of sprinkler treatment at the fall trial. Variable leaf chlorophyll content was observed over the plant development in each of the treatments, which corresponds to the plant N accumulation over the growing seasons.

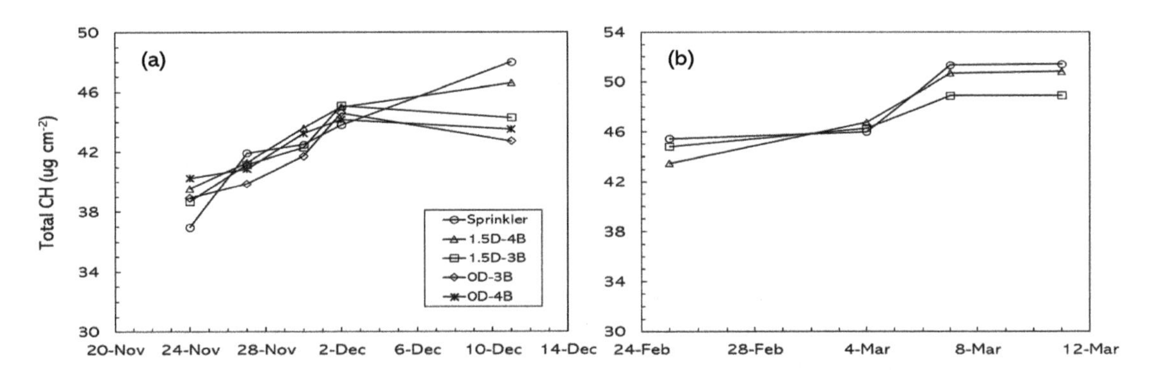

Figure 8. The trends of average leaf chlorophyll content of spinach at different irrigation treatments over (**a**) the fall crop season and (**b**) winter crop season.

Figure 9 shows the average day NDVI values for the drip treatment with four driplines (the 3.8D-4B treatment) and the sprinkler treatment for the fall season experiment. The results indicated that the NDVI values varied from 0.2 (two weeks after planting) to 0.84 (the day before final harvest). By late November, the NDVI values were very similar in both irrigation treatments, where this index had higher values in the drip treatment even over a short period. The results demonstrated lower NDVI values in the drip treatment than sprinkler treatment over the last two weeks of the crop season.

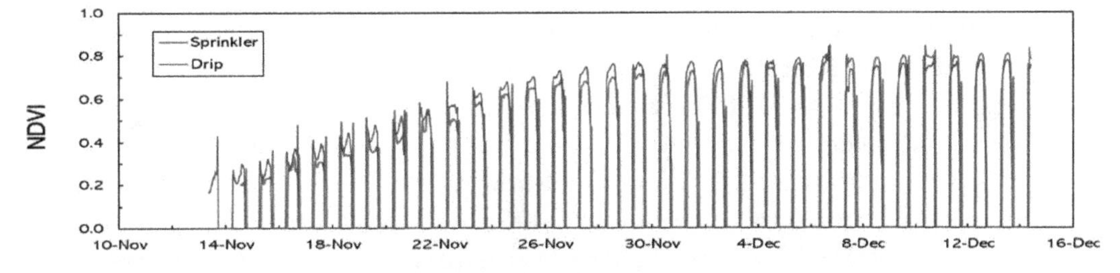

Figure 9. NDVI values of sprinkler and drip (3.8D-4B) treatments over the fall crop season. The daily values vary between a maximum during daytime and zero during nighttime.

The values of total plant nitrogen content, leaf chlorophyll content of spinach, and NDVI confirmed that N uptake at the drip treatments was not entirely as effective as the sprinkler treatment, particularly in the fall experiment. The nutrient management issue in spinach drip irrigation in combination with water management is likely a critical issue that we need to learn more about, since it may affect the adoption and viability of the drip for spinach production. Spinach is a very fast short-season crop, and hence, water and nitrogen management may significantly influence the leaf as the major organ for important physiological processes and essential indicator used to measure the growth and yield of the crop. While the leaf area and shoot biomass of spinach are greatly affected by water and nitrogen levels [4], this dependency is likely more critical in organic spinach.

3.5. Leaf Wetness

Figure 10 shows the probe output of the leaf wetness sensors placed in the sprinkler treatment and drip treatment (3.8D-4B) for a period of 12 days during the fall season experiment. At this period, there were two irrigation events in each of the treatments, and two rainy days. The sensor output from dew was typically lower (less than 700 counts) than that from the rain or the irrigation event. The results revealed that sprinkler-irrigated crop canopies remained wet 24.3% (= (70 h/288 h) × 100) times longer during this period than the crop canopy irrigated with drip treatment.

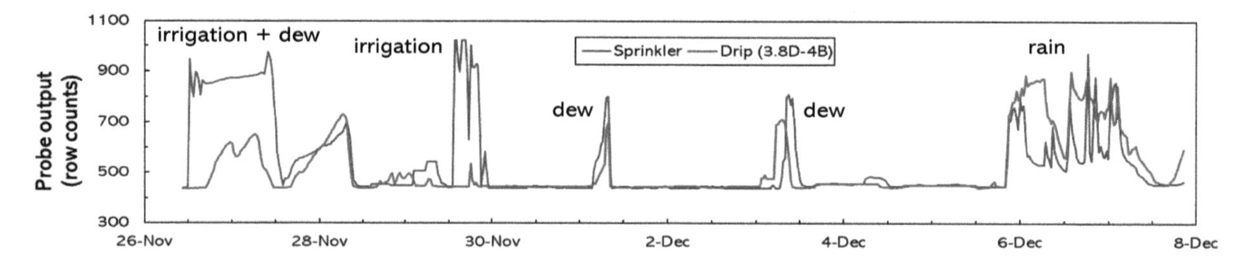

Figure 10. The row counts of leaf wetness sensors at the sprinkler and drip (3.8D-4B) treatments over a 12-day period in the fall crop season.

Spinach downy mildew requires a cool environment with long periods of leaf wetness or high humidity. Wet foliage is especially favorable. Considering the above analysis, and in the case where the weather and farming conditions are similar, there is higher risk for infection and downy mildew disease development in spinach irrigated by sprinklers in comparison with spinach irrigated by drips. The air temperature and relative humidity pattern (Figure 2) indicate that there was a desirable weather condition and more possibility for downy mildew disease in mid-December, but the fall experiment was just terminated at the time. The next desirable period was mid-February when temperatures had cooled to the range believed to be optimal for downy mildew, days became windier, and there was a period of leaf surface wetness caused by sprinkler irrigation, rainfall, or high relative humidity.

3.6. Shoot Biomass

The effects of various irrigation treatments on spinach shoot biomass yield over the two experimental seasons are summarized in Table 2 and Figure 11. In the fall trial, the mean biomass yield in the sprinkler treatment was 13,905 kg ha^{-1}, approximately 9% more than the 3.8D-4B treatment. The lowest mean yield (11,136 kg ha^{-1}) was observed in the 0D-3B treatment. Statistical analysis indicated very strong evidence ($p = 0.001$) for an overall effect of the irrigation system on spinach yield. A significant difference between the individual treatments was investigated using the Tukey-HSD analysis. The results demonstrated a significant yield difference between the sprinkler irrigation and each of the drip irrigation treatments (p values of 0.0001 to 0.0009). Even though no significant yield difference was obtained between the surface drip and sub-surface drip (driplines at 3.8 cm depth) with the same dripline number in bed (p value of 0.8276 for the three-dripline and 0.1995 for the four-dripline), the number of driplines in bed had a very significant impact on spinach biomass yield (p values of 0.0001 to 0.0009).

In the winter trial, the mean biomass yield in the sprinkler treatment was 14,886 kg ha^{-1}, approximately 7% more than the 3.8D-4B treatment. Statistical analysis indicated strong evidence ($p = 0.0424$) for an overall effect of irrigation system on spinach fresh yield. While we could not find a significant difference between the impact of the sprinkler and the 3.8D-4B irrigation treatments on spinach yield ($p = 0.1161$), there was a statistically significant yield difference between the sprinkler and the 3.8D-3B irrigation treatments ($p = 0.04147$).

Table 2. Mean spinach fresh yield values of each irrigation treatment in each of the fall and winter experiments. Yields with different letters significantly differ ($p < 0.05$) by Tukey's test.

Fall 2018		Winter 2019	
Irrigation Treatment	**Fresh Yield (kg ha^{-1})**	**Irrigation Treatment**	**Fresh Yield (kg ha^{-1})**
Sprinkler	13,905 [a]	Sprinkler	14,886 [a]
3.8D-4B	12,753 [b]	3.8D-4B	13,914 [ab]
0D-4B	12,273 [b]	3.8D-3B	13,580 [b]
0D-3B	11,136 [c]	-	-
3.8D-3B	11,351 [c]	-	-

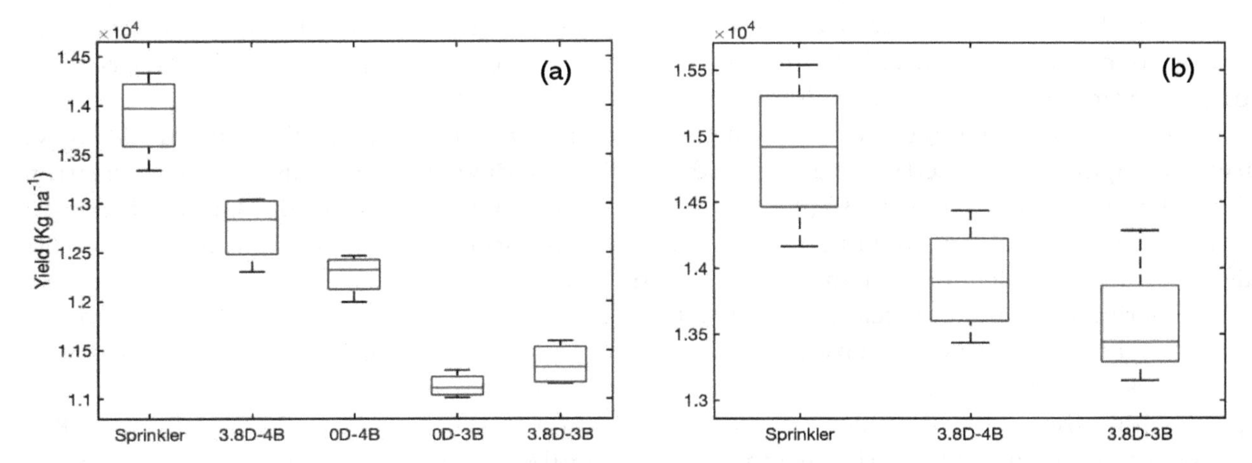

Figure 11. Mean yield data of the sample beds in each of the irrigation treatments at the fall experiment and (**a**) and the winter experiment (**b**). Horizontal red lines indicate the average for each treatment.

The yield reduction in drip irrigation treatments compared to the sprinkler irrigation ranged between 7% (the 3.8D-4B treatment against the sprinkler treatment in the winter trial) and 24.8% (the 0D-3B treatment against the sprinkler treatment in the fall trial). The yield difference may have likely been caused by suboptimal irrigation and nutrient management conditions of the drip treatments. Since drip irrigation was tested for the first time for spinach in this study, subsequent trials need to plan for irrigation and nutrient improvements and be conducted in different aspects. These practices had to be adjusted in real time as the study progressed. The biomass yield reported here are related to a final harvest, which was on the same day for all the treatments. Since the findings showed that the leaf density of drip irrigation treatments was more behind than sprinkler treatment, scheduling a different harvest time for the drip treatments may compensate for some of the yield reductions against the sprinkler treatment.

Several factors influence appropriate drip irrigation management, including system design, soil characteristics, and environmental conditions. The influences of these factors can be integrated into a practical and efficient system which determine the quantity and timing of drip irrigation. Drip irrigation offers the potential for precise water management, and also provides the ideal vehicle to deliver nutrients in a timely and efficient manner. However, achieving high water and nutrient use efficiency while maximizing crop productivity requires intensive and proper management.

However, the 7% through 13% shoot biomass difference between the four dripline treatments and the sprinkler treatment demonstrates the potential of sub-surface drip irrigation for profitable spinach production. This yield difference could be reduced through optimal system design and better irrigation and nutrient management practices for the drip system.

Optimizing water and nitrogen management had not been an objective for this study, therefore, evaluating water use efficiency of the treatments may not be an interest to be discussed. With the 7% lower biomass yield and 11% less applied water in the 3.8D-4B than the sprinkler treatment, we may simply conclude that drip irrigation may have the potential to enhance the efficiency of water use in spinach production or at least at this point, there is no major difference between the two systems regarding this indicator. High water use efficiency for sub-surface drip irrigation was reported by researchers for multiple crops [12,13,15,21].

3.7. Downy Mildew Incidence

Downy mildew was not observed in the fall experiment. In the winter experiment, downy mildew activity was first confirmed in the study area on 5 March. Disease incidence was rated on 11 March. Downy mildew incidence was low on this date, with only two beds (0.12% and 0.20%) exhibiting incidence values above 0.1% (Figure 12). Mean downy mildew incidence in plots irrigated with sprinklers following emergence was 0.08%, approximately 4 to 5× higher than treatments irrigated with drip following emergence. Statistical analysis indicated evidence ($p = 0.0461$) for an overall effect of irrigation treatment on downy mildew.

Analysis of means suggested that all three treatments were statistically similar. However, a pairwise comparison revealed some evidence that downy mildew incidence was lower in plots irrigated with 3.8D-4B ($p = 0.0671$) or 3.8D-3B ($p = 0.1139$) following emergence when compared to the sprinkler.

The likely mechanism causing this effect was a reduction under drip irrigation of leaf wetness, which is critical for infection and sporulation by the downy mildew pathogen. Additional repetitions of this experiment in higher disease pressure situations are needed for further evaluation of the ability of drip irrigation to reduce downy mildew. Another mechanism that could partially account for the observed differences among the treatments is that the leaf density in drip-irrigated plots was slightly behind that of the sprinkler irrigated plots in time. A less dense canopy could reduce the leaf wetness potential, and in turn, disease incidence potential. However, it is unclear if the magnitude of differences in density could account for the magnitude in differences in downy mildew incidence between sprinkler- and drip-irrigated treatments.

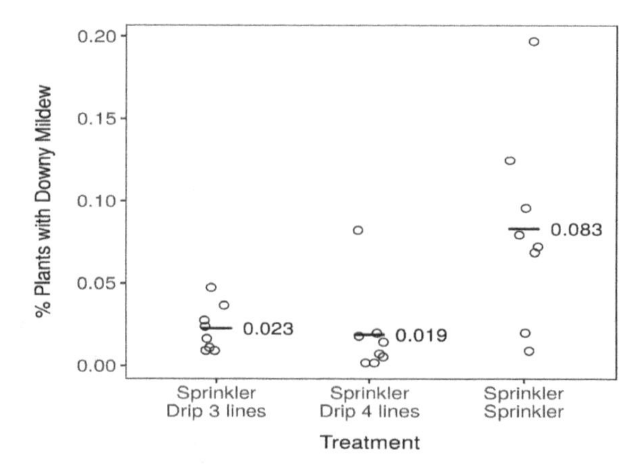

Figure 12. Raw data (all plots and beds within plots) of downy mildew incidence. Horizontal bars and labels indicate the mean for each treatment. Points are jittered to reduce overlap. The top line of *x*-axis labels indicate what the plot was germinated with, and the bottom row indicates the irrigation used following emergence.

Perhaps the most important benefit of drip for spinach production could be less yield loss as a result of downy mildew management. Spinach is a high-value crop—for instance, the value of spinach per kg was about $2 in 2016 [22]. While more data are needed to conduct an economic feasibility analysis of drip irrigation for spinach production, the results of this study demonstrated a positive impression. The initial cost associated with a drip system for spinach fields is estimated at about $3000 to $4000 per hectare in the region.

4. Conclusions

This study demonstrated the potential for drip irrigation to be used to produce organic spinach, conserve water, enhance the efficiency of water use, and reduce incidence of downy mildew. Further work is needed to comprehensively evaluate the viability of utilizing drips—specifically, the optimal system design, the impacts of irrigation and nitrogen management practices in various soil types and climates, and strategies to maintain productivity and economic viability at spinach. Assessing drip irrigation for the entire crop season, including germination, could be another research interest since spinach is a short-season crop and combining the sprinkler for crop germination and drip for such a short period might cause some practical issues.

Author Contributions: Conceptualization, A.M. and M.C.; Data curation, A.M.; Formal analysis, A.M.; Investigation, A.M. and A.P.; Methodology, A.M., M.C. and A.P.; Project administration, A.M.; Supervision, A.M.; Writing—original draft, A.M.; Writing—review & editing, A.M., M.C. and A.P.

References

1. California Department of Food and Agriculture (CDFA). *California Agricultural Statistics Review*; Agricultural Statistical Overview: Sacramento, CA, USA, 2018.
2. Ortiz, C. *Office of the Agricultural Commissioner Sealer of Weights and Measures-Imperial County*; Annual Crop and Livestock Report: Imperial County, CA, USA, 2018.
3. Birdsall, S. *Office of the Agricultural Commissioner Sealer of Weights and Measures-Imperial County*; Annual Crop and Livestock Report: Imperial County, CA, USA, 2009.
4. Zhang, J.; Sha, Z.; Zhang, Y.; Bei, Z.; Cao, L. The effects of different water and nitrogen levels on yield, water and nitrogen utilization efficiencies of spinach. *Can. J. Plant Sci.* **2015**, *95*, 671–679. [CrossRef]
5. Cantliffe, D.J. Nitrate accumulation in vegetable crops as affected by photoperiod and light duration. *J. Amer. Soc. Hort. Sci.* **1972**, *97*, 414–418.
6. Heinrich, A.; Smith, R.; Cahn, M. Nutrient and Water Use of Fresh Market Spinach. *HortTechnology* **2013**, *23*, 325–333. [CrossRef]
7. Zhang, J.; Bei, Z.; Zhang, Y.; Cao, L. Growth Characteristics, Water and Nitrogen Use Efficiencies of Spinach in Different Water and Nitrogen Levels. *Sains Malays.* **2014**, *43*, 1665–1671.
8. Choudhury, R.A.; Koike, S.T.; Fox, A.D.; Anchieta, A.; Subbarao, K.V.; Klosterman, S.J.; McRoberts, N. Season-Long Dynamics of Spinach Downy Mildew Determined by Spore Trapping and Disease Incidence. *Phytopathology* **2016**, *106*, 1311–1318. [CrossRef] [PubMed]
9. Correll, J.C.; Bluhm, B.H.; Feng, C.; Lamour, K.; Du Toit, L.J.; Koike, S.T. Spinach: Better management of downy mildew and white rust through genomics. *Eur. J. Plant Pathol.* **2011**, *129*, 193–205. [CrossRef]
10. Correll, J.C.; Feng, C.; Matheron, M.E.; Porchas, M.; Koike, S.T. Evaluation of spinach varieties for downy mildew resistance. *Plant Dis. Manag. Rep.* **2017**, *11*, 122.
11. Koike, S.; Cahn, M.; Cantwell, M.; Fennimore, S.; Natwick, E.; Smit, R.; Takele, E. *Spinach Production in California*; University of California Agriculture and Natural Resources: Auckland, CA, USA, 2011; ISBN 978-1-60107-719-6.
12. Montazar, A.; Bali, K.; Zaccaria, D.; Putnam, D. Viability of subsurface drip irrigation for alfalfa production in the low desert of California. In Proceedings of the ASABE International Meeting, Detroit, MI, USA, 29 July–1 August 2018.

13. Hartz, T.K.; Bottoms, T.G. Nitrogen requirements of drip-irrigated processing tomatoes. *HortScience* **2009**, *44*, 1988–1993. [CrossRef]
14. Lamm, F.R.; Camp, C.R. Subsurface drip irrigation. In *Microirrigation for Crop Production—Design, Operation and Management*; Lamm, F.R., Ayars, J.E., Nakayama, F.S., Eds.; Elsevier Publications: Amsterdam, The Netherlands, 2017; Chapter 13; pp. 473–551.
15. Lamm, F.R.; Bordovsky, J.P.; Schwankl, L.J.; Grabow, G.L.; Enciso-Medina, J.; Peters, R.T.; Colaizzi, P.D.; Trooien, T.P.; Porter, D.O. Subsurface Drip Irrigation: Status of the Technology. In *2010 5th National Decennial Irrigation Conference*; Phoenix Convention Center: Phoenix, AZ, USA, 5–8 December 2010.
16. Montazar, A.; Putnam, D.; Bali, K.; Zaccaria, D. A model to assess the economic viability of alfalfa production under subsurface Drip Irrigation in California. *Irrig. Drain.* **2017**, *66*, 90–102. [CrossRef]
17. Luque Quispe, M.R. Evaluation of drip and Microasperation Irrigation Methods in Spinach (Spinaca oleracea) and Swiss Lettuce (Valerianella locusta) in Walipinis. Brigham Young University. Available online: https://scholarsarchive.byu.edu/etd (accessed on 13 June 2019).
18. Takebe, M.; Okazaki, K.; Kagishita, K.; Karasawa, T. Effect of drip fertigation system on nitrate control in spinach *Spinacea oleracea L.. Soil Sci. Plant Nutr.* **2006**, *52*, 251. [CrossRef]
19. ASAE Standards. *Procedure for Sprinkler Testing and Performance Reporting*; ASAE: St. Joseph, MI, USA, 1985.
20. Piccinni, G.; Ko, J.; Marek, T.; Leskovar, D.I. Crop Coefficients Specific to Multiple Phenological Stages for Evapotranspiration-based Irrigation Management of Onion and Spinach. *HotrtScience* **2009**, *44*, 421–425. [CrossRef]
21. Montazar, A.; Radawich, J.; Zaccaria, D.; Bali, K.; Putnam, D. Increasing water use efficiency in alfalfa production through deficit irrigation strategies under subsurface drip irrigation. Water shortage and drought: From challenges to solution. In Proceedings of the USCID Water Management Conference, San Diego, CA, USA, 17–20 May 2016.
22. Ortiz, C. *Office of the Agricultural Commissioner Sealer of Weights and Measures–Imperial County*; Annual Crop and Livestock Report: Imperial County, CA, USA, 2017.

7

Modeling the Effects of Irrigation Water Salinity on Growth, Yield and Water Productivity of Barley in Three Contrasted Environments

Zied Hammami [1,*], Asad S. Qureshi [1], Ali Sahli [2], Arnaud Gauffreteau [3], Zoubeir Chamekh [2,4], Fatma Ezzahra Ben Azaiez [2], Sawsen Ayadi [2] and Youssef Trifa [2]

1 International Center for Biosaline Agriculture (ICBA), Dubai P.O. Box 14660, UAE; a.qureshi@biosaline.org.ae
2 Laboratory of Genetics and Cereal Breeding, National Agronomic Institute of Tunisia, Carthage University, 43 Avenue Charles Nicole, 1082 Tunis, Tunisia; sahli_inat_tn@yahoo.fr (A.S.); zoubeirchamek@gmail.com (Z.C.); zahra.azaiez@gmail.com (F.E.B.A.); sawsen.ayadi@gmail.com (S.A.); youssef.trifa@gmail.com (Y.T.)
3 INRA–INA-PG–AgroParisTech, UMR 0211, Avenue Lucien Brétignières, F-78850 Thiverval Grignon, France; arnaud.gauffreteau@inrae.fr
4 Carthage University, National Agronomic Research Institute of Tunisia, LR16INRAT02, Hédi Karray, 1082 Tunis, Tunisia
* Correspondence: z.hammami@biosaline.org.ae

Abstract: Freshwater scarcity and other abiotic factors, such as climate and soil salinity in the Near East and North Africa (NENA) region, are affecting crop production. Therefore, farmers are looking for salt-tolerant crops that can successfully be grown in these harsh environments using poor-quality groundwater. Barley is the main staple food crop for most of the countries of this region, including Tunisia. In this study, the AquaCrop model with a salinity module was used to evaluate the performance of two barley varieties contrasted for their resistance to salinity in three contrasted agro-climatic areas in Tunisia. These zones represent sub-humid, semi-arid, and arid climates. The model was calibrated and evaluated using field data collected from two cropping seasons (2012–14), then the calibrated model was used to develop different scenarios under irrigation with saline water from 5, 10 to 15 dS m^{-1}. The scenario results indicate that biomass and yield were reduced by 40% and 27% in the semi-arid region (KAI) by increasing the irrigation water salinity from 5 to 15 dS m^{-1}, respectively. For the salt-sensitive variety, the reductions in biomass and grain yield were about 70%, respectively, although overall biomass and yield in the arid region (MED) were lower than in the KAI area, mainly with increasing salinity levels. Under the same environmental conditions, biomass and yield reductions for the salt-tolerant barley variety were only 16% and 8%. For the salt-sensitive variety, the biomass and grain yield reductions in the MED area were about 12% and 43%, respectively, with a similar increase in the salinity levels. Similar trends were visible in water productivities. Interestingly, biomass, grain yield, and water productivity values for both barley varieties were comparable in the sub-humid region (BEJ) that does not suffer from salt stress. However, the results confirm the interest of cultivating a variety tolerant to salinity in environments subjected to salt stress. Therefore, farmers can grow both varieties in the rainfed of BEJ; however, in KAI and MED areas where irrigation is necessary for crop growth, the salt-tolerant barley variety should be preferred. Indeed, the water cost will be reduced by 49% through growing a tolerant variety irrigated with saline water of 15 dS m^{-1}.

Keywords: salinity; environments; AquaCrop model; water productivity; scenarios; tolerant

1. Introduction

The world food supply is affected by environmental abiotic stresses, which damages up to 70% of food crop yields [1–3]. In the Near East and North Africa (NENA) region, physical water scarcity is already affecting food production [4]. The NENA region is characterized by an arid climate with a total annual rainfall much lower than the evapotranspiration of the field crops. In the Arab World, more than 85% of the available water resources are used for agriculture [5]. Despite this high-water allocation for the agriculture sector, about 50% of food requirements are imported [4]. Crop irrigation uses poor quality groundwater, which is saline in nature. The uninterrupted application of groundwater for irrigation is replete, which leads to a severe increase in soil salinity and reduction in crop yields. Climate changes, namely the increase in global temperatures and the decline in rainfall, exacerbate soil salinization, resulting in loss of production in arable lands [6]. According to recent estimates, one-fifth of the irrigated lands in the world are affected by salinity. Every day, on average, 2000 ha of irrigated land in arid and semi-arid areas is adversely affected by salinity problems [7]. The annual economic loss due to these increases in soil salinity is about USD 27.3 billion [8].

Cereals are the main crops in the Mediterranean and NENA regions, contributing to food security and social stability. Barley is one of these staple crops in the area. However, its production is constrained by abiotic factors, such as the arid climate, low and erratic rainfall, and soil and water salinity. The anticipated climate changes will further increase the negative impacts of these factors in the future [9]. Barley (*Hordeum vulgare* L.) is a drought- and salt-tolerant crop with considerable economic importance in Mediterranean and NENA regions since it is a source of stable farm income [10]. Indeed, barley is a staple food for over 106 countries in the world [11]. Barley is characterized by its high adaptability from humid to arid and even Saharan environments. Barley is grown in many areas of the world and is used for feed, food, and malt production [2,12].

To improve barley production in these regions, plant scientists have adopted a strategy to identify tolerant genotypes for maintaining reasonable yield on salt-affected soils [13]. Crops physiologists and breeders are working to assess how efficient a genotype is in converting water into biomass or yield. To do so, they use production parameters, with which measurement in field experiments is difficult and time-consuming. However, these complex parameters can be determined with the help of crop growth simulation models [6,13]. Dynamic simulation models describe the growth and development of crops based on the interaction with soil, water, and climate parameters. Models can be used to simulate soil and water salinity and crop management practices on the growth and yield of crops under different agro-climatic conditions [6].

Models were used to test the impact of salinity on crops under different environmental conditions and different fertilization practices [14,15].

AquaCrop is a water-driven dynamic model (Vanuytrecht et al., 2014). AquaCrop is a simulation model to study crops' water productivity. As crop-water-productivity is affected by climatic conditions, it is crucial to understand water productivity's response to changing rainfall and temperature patterns [9].

Among the available models, AquaCrop is preferred due to its robustness, precision, and the limited number of variables to be introduced [16]. It uses a small number of explicit and intuitive parameters that require simple calculation [16]. AquaCrop is a software system developed by the Land and Water Division of FAO to estimate water use efficiency and improve agricultural systems' irrigation management practices [17,18].

Water productivity (WP) can be described as the ratio of crops' net benefits, including both rain and irrigation.

According to [19], irrigation management organizations are interested in the yield per unit of irrigation water applied, as they have to improve the yield through human-induced irrigation processes. However, the downside is that not all irrigation water is used to generate crop production. Therefore, FAO defines water productivity as a ratio between a unit of output and a unit of input. Here, water productivity is used exclusively to indicate the amount or value of the product over the volume or value of water that is depleted or diverted [20].

This model was developed by the Food and Agriculture Organization (FAO) [16,21]. AquaCrop simulates the response of crop yield to water and is particularly suited to regions where water is the main limiting factor for agricultural production. The model is based on the concepts of crops' yield response to water developed by Doorenbos and Kassam [22]. The AquaCrop model (v4.0) published in 2012 can estimate yield under salt stress conditions.

The AquaCrop model has been used to predict crop yields under salt stress conditions in different parts of the world [23,24]. Kumar et al. [23] successfully used the AquaCrop model to predict the water productivity of winter wheat under different salinity irrigation water regimes. Mondal et al. [24] used AquaCrop to evaluate the potential impacts of water, soil salinity, and climatic parameters on rice yield in the coastal region of Bangladesh. The AquaCrop model has also been widely used to simulate yields of various crops under diverse environments. For example, barley (*Hordeum vulgare* L.) [5,25,26], teff (*Eragrostis teff* L.) [5], cotton (*Gossypium hirsutum* L.) [27], maize (*Zea mays* L.) [28] wheat (*Triticum aestivum* L.) [3].

In this study, the AquaCrop model (v4.0) is used to assess the performance of two barley genotypes under three contrasted agro-ecosystems (soil, salinity, and climate). In these areas, groundwater is primarily used for irrigation. The salinity of irrigation water ranges from 3 to 15 dS m^{-1}. Farmers do not know which barley variety is most tolerant to producing a reasonable yield under these saline environments. Furthermore, model simulations were also performed to evaluate the impact of three irrigation water salinity levels (5, 10, and 15 dS m^{-1}) on the barley yield. A cost–benefit analysis was performed to determine the economic returns of each level of salinity water irrigation and genotype tolerance based on model simulation results. Those results should help recommend the farmers of saline areas to enhance barley yield and economic return.

2. Materials and Methods

Description of Field Trial Sites

Field experiments were conducted during the 2012–2014 period in three contrasting locations (Beja, Kairouan, Medenine) of Tunisia. The Beja site (36°44′01.13″ N; 9°08′14.30″ E) is sub-humid, Kairouan (35°34′34.97″ N; 10°02′50.88″ E) is located in the semi-arid area of central Tunisia, and Medenine (33°26′54″ N, 10°56′31″ E) is part of the South East arid region of Tunisia (Figure 1). Two barley varieties (Konouz from Tunisia and Batini 100/1 B from Oman) were used for field experiments. The Konouz variety is salt-sensitive [29,30], whereas Batini 100/1 B is salt-tolerant [29,31].

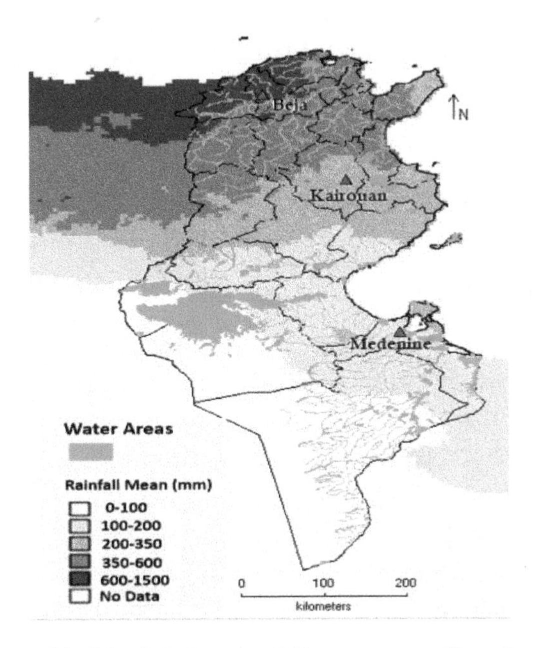

Figure 1. Location of field trial sites in different agro-climatic zones of Tunisia.

In Kairouan (KAI) and Medenine (MED) field trial sites were divided into two sub-plots. Each subplot was irrigated by one water salinity treatment (EC = 2 and 13 dS m^{-1}). Three blocks were defined perpendicularly to the sub-plots so that both treatments were observed in each block. As Beja is located in the rainfed cereal growing area of Tunisia, no irrigation was applied.

The weather data characterize the trials sites related to temperature, and rain was described by [29]. The irrigation water applied and reference evapotranspiration (ET_o) registered in the trials during the two growing seasons are presented in Table 1. The collected data from each site were used to estimate the reference evapotranspiration (ET) according to the Penman-Monteith Evapotranspiration FAO-56 Method, and then the total water supplied was determined for each site to obtain the water barley requirement. Irrigation was applied using a drip system. To ensure water supply homogeneity, line source emitters were installed at each planting row and 33-cm spacing between emitters on the same row.

Table 1. Rainfall, irrigation water applied and evapotranspiration (ET_o) in three trial sites.

Growing Season	Rainfall (mm)			Irrigation Water Applied (mm)			ET_o (mm)		
	Sites			Sites			Sites		
	Beja	KAI	MED	Beja	KAI	MED	Beja	KAI	MED
2012–2013	472.2	151.9	81.1	0	360	455	393.8	364.7	327.6
2013–2014	413.5	180.0	156.1	0	360	405	390.1	363.7	328.4

Soil samples were taken from the trial sites, and physico-chemical analyses were performed. The site's soil characteristics are diverse, from soil rich in clay and organic matter in BEJ to sandy soil with impoverished organic matter continent in MED (Table 2).

Table 2. Soil properties in three field trial sites.

Site	Sand (%)	Clay (%)	Silt (%)	OM (%)	Na^+ Content (ppm)	K^+ Content (ppm)	Ca^{2+} Content (ppm)	PWP (% vol)	FC (% vol)
Beja	15.0	57.5	27.5	4.7	10–20	250–300	100–110	32.0	50.0
KAI	14.8	45.1	40.1	4.0	230–270	390–550	90–140	23.0	39.0
MED	55.5	20.5	24.0	0.9	120–200	30–70	30–55	6.0	13.0

(OM: organic matter, PWP: permanent wilting point; FC: field capacity).

Crops were sown during the last week of November. Seeds were hand sown at the rate of 200 viable grains per m^2. Nitrogen, potassium and phosphorus were applied separately at 85, 50, and 50 kg/ha rates, respectively.

At the five different stages, plants for each genotype, from three small areas (25 × 25 cm) were taken from each experimental unit and used to determine the biomass. At a final harvest stage, plot (1 × 2 m) was used for biomass and grain yield assessment. Water productivity (WP) was calculated as the ration between the collected yield expressed in kg ha^{-1} and the daily transpiration simulated by the model.

3. Description of the AquaCrop Model

The model describes soil, water, crop, and atmosphere interactions through four sub-model components: (i) the soil with its water balance; (ii) the crop (development, growth, and yield); (iii) the atmosphere (temperature, evapotranspiration, and rainfall), and carbon dioxide (CO_2) concentration; and (iv) the management, such as irrigation and crop fertilization soil fertility.

The AquaCrop model is based on the relationship between the relative yield and the relative evapotranspiration [22] as follows

$$\frac{Y_x - Y_a}{Y_x} = K_y\left(\frac{ET_x - ET_a}{ET_x}\right) \quad (1)$$

where Y_x is the maximum yield, Y_a is the actual yield, ET_x is the maximum evapotranspiration, ET_a is the actual evapotranspiration, and K_y is the yield response factor between the decrease in the relative yield and the relative reduction in evapotranspiration.

The AquaCrop model does not take into account the non-productive use of water for separating evapotranspiration (ET) into crop transpiration (T) and soil evaporation (E)

$$ET = E + Tr \tag{2}$$

where ET = actual evapotranspiration, E = soil evaporation and Tr = the sweating of crop.

At a daily time step, the model successively simulates the following processes: (i) groundwater balance; (ii) development of green canopy (CC); (iii) crop transpiration; (iv) biomass (B); and (v) conversion of biomass (B) to crop yield (Y). Therefore, through the daily potential evapotranspiration (ET_o) and productivity of water (WP*), the daily transpiration (Tr) is converted into vegetal biomass as follows

$$B_i = WP^*\left(\frac{T_{ri}}{ETo_i}\right) \tag{3}$$

where WP* is the normalized water productivity [32,33] relative to Tr. After the normalization of water productivity for different climatic conditions, its value can be converted into a fixed parameter [34]. The estimation and prediction of performance are based on the final biomass (B) and harvest index (HI). This allows a clear distinction between impact of stress on B and HI, in response to the environmental conditions

$$Y = HI * (B) \tag{4}$$

where: Y = final yield; B = biomass; HI = harvest index.

During the calibration and testing of the model, we calculated water productivity (WP) as presented by Araya et al. [5]

$$WP = \left[\frac{Y}{\sum Tr}\right] \tag{5}$$

where Y is the yield expressed in kg ha^{-1} and Tr is the daily transpiration simulated by the model.

3.1. Crop Response to Soil Salinity Stress

The electrical conductivity of saturation soil-past extracts from the root zone (ECe) is commonly used as an indicator of the soil salinity stress to determine the total reduction in biomass production, determines the value for soil salinity stress coefficient ($K_{s,\ salt}$).

The coefficient of soil salinity stress (Ks_{salt}) varied between 0 (full effect of stress of soil salinity) and 1 (no effect). The following equation determined the reduction in biomass

$$B_{rel} = 100\ (1 - Ks_{salt}) \tag{6}$$

B_{rel} represents the expected biomass production under given salinity stress relative to the biomass produced in the absence of salt stress. The coefficient is adjusted daily to the average ECe in the root zone [35].

Then, the thresholds values are given for the sensitive and tolerant barley genotype and expressed in dS m^{-1}. This allows the estimation of the lower limit (EC_{en}) to which the soil salinity stress begins to affect the production of biomass and the upper threshold (EC_{ex}), in which soil salinity stress has reached its maximum effect.

3.2. Soil Salinity Calculation

AquaCrop adopts the calculation procedure presented in BUDGET [36] to simulate the movement and retention of salt in the soil profile. The salts enter the soil profile as solutes after irrigation with saline water or through capillary rise from a shallow groundwater table (vertical downward and upward salt movement). The average ECe in the compartments of the effective rooting depth determines the effects of soil salinity on biomass production.

To explain the movement and retention of soil water and salts in the soil profile, AquaCrop divides the soil profile into 2 to 11 soil compartments called "cells", depending on the type of soil in each horizon (clay, sandy horizon) and its saturated hydraulic conductivity (Ksat in mm/day). The salt diffusion between two adjacent cells (cell j and cell j+1) is determined by the differences in salt concentration and expressed by the electrical conductivity (EC) of soil water.

AquaCrop determines the vertical salt movement in response to soil evaporation, considering the amount of water extracted from the soil profile by evaporation and the wetness of the upper soil layer. The relative soil water content of the topsoil layer determines the fraction of the dissolved salts that moves with the evaporating water.

AquaCrop determines the vertical salt movement because of the capillary rise. Finally, the salt content of a cell is determined by

$$\mathrm{Salt_{cell}} = 0.64\ \mathrm{W_{cell}EC_{cell}} \tag{7}$$

$\mathrm{Salt_{cell}}$ is the salt content expressed in grams salts per m^2 soil surface, Wcell its volume expressed in liter per m^2 (1 mm = 1 L/m^2), and 0.64 a global conversion factor used in AquaCrop to convert dS/m to g/L. The electrical conductivity of the soil water (ECsw) and of the electrical conductivity of saturation soil-past extract (ECe) at a particular soil depth (soil compartment) is calculated as

$$\mathrm{EC_{SW}} = \frac{\sum_{j=1}^{n} \mathrm{Salt_{cell.j}}}{0.64\ (1000\ \theta\Delta_z)\left\{1 - \frac{\mathrm{Vol\%_{gravel}}}{100}\right\}} \tag{8}$$

$$\mathrm{EC_e} = \frac{\sum_{j=1}^{n} \mathrm{Salt_{cell.j}}}{0.64\ (1000\ \theta_{sat}\Delta_z)\left\{1 - \frac{\mathrm{Vol\%_{gravel}}}{100}\right\}} \tag{9}$$

where n is the number of cells in each soil compartment; θ is the soil water content (m^3/m^3); θ_{sat} is the soil water content (m^3/m^3) at saturation; Δz (m) is the thickness of the soil compartment and Vol% gravel is the volume percentage of the gravel in the soil horizon of each compartment.

4. Model Calibration

4.1. Input and Output Variables of the Model

The model was calibrated using data from the growing season of 2012–2013 and evaluated using data from 2013–2014. Determining parameters for crop development and production, as well as water and salinity stress, was fundamental for calibrating the AquaCrop model. The parameters of climate, soil, and crop management used for the model calibration are presented in Table 3.

Table 3. Climate, soil and crop parameters used for the simulation model AquaCrop.

Climate		- Daily rainfall, daily ET_o, daily temperatures - CO_2 concentration
Crop	Limited set	Crop development and production parameters which include phenology and life cycle
	Crop parameters	- Harvest index - Root zone threshold at the end of the canopy expansion - Threshold root zone depletion for early senescence - Time for the maximum canopy cover - Maximum vegetation - Flowering time - Initial vegetative cover - Depletion threshold root zone for stomata closure - Extraction of water

Table 3. *Cont.*

	Field	- Soil fertility, mulch - Field practices (surface runoff presence, ground bond)
Soil	Soil profile	Characteristics of soil horizon (no of soil horizon, thickness, Permanent Wilting Point (PWP), Field Capacity (FC), Soil saturation (SAT), Ksat); soil surface (runoff, evaporation); Restrictive soil layer capillary rise).
	Soil water and groundwater	Constant depth; variable depth; water quality.

4.2. Statistical Parameters Used for the Calibration and Evaluation of Model

Several statistical indices were used to evaluate the performance of the model on the field measured data. These include Percentage Error (PE), Root Mean Square Error (RMSE), Model Efficiency (ME) and Coefficient of Determination (R^2).

Percentage Error (PE) was determined using the following equation

$$EP = \frac{(S_i - O_i)}{O_i} \times 100 \quad (10)$$

where S_i and O_i are simulated and observed values, respectively.

The root means square error (RMSE) [37] is presented by the following equation

$$RMSE = \sqrt{\frac{1}{n}\sum_{i=1}^{n}(S_i - O_i)^2} \quad (11)$$

with the values of RMSE close to zero indicate the best model fit.

The model efficiency (ME) [38] was applied to assess the effectiveness of the model. The ME indicator compares the variability of prediction errors by the model to those of collected data from the field. If the prediction errors are greater than the data error, then the indicator becomes negative. The upper ME bound is at 1.

$$ME = \frac{\sum_{i=1}^{n}(O_i - MO)^2 - \sum_{i=1}^{n}(S_i - O_i)^2}{\sum_{i=1}^{n}(O_i - MO)^2} \quad (12)$$

The coefficient of determination (R^2), as a result of regression analysis, is the proportion of the variance in the dependent variable (predict value) that is predictable from the independent variable (observed value) and is computed according to [35]

$$R^2 = \left\{\frac{\sum_{i=1}^{n}(O_i - \overline{O})(S_i - \overline{S})}{\left[\sum_{i=1}^{n}(O_i - \overline{O})^2\right]^{0.5}\left[\sum_{i=1}^{n}(S_i - \overline{S})^2\right]^{0.5}}\right\}^2 \quad (13)$$

R^2 is between 0 and 1.

4.3. Parameters Used for Model Calibration

In total, 26 input parameters were used for the model calibration (Table 4). Out of these, 14 parameters were considered as "conservative" because they do not change with salinity and are independent of limiting or non-limiting conditions. These parameters include normalized crop water productivity and crop transpiration coefficient. The remaining 12 are site-specific (climate, water, and soil salinity) and crop-specific (tolerant or sensitive). These input parameters were adjusted during the calibration process to obtain better adequacy between the measured and simulation values.

Table 4. Final values of different model input parameters obtained after calibration for two genotypes under different salinity levels (S1 = 2 dS m^{-1}; S2 = 13 dS m^{-1}).

		Batini-100/1 B (Salt-Tolerant)			Konouz (Salt-Sensitive)			Remarks
		BEJ	KAI	MED	BEJ	KAI	MED	
Base temperature (°C)	S1	0	0	0	0	0	0	Conservative
	S2	-	0	0	-	0	0	
Upper temperature (°C)	S1	30	30	30	30	30	30	Conservative
	S2	-	30	30	-	30	30	
Initial canopy cover, CC0 (%)	S1	1.5	1.5	1.5	1.5	1.5	1.5	Conservative
	S2	-	1.5	1.5	-	1.5	1.5	
Canopy cover per seeding (cm^2/plant)	S1	0.75	0.75	0.75	0.75	0.75	0.75	Conservative
	S2	-	0.75	0.75	-	0.75	0.75	
Maximum coefficient for transpiration, KcTr, x	S1	0.90	0.90	0.90	0.90	0.90	0.90	Conservative
	S2	-	0.90	0.90	-	0.90	0.90	
Maximum coefficient for soil evaporation, Kex	S1	0.4	0.4	0.4	0.4	0.4	0.4	Conservative
	S2	-	0.4	0.4	-	0.4	0.4	
Upper threshold for canopy expansion, Pexp, upper	S1	0.30	0.30	0.30	0.20	0.20	0.20	Varietal effect
	S2	-	0.30	0.30	-	0.20	0.20	
Lower threshold for canopy expansion, Pexp, lower	S1	0.65	0.65	0.65	0.55	0.55	0.55	Varietal effect
	S2	-	0.65	0.65	-	0.55	0.55	
Leaf expansion stress coefficient curve shape	S1	4.5	4.5	4.5	4.5	4.5	4.5	Conservative
	S2	4.5	4.5	4.5	4.5	4.5	4.5	
Upper threshold for stomatal closure, Psto, upper	S1	0.6	0.6	0.6	0.55	0.55	0.55	Varietal effect
	S2	-	0.6	0.6	-	0.55	0.55	
Leaf expansion stress coefficient curve shape	S1	4.5	4.5	4.5	4.5	4.5	4.5	Conservative
	S2	4.5	4.5	4.5	4.5	4.5	4.5	
Canopy senescence stress coefficient, Psen, upper	S1	0.65	0.65	0.65	0.55	0.45	0.45	Varietal effect and site effect for the sensitive
	S2	-	0.65	0.65	-	0.45	0.45	
Senescence stress coefficient curve shape	S1	4.5	4.5	4.5	4.5	4.5	4.5	Conservative
	S2	4.5	4.5	4.5	-	4.5	4.5	

Table 4. *Cont.*

		Batini-100/1 B (Salt-Tolerant)			Konouz (Salt-Sensitive)			Remarks
		BEJ	KAI	MED	BEJ	KAI	MED	
Reference harvest index, HI0 (%)	S1	40	40	41	41	42	45	Varietal and salt stress effect
	S2		40	45		41	41	
Normalized crop water productivity, WP* (g/m^2)	S1	14	14	14	14	14	14	Conservative
	S2	14	14	14	14	14	14	
Time from sowing to emergence (day)	S1	7	7	7	7	7	7	Conservative
	S2	7	7	7	7	7	7	
Time from sowing to maximum CC (jours)	S1	60	60	60	62	60	57	Varietal and salt stress effect
	S2	-	60	58	-	59	55	
Time from sowing to maximum CC (day)	S1	145	145	145	145	145	145	Conservative
	S2	-	145	145	145	145	145	
Time from sowing to maturity (day)	S1	178	157	157	178	157	157	Varietal and salt stress effect
	S2	-	157	157	-	157	157	
Maximum canopy cover, CCx (%)	S1	87	87	87	87	75	63	Varietal and salt stress effect
	S2	-	87	70	-	60	40	
Canopy growth coefficient, CGC (%/day)	S1	12.5	12.5	12.5	12	12	12	Varietal effect
	S2		12.5	12.5		12	12	
Canopy decline coefficient, CDC (%/day)	S1	6	6	6	6	6	6	Conservative
	S2	-	6	6	-	6	6	
Maximum effective rooting depth, Zx (m)	S1	0,9	0.75	0.75	0,9	0.75	0.75	Site effect
	S2	-	0.75	0.75	-	0.75	0.75	
Salinity stress, lower threshold, ECen (dS m^{-1})	S1	3	3	3	1	1	1	Varietal effect
	S2	3	3	3	1	1	1	
Salinity stress, upper threshold, ECex (dS m^{-1})	S1	22	22	22	18	18	18	Varietal effect
	S2	22	22	22	18	18	18	
Shape factor for salinity stress coefficient curve	S1	1	1	1	1	1	1	Conservative
	S2	1	1	1	1	1	1	

For Bej only rainfall; for Kai, two levels of water salinity (S1 = 1.2 dS m^{-1}; S2 = 13 dS m^{-1} (S2)); for Med, two levels of water salinity (S1 = 2 dS m^{-1}; S2 = 13 dS m^{-1} (S2)).

4.4. Development of Different Scenarios

After calibration and evaluation, the model was used to assess the performance of two barley varieties under three water salinity conditions scenarios i.e., 5, 10, and 15 dS m^{-1}, using the weather data for the growing season 2013–2014.

4.5. The Economic Gain from the Use of a Unit of Water Consumed in the Tow Barley Varieties under Different Climatic Conditions

The economic productivity of two barley varieties was estimated using the average unit cost of one water cubic meter in Tunisia and the water use predicted by AquaCrop. The crop water economic productivity of the tolerant and the sensitive barley varieties as the measure of the biophysical and then economic gain from the use of a unit of water consumed were estimated by AquaCrop model in grain yield production [20]. This is expressed in productive crop units of kg/m^3 and money unit/m^3.

5. Results

5.1. Biophysical Environments Variability of Experimental Sites

The experiments are conducted in adaptability trials set up in three contrasting biophysical environments (from the sub-humid to the arid interior). These sites, namely Beja, Kairouan and Medenine, were selected on a North–South transect (Figure 1). The soils of the trial sites are very diverse, from soil rich in clay and poor in organic matter in BEJ to sandy soil with poor organic matter continent in MED (Table 2). Beja's sub-humid site received annual rainfall of 472 and 413 mm respectively during the two cropping seasons. However, in the semi-arid and arid sites, low rainfall was registered. The arid site of MED received an annual rainfall of 81 mm during the first cropping season and 156 mm during the second. At Kairouan, the rainfall for the 2012/2013 and 2013/2014 seasons was 152 and 180 mm, respectively (Table 1). As Beja is located in the rainfed cereal-growing area of Tunisia, no irrigation was applied. KAI and MED field trial sites, two different salinities (EC = 2 and 13 dS m^{-1}) of water were used for irrigation.

Soil calcium and potassium content was higher in KAI as compared to MED. Soil sodium content changes during the different experimentation period following irrigation with saline water in KAI and MED, where sodium is the dominant element present in the saline irrigation water. The variation between sites might be explained by the variation in the cationic exchange capacity of the sandy soil and torrential character of the rainfall in this area (Table 2).

5.2. Biomass, Grain Yield, and Water Use Efficiency

The correlation between grain yield, biomass, and water productivity values for two barley genotypes showed that the observed and simulated values are closely co-related, as evidenced by the high R^2 values, i.e., 0.91, 0.93, and 0.89 for grain yield, biomass, and water productivity, respectively (Figure 2).

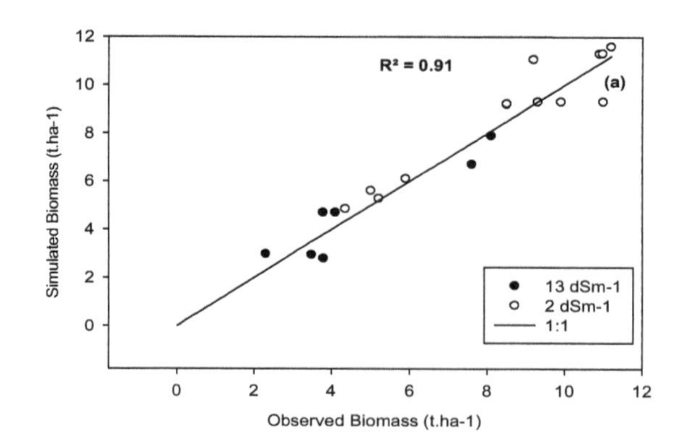

Figure 2. *Cont.*

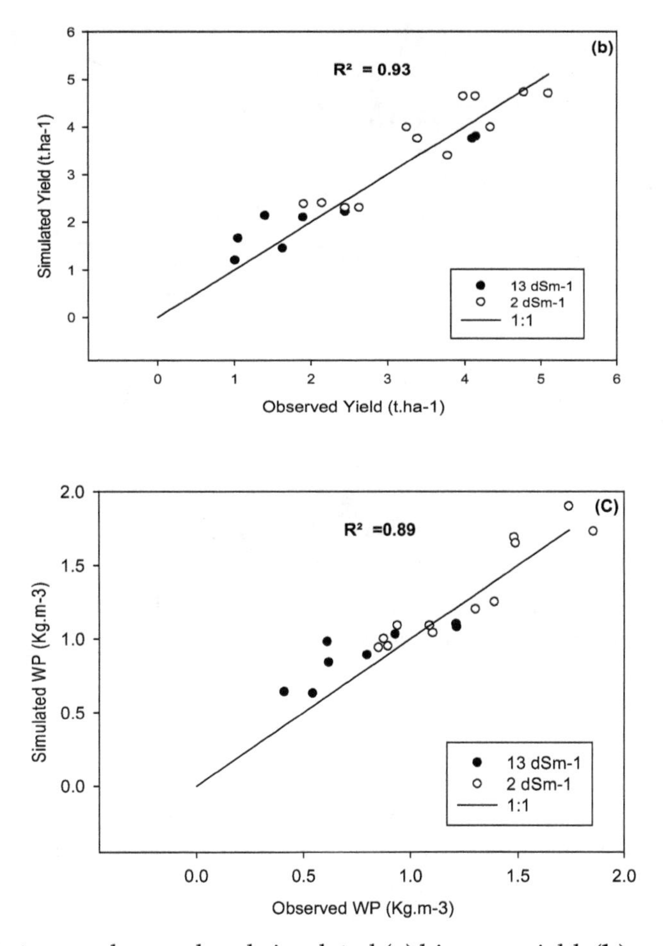

Figure 2. Correlation between observed and simulated (**a**) biomass yield; (**b**) grain yield; and (**c**) water productivity compared with 1:1 line.

The correlation between observed and simulated values of biomass yield for two barley genotypes at three locations showed proximity (Figure 3), which indicates the excellent ability of the AquaCrop model to predict biomass yield under different agro-climatic conditions. The results also show that the sensitive barley variety at MED produces the lowest biomass for both irrigation water qualities. Similar trends were observed for grain yield, where the tolerant barley variety performed better than the sensitive variety regardless of the location and the quality of irrigation water.

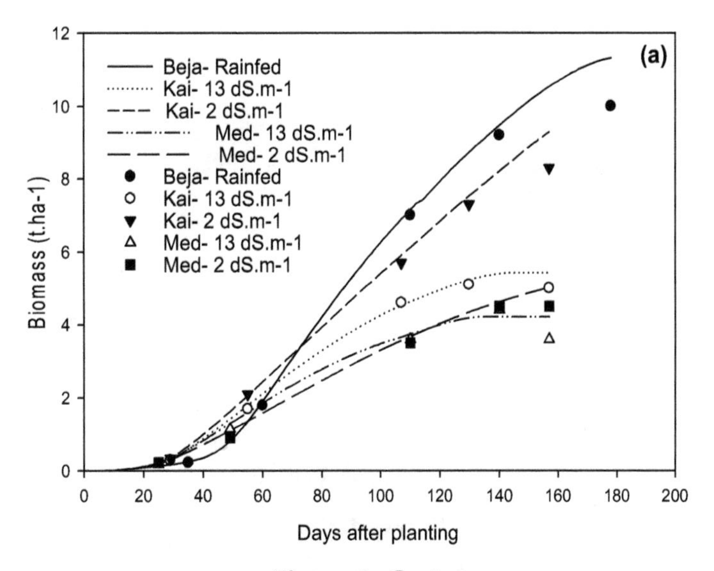

Figure 3. *Cont.*

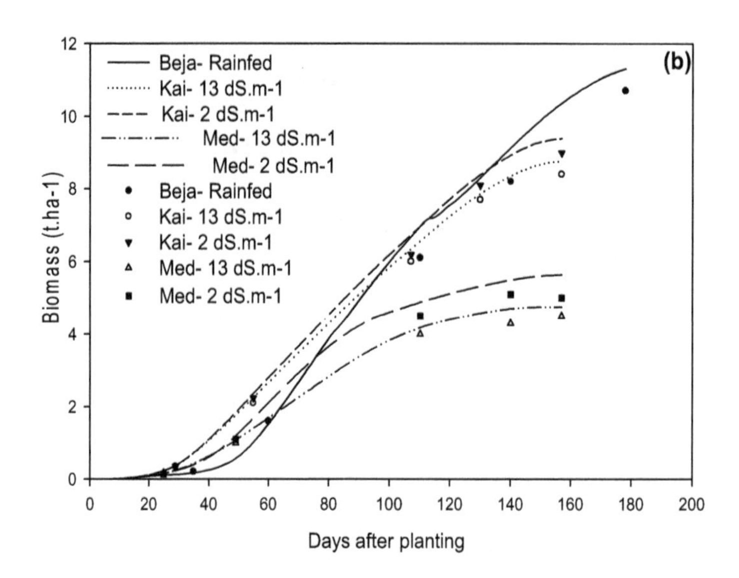

Figure 3. Simulated and observed biomass of (**a**) tolerant and (**b**) sensitive barley genotypes (dots represent observations; simulations are represented by lines).

5.3. Canopy Cover (CC)

The maximum and minimum CC were 85% and 30% in the sub-humid and arid areas, respectively. The salinity induces a 10% reduction in the CC in the sub-humid environment and 5–30% in the dry climate of MED. CC reduction under saline irrigation water is less noticeable in the tolerant variety than the sensitive variety for both salinity levels. However, in the rainfed area of Beja, the growth of both varieties was comparable.

Figure 4 shows a strong correlation between measured and simulated CC values for both varieties of barley ($R^2 = 0.91$ and $R^2 = 0.93$). In general, a good match between the observed and the simulated CC was observed in all three locations. However, the model somewhat over-estimated CC in the rainfed environment of Beja and slightly under-estimated it in the other two situations.

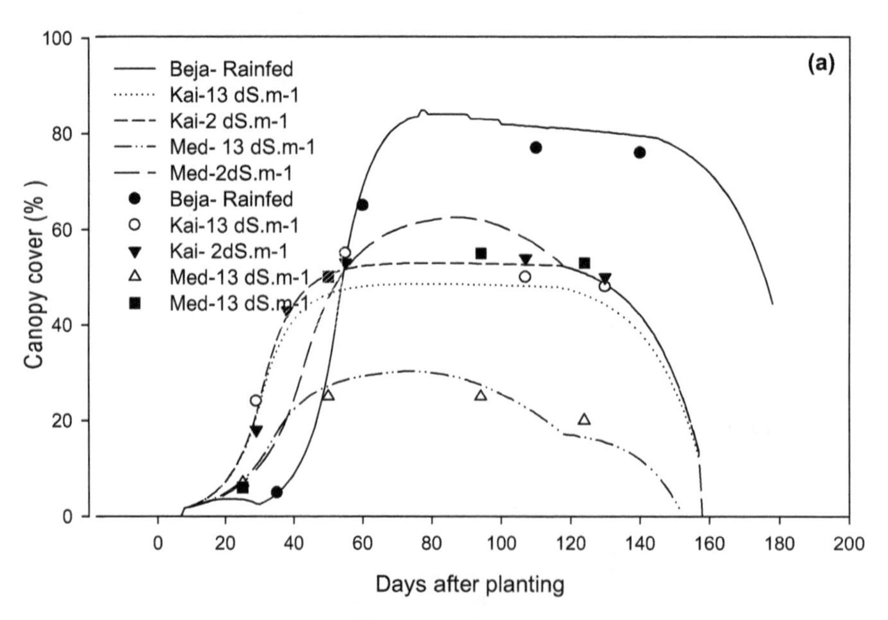

Figure 4. *Cont.*

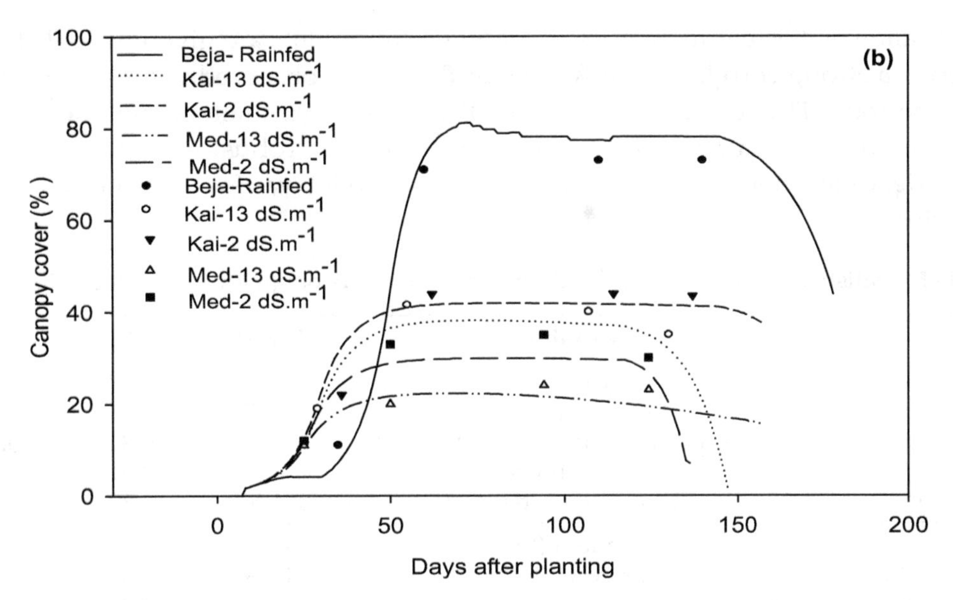

Figure 4. Simulated and observed canopy cover for (**a**) tolerant and (**b**) sensitive barley varieties.

5.4. Effects of Soil Salinity

The maximum soil salinity was in the arid and semi-arid areas irrigated with saline water, respectively. The soil salinisation dynamic depends on the salinity of irrigation water. However, in the rainfed area of Beja, we noted the absence of any salty issue.

Figure 5 shows that the simulated soil salinity trend in the root zone (up to a depth of 0.7 m) corresponds very well with the measured values under different saline water regimes across different environments throughout the growing season. The observed and modeled soil salinity correlated well, with an R^2 of 0.96. Figure 5 shows that the model reliably simulated average root zone salinity when the crop is irrigated with low-salinity water (2 dS m^{-1}). However, it slightly underestimated soil salinity under higher saline water conditions (13 dS m^{-1}), particularly for the late growing season.

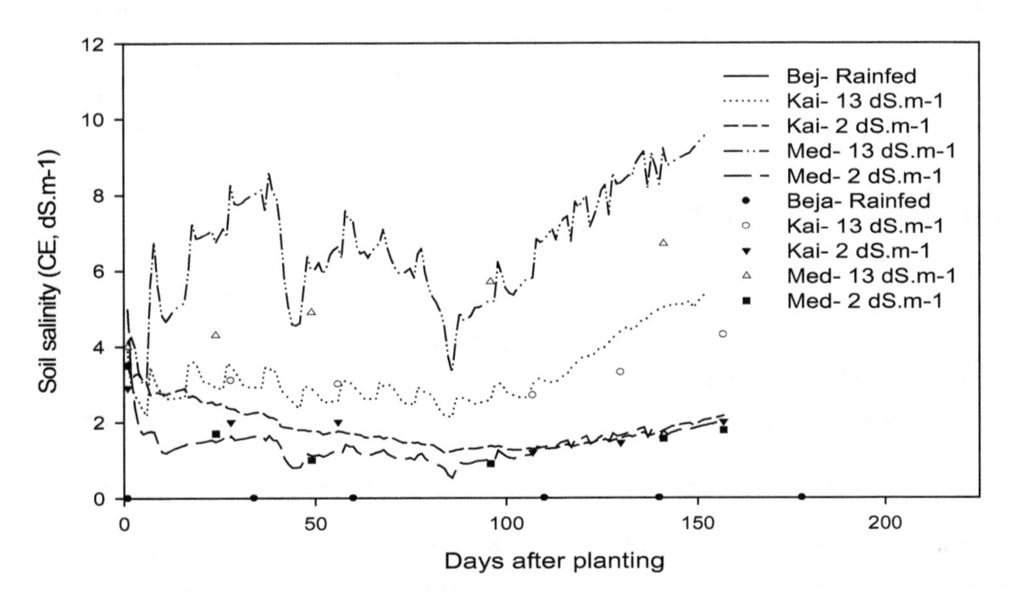

Figure 5. Simulated and observed soil salinity in the testing-cropping season under different saline water regimes and across different environments.

5.5. Statistical Indices for AquaCrop Model Evaluation

The statistical indices derived for evaluating the AquaCrop model's performance in predicting soil water content, yield, canopy cover percent, biomass, and water productivity (*WP*) of barley genotypes under different saline water regimes across different environments are given in Table 5. All statistical parameters depict a strong correlation between simulated and observed values for model calibration and evaluation periods. The correlation between all statistical parameters remained almost the same for the calibration and evaluation period, which indicates the robustness of the model prediction. Based on the model calibration and evaluation results, the model was found robust enough to calculate different scenarios.

Table 5. Statistical indices values for different parameters obtained for model calibration.

	Variable	RMSE	ME	R^2
Calibration	Grain yield (t ha^{-1})	0.40	0.89	0.91
	Biomass (t ha^{-1})	0.87	0.96	0.93
	water productivity (kg ha^{-1} mm^{-1})	0.15	0.84	0.89
	Soil salinity	0.34	0.91	0.95
	Canopy cover percent	1.5	0.89	0.91
Evaluation	Grain yield (t ha^{-1})	0.45	0.87	0.89
	Biomass (t ha^{-1})	0.89	0.86	0.87
	water productivity (kg ha^{-1} mm^{-1})	0.13	0.91	0.84
	Soil salinity	1.25	0.91	0.96
	Canopy cover percent	2.25	0.89	0.91

6. Development of Different Scenarios

Due to a shortage of surface water, farmers of KAI and MED regions have no option than to use groundwater for irrigation. The quality of groundwater ranges from 4 to15 dS m^{-1} in these two regions. Farmers are interested to know which barley varieties would be most suitable to grow under these groundwater quality conditions. The calibrated and evaluated model was used to assess the performance of two barley varieties under three water salinity conditions i.e., 5, 10, and 15 dS m^{-1}, and the results are presented in Table 6.

Table 6. Predicted values of biomass, yield, and water productivity of two barley varieties for different scenarios.

	BEJ	KAI			MED		
	Rainfed	5 dS m^{-1}	10 dS m^{-1}	15 dS m^{-1}	5 dS m^{-1}	10 dS m^{-1}	15 dS m^{-1}
			Tolerant genotype				
Biomass (t ha^{-1})	11.30	9.07	8.36	5.48	5.60	4.74	4.70
Yield (t ha^{-1})	4.70	3.65	3.44	2.20	2.29	2.13	2.10
WP (kg m^{-3})	1.73	1.29	1.19	0.85	1.25	1.18	1.00
			Sensitive genotype				
Biomass (t ha^{-1})	11.33	6.62	4.60	1.90	3.18	3.03	2.80
Yield (t ha^{-1})	4.64	2.70	1.90	0.80	1.40	1.30	0.80
WP (kg m^{-3})	1.65	1.12	0.85	0.45	0.74	0.72	0.51

The performance of both barley varieties in the KAI area is predicted to be much higher than MED area under all salinity levels due to prevailing climatic conditions. In the KAI area, biomass and grain yield reductions are much higher with the increasing water salinity for both varieties. For example, the biomass and yield reductions in the KAI area were about 40%with an increase in salinity from 5 to 10 and 15 dS m^{-1}. For the sensitive genotype, the biomass and yield reductions in the KAI area would be above 72% with a similar increase in the salinity levels. Although overall biomass and grain yields in the MED area were lower than in the KAI area, biomass and yield reductions for the salt-tolerant barley variety were only 16% and 8%, with an increase in salinity from 5 to 15 dSm^{-1}, respectively.

However, for the sensitive genotype, reductions in biomass and yield were 12% and 43%, respectively, with a similar increase in salinity levels. Similar trends are obtained for water productivities.

Without salt stress, both varieties have the same performance. However, the tolerant variety performs better than the sensitive variety under salt stress. This is because it has better potential. Therefore, farmers can grow both varieties in the rainfed areas of BEJ, while, in KAI and MED areas where irrigation is necessary for crop growth, the salt-tolerant barley variety should be preferred. The cultivation of the salt-sensitive barley variety in the MED area will be risky, as the yields will be low, and the development of soil salinity over time will remain a challenge. This situation will be very critical for long-term sustainable crop production in the area.

7. Economic Productivity of Barley Varieties under Different Climatic Conditions

The economic productivity of two barley varieties was estimated using the average unit cost of one water cubic meter in Tunisia and the water use predicted by AquaCrop. The results show that the production cost of 1 kg of barley is lowest in the BEJ area compared to those areas where it is irrigated with saline water.

In the KAI region, the cost will be reduced by 13.28% 28.72% and 47.19% by growing the tolerant variety irrigated with saline water of 5, 10, and 15 dS m^{-1}, respectively. In the arid region of MED, the benefit will be reduced by 40%, 38%, and 49% by growing the tolerant barley variety by irrigating with saline water of 5, 10 and 15 dS m^{-1}, respectively (Figure 6). However, in the sub-humid region of BEJ, there is no significant difference between susceptible and tolerant genotypes. The results show the economic interest for arid region farmers to grow the tolerant barley variety. This stresses the need for appropriate breeding programs for the saline environments for optimizing crop production instead of targeting potential yields.

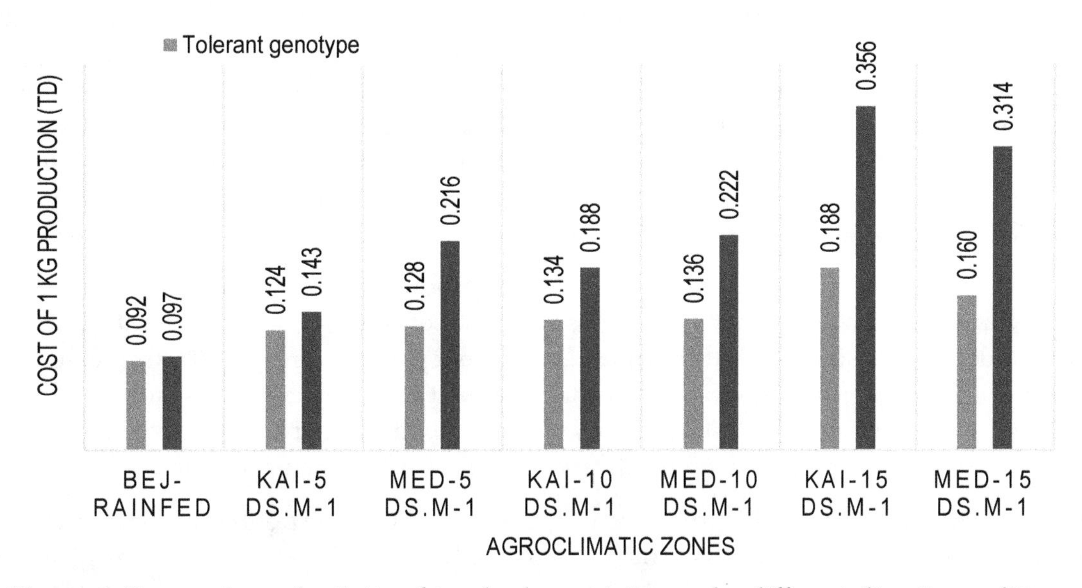

Figure 6. Economic productivity of two barley varieties under different climatic conditions.

8. Discussion

We evaluated the AquaCrop model for two barley varieties under contrasting environments and different water salinity levels. The simulated model values were close to the field measurements concerning biomass, yield and soil salinity. ME and R^2 parameters were close to 0.9, showing the model's ability to simulate the behavior of sensitive and resistant cultivars in contrasting environments and irrigation practices. Araya et al. [5] reported R^2 values of 0.80 when simulating barley biomass and grain yield using AquaCrop. El Mokh et al. [25] reported R^2 values of 0.88 when simulating barley yield under different irrigation regimes in a dry environment using AquaCrop. Mondal et al. [24] reported a 0.12 t ha^{-1} root mean square error after simulating the yield response of rice to salinity stress with the AquaCrop model. Our results also show a correct prediction with an RMSE of 0.45 t ha^{-1} (Table 5). This shows that the AquaCrop model simulates biomass production for all environments with an acceptable accuracy level.

AquaCrop model produces consistent simulation results for CC with an R^2 of 0.89 and RMSE of 2.25 (Table 5). The model also simulated soil salinity satisfactorily for all environments (R^2 = 0.96) for all situations. The R^2 values exceeding 0.8 are considered excellent for model performance [39]. The ability of AquaCrop to predict yield depends on the appropriate calibration of the canopy cover curve [1,40]. Indeed, after simulation of soil water balance at a daily time step, the model simulates CC and then simulates the transpiration of a crop, biomass above the soil, and converts biomass into yield. Therefore, it is essential to make accurate predictions of the canopy cover by the proper calibration of crop traits.

Therefore, through proper calibration, models can be used for additional solutions for the quantification of salinity build-up in the root zone [41].

We also noted the overestimation of the soil salinity at the end of the growing season when saline water is used for irrigation (Figure 5). This could be due to the excessive leaching of salts from the soil profile through irrigation, as reported by Mohammadi et al. [42]. Over- or underestimation at the end of the season could be the simplification of soil salt transport calculations in the model based on some empirical functions, including the parameters of Ks and the drainage coefficient for vertical downward salt movement. Furthermore, the occasional leaching of salts from the root zone using relatively better-quality water is also recommended. Changing cropping patterns is also a useful strategy for the rehabilitation and management of saline soils, especially when only saline water is available for irrigation.

The AquaCrop model was also capable of predicting water productivity under sub-humid, semi-arid, and arid environments and the effect of salinity. Plants subjected to salinity stress show a varying response in *WP*. The sensitive genotype was more exposed to varying responses in *WP*. Besides, heat stress induced by increased temperatures and the water deficit also decreases productivity, as demonstrated by Hatfield [43]. The observed and predicted water productivities were directly affected by climate aridity and the salinity of the irrigation water. However, the tolerant barley variety was less affected by these factors. These results are in agreement with the earlier studies [16,44].

Water scarcity is already hampering agricultural production in the MENA region. Therefore, the adoption of integrated management strategies will be useful for growing tolerant genotypes under saline water conditions and increasing the water use efficiency. For the sustainable management of crop growth in saline environments, soil-crop-water management interventions consistent with site-specific conditions need to be adopted [41]. These may include cyclic or conjunctive saline water use and freshwater through proper irrigation scheduling to avoid salinity development.

There are several traits available for screening genetic material for enhanced production and *WP* under different climate scenarios. This study shows that, under different water salinity conditions, sensitive barley genotype is more affected by the increasing water salinity than the tolerant barley genotype. The crop yields for both genotypes under all water salinity levels were higher in KAI area compared to the MED area. Therefore, this study recommends that farmers with higher salinity water for irrigation should grow tolerant barley genotypes, allowing them to reduce the cost, on average, by 30% (Figure 6). However, from a sustainability point of view, irrigation amounts should be kept to a minimum to optimize crop yields instead of targeting potential yields [45]. This exercise will help there be less accumulation of salts in the root zone. Besides, the occasional leaching of salts from the root zone using relatively better-quality water is also recommended. Changing cropping patterns is also regarded as a useful strategy for the rehabilitation and management of saline soils, especially when only saline water is available for irrigation [46,47].

9. Conclusions

The AquaCrop model with a salinity module was used to evaluate the agronomic performance of two barley varieties for the three different agro-climatic zones in Tunisia. These zones represent sub-humid, semi-arid, and arid climates. The model was calibrated and evaluated using field data from two years (2012 and 2014). The excellent correlation between the simulated and measured data of biomass, yield, and soil salinity confirms the ability of AquaCrop model to simulate crop growth under different climatic conditions. The scenario results using the calibrated model indicate that farmers with higher salinity water for irrigation should grow tolerant barley genotypes. However, from a sustainability point of view, irrigation amounts should be kept to a minimum to optimize crop yields instead of targeting potential yields.

Author Contributions: Conceptualization, Z.H.; Data curation, F.E.B.A. and S.A.; Formal analysis, Z.H.; Methodology, Z.H., A.S. and Z.C.; Supervision, Y.T.; Writing—Original draft, Z.H.; Writing—Review and editing, A.S.Q. and A.G. All authors have read and agreed to the published version of the manuscript.

References

1. Abi Saab, M.T.; Albrizio, R.; Nangia, V.; Karam, F.; Rouphael, Y. Developing scenarios to assess sunflower and soybean yield under different sowing dates and water regimes in the Bekaa valley (Lebanon): Simulations with Aquacrop. *Int. J. Plant Prod.* **2014**, *8*, 457–482.
2. Ahmed, M.; Goyal, M.; Asif, M. Silicon the non-essential beneficial plant nutrient to enhanced drought tolereance in wheat. In *Crop Plant*; Goyal, A., Ed.; Intech Publication House: London, UK, 2012; pp. 31–48.
3. Andarzian, B.; Bannayan, M.; Steduto, P.; Mazraeh, H.; Barati, M.E.; Barati, M.A.; Rahnama, A. Validation and testing of the AquaCrop model under fulland deficit irrigated wheat production in Iran. *Agric. Water Manag.* **2011**, *100*, 1–8. [CrossRef]
4. Araya, A.; Habtub, S.; Hadguc, K.; Kebedea, A.; Dejened, T. Test of AquaCrop model in simulating biomass and yield of water deficient and irrigated barley (Hordeum vulgare). *Agric. Water Manag.* **2010**, *97*, 1838–1846. [CrossRef]
5. Araya, A.; Keesstra, S.D.; Stroosnijder, L. Simulating yield response to water of Tef (Eragrostistef) with FAO's AquaCrop model. *Field Crop. Res.* **2010**, *116*, 1996–2204. [CrossRef]
6. Chauhdarya, J.N.; Bakhsh, A.; Ragab, R.; Khaliq, A.; Bernard, A.; Engeld, M.R.; Shahid, M.N.; Nawaz, Q. Modeling corn growth and root zone salinity dynamics to improve irrigation and fertigation management under semi-arid conditions. *Agric. Water Manag.* **2020**, *230*, 105952. [CrossRef]
7. FAO (Food and Agriculture Organization of the United Nations). *Advances in the Assessment and Mmonitoring of Salinization and Status of Biosaline Agriculture*; FAO: Rome, Italy, 2010.
8. Qadir, M.; Quillérou, E.; Nangia, V.; Murtaza, G.; Singh, M.; Thomas, R.J.; Drechsel, P.; Noble, A.D. Economics of salt-induced land degradation and restoration. *Nat. Resour. Forum* **2014**, *38*, 282–295. [CrossRef]
9. Hatfield, J.L.; Dold, C. Water-Use efficiency: Advances and challenges in a changing climate. *Front. Plant Sci.* **2019**. [CrossRef]
10. Zhou, G.; Johnson, P.; Ryan, P.R. Quantitative trait loci for salinity tolerance in barley (*Hordeum vulgare* L.). *Mol. Breed.* **2012**, *29*, 427–436. [CrossRef]
11. Newton, A.C.; Flavell, A.J.; George, T.S.; Leat, P.; Mullholland, B.; Ramsay, L.; RevoredoGiha, C.; Russell, J.; Steffenson, B.J.; Swanston, J.S.; et al. Crops that feed the world 4. Barley: A resilient crop? Strengths and weaknesses in the context of food security. *Food Secur.* **2011**, *3*, 141–178. [CrossRef]
12. Zhang, H.; Han, B.; Wang, T.; Chen, S.; Li, H.; Zhang, Y.; Dai, S. Mechanisms of Plant Salt Response: Insights from Proteomics. *J. Proteome Res.* **2012**, *11*, 49–67. [CrossRef]
13. Negrao, S.; Schmockel, S.M.; Tester, M. Evaluating physiological responses of plants to salinity stress. *Ann. Bot.* **2017**, *119*, 1–11. [CrossRef] [PubMed]
14. Heng, L.K.; Hsiao, T.; Evett, S.; Howell, T.; Steduto, P. Validating the FAO AquaCrop model for irrigated and water deficient field maize. *Agron. J.* **2009**, *101*, 488–498. [CrossRef]
15. Verma, A.K.; Gupta, S.K.; Isaac, R.K. Use of saline water for irrigation in monsoon climate and deep water table regions: Simulation modelling with SWAP. *Agric. Water Manag.* **2012**, *115*, 186–193. [CrossRef]

14. Heng, L.K.; Hsiao, T.; Evett, S.; Howell, T.; Steduto, P. Validating the FAO AquaCrop model for irrigated and water deficient field maize. *Agron. J.* **2009**, *101*, 488–498. [CrossRef]
15. Verma, A.K.; Gupta, S.K.; Isaac, R.K. Use of saline water for irrigation in monsoon climate and deep water table regions: Simulation modelling with SWAP. *Agric. Water Manag.* **2012**, *115*, 186–193. [CrossRef]
16. Soothar, R.K.; Wenying, Z.; Yanqing, Z.; Moussa, T.; Uris, M.; Wang, Y. Evaluating the performance of SALTMED model under alternate irrigation using saline and fresh water strategies to winter wheat in the North China Plain. *Environ. Sci. Pollut. Res.* **2019**, *26*, 34499–34509. [CrossRef] [PubMed]
17. Steduto, P.; Hsiao, T.C.; Raes, D.; Fereres, E. AquaCrop—The FAO crop model to simulate yield response to water. I. Concepts and underlying principles. *Agron. J.* **2009**, *101*, 426–437. [CrossRef]
18. Van Gaelen, H. *AquaCrop Training Handbooks–Book II Running AquaCrop*; Food and Agriculture Organization of the United Nations: Rome, Italy, 2016.
19. Doorenbos, J.; Kassam, A.H. *Yield Response to Water*; FAO Irrigation and Drainage Paper No. 33; FAO: Rome, Italy, 1979.
20. Kumar, A.; Sarangi, D.K.; Singh, R.; Parihar, S.S. Evaluation of aquacrop model in predicting wheat yield and water productivity under irrigated saline regimes. *Irrig. Drain.* **2014**, *63*, 474–487. [CrossRef]
21. Mondal, M.S.; Fazal, M.A.; Saleh, M.D.; Akanda, A.R.; Biswas, S.K.; Moslehuddin, Z.; Sinora, Z.; Attila, N. Simulating yield response of rice to salinity stress with the AquaCrop model. *Environ. Sci. Process. Impacts* **2015**, *17*, 1118–1126. [CrossRef]
22. El Mokh, F.; Nagaz, K.; Masmoudi, M.M.; Mechlia, N.B.; Fereres, E. *Calibration of AquaCrop Salinity Stress Parameters for Barley under Different Irrigation Regimes in a Dry Environment*; Springer: Cham, Germany, 2017. [CrossRef]
23. Hellal, F.; Mansour, H.; Mohamed, A.H.; El-Sayed, S.; Abdelly, C. Assessment water productivity of barley varieties under water stress by AquaCrop model. *AIMS Agric. Food* **2019**, *4*, 501–517. [CrossRef]
24. Tan, S.; Wang, Q.; Zhang, J.; Chen, Y.; Shan, Y.; Xu, D. Performance of AquaCrop model for cotton growth simulation under film-mulched drip irrigation in southern Xinjiang, China. *Agric. Water Manag.* **2018**, *196*, 99–113. [CrossRef]
25. Hammami, Z.; Gauffreteau, A.; BelhajFraj, M.; Sahlia, A.; Jeuffroy, M.H.; Rezgui, S.; Bergaoui, K.; McDonnell, R.; Trifa, Y. Predicting yield reduction in improved barley (*Hordeum vulgare* L.) varieties and landraces under salinity using selected tolerance traits. *Field Crop. Res.* **2017**, *211*, 10–18. [CrossRef]
26. Sbei, H.; Sato, K.; Shehzad, T.; Harrabi, M.; Okuno, K. Detection of QTLs for salt tolerance in Asian barley (*Hordeum vulgare* L.) by association analysis with SNP markers. *Breed. Sci.* **2014**, *64*, 378–388. [CrossRef] [PubMed]
27. Jaradat, A.A.; Shahid, M.; Al-Maskri, A.Y. Genetic diversity in the Batini barley landrace from Oman: Spike and seed quantitative and qualitative traits. *Crop. Sci.* **2014**, *44*, 304–315. [CrossRef]
28. Raes, D.; Steduto, P.; Hsiao, T.C.; Fereres, E. *Crop Water Productivity. Calculation Procedures and Calibration Guidance. AquaCrop Version 3.0. FAO*; Land and Water Development Division: Rome, Italy, 2009.
29. Trombetta, A.; Iacobellis, V.; Tarantino, E.; Gentile, F. Calibration of the AquaCrop model for winter wheat using MODIS LAI images. *Agric. Water Manag.* **2015**, *164*. [CrossRef]
30. Hanks, R.J. Yield and water-use relationships. In *Limitations to Efficient Water Use in Crop Production*; Taylor, H.M., Jordan, W.R., Sinclair, T.R., Eds.; ASA, CSSA, and SSSA: Madison, WI, USA, 1983; pp. 393–411.
31. Tanner, C.B.; Sinclair, T.R. Efficient water use in crop production: Research or re-search? In *Limitations to Efficient Water Use in Crop Production*; Taylor, H.M., Jordan, W.R., Sinclair, T.R., Eds.; ASA, CSSA, and SSSA: Madison, WI, USA, 1983; pp. 1–27.
32. Steduto, P.; Hsiao, T.C.; Fereres, E. On the conservative behavior of biomass water productivity. *Irrig. Sci.* **2007**, *25*, 189–207. [CrossRef]
33. Raes, D.; Steduto, P.; Hsiao, T.C.; Fereres, E. *AquaCrop Version 5.0 Reference Manual*; Food and Agriculture Organization of the United Nations: Rome, Italy, 2016.
34. Loague, K.; Green, R.E. Statistical and graphical methods for evaluating solute transport models: Overview and application. *J. Contam. Hydrol.* **1991**, *7*, 51–73. [CrossRef]
35. Minhas, P.S.; Tiago, B.; AlonBen-Gal, R.; Pereira, L.S. Coping with salinity in irrigated agriculture: Crop evapotranspiration and water management issues. *Agric. Water Manag.* **2020**, *227*, 105832. [CrossRef]
36. Pereira, L.S.; Paredes, P.; Rodrigues, G.C.; Neves, M. Modeling barley water use and evapotranspiration

partitioning in two contrasting rainfall years. Assessing SIMDualKc and AquaCrop models. *Agric. Water Manag.* **2015**, *159*, 239–254. [CrossRef]

37. Iqbal, M.A.; Shen, Y.; Stricevic, R.; Pei, H.; Sun, H.; Amiri, E.; Rio, S. Evaluation of FAO Aquacrop model for winter wheat on the North China plain under deficit from field experiment to regional yield simulation. *Agric. Water Manag.* **2014**, *135*, 61–72. [CrossRef]
38. Mohammadi, M.; Ghahraman, B.; Davary, K.; Ansari, H.; Shahidi, A.; Bannayan, M. Nested validation of AquaCrop model for simulation of winter wheat grain yield soil moisture and salinity profiles under simultaneous salinity and water stress. *Irrig. Drain.* **2016**, *65*, 112–128. [CrossRef]
39. Hatfield, J.L. Increased temperatures have dramatic effects on growth and grain yield of three maize hybrids. *Agric. Environ. Lett.* **2016**, *1*, 1–5. [CrossRef]
40. Hsiao, T.C.; Heng, L.; Steduto, P.; Roja-Lara, B.; Raes, D.; Fereres, E. AquaCrop—The FAO model to simulate yield response to water: Parametrization and testing for maize. *Agron. J.* **2009**, *101*, 448–459. [CrossRef]
41. Wiegmanna, M.; William, T.B.; Thomasb, H.J.; Bullb, I.; Andrew, J.; Flavellc, J.; Annette, Z.; Edgar, P.; Klaus, P.; Andreas, M. Wild barley serves as a source for biofortification of barley grains. *Plant Sci.* **2019**, *283*, 83–94. [CrossRef] [PubMed]
42. Roberts, D.P.; Mattoo, A.K. Sustainable agriculture—Enhancing environmental benefits, food nutritional quality and building crop resilience to abiotic and biotic stresses. *Agriculture* **2018**, *8*, 8. [CrossRef]
43. Tavakoli, A.R.; Moghadam, M.M.; Sepaskhah, A.R. Evaluation of the AquaCrop model for barley production under deficit irrigation and rainfed condition in Iran. *Agric. Water Manag.* **2015**, *161*, 136–146. [CrossRef]
44. Eisenhauer, J.G. Regression through the origin. *Teach. Stat.* **2003**, *25*, 76–80. [CrossRef]
45. Teixeira, A.D.C.; Bassoi, L.H. HBassoi Crop Water Productivity in Semi-arid Regions: From Field to Large Scales. *Ann. Arid Zone* **2009**, *48*, 1–13.
46. FAO. The Irrigation Challenge. In *Increasing Irrigation Contribution To Food Security through Higher Water Productivity from Canal Irrigation Systems*; Issue paper; FAO: Rome, Italy, 2003.
47. Barnston, A. Correspondence among the Correlation [root mean square error] and Heidke Verification Measures; Refinement of the Heidke Score. *Notes Corresp. Clim. Anal. Cent.* **1992**, *7*, 699–709.

8

Soil Water Infiltration Model for Sprinkler Irrigation Control Strategy: A Case for Tea Plantation in Yangtze River Region

Yong-zong Lu [1,2,3], Peng-fei Liu [1], Aliasghar Montazar [2], Kyaw-Tha Paw U [3] and Yong-guang Hu [1,*]

1 Key Laboratory of Modern Agricultural Equipment and Technology, Ministry of Education Jiangsu Province, Jiangsu University, Zhenjiang 212013, China; luyongzong@126.com (Y.-z.L.); 18252585090@163.com (P.-f.L.)
2 Division of Agriculture and Natural Resources, UC Cooperative Extension, University of California, Imperial County, Holtville, CA 92250, USA; amontazar@ucanr.edu
3 Department of Land, Air and Water Resources, University of California, Davis, CA 95616, USA; ktpawu@ucdavis.edu
* Correspondence: deerhu@ujs.edu.cn.

Abstract: The sprinkler irrigation method is widely applied in tea farms in the Yangtze River region, China, which is the most famous tea production area. Knowledge of the optimal irrigation time for the sprinkler irrigation system is vital for making the soil moisture range consistent with the root boundary to attain higher yield and water use efficiency. In this study, we investigated the characteristics of soil water infiltration and redistribution under the irrigation water applications rates of 4 mm/h, 6 mm/h, and 8 mm/h, and the slope gradients of 0°, 5°, and 15°. A new soil water infiltration model was established based on water application rate and slope gradient. Infiltration experimental results showed that soil water infiltration rate increased with the application rate when the slope gradient remained constant. Meanwhile, it decreased with the increase in slope gradient at a constant water application rate. In the process of water redistribution, the increment of volumetric water content (VWC) increased at a depth of 10 cm as the water application rate increased, which affected the ultimate infiltration depth. When the slope gradient was constant, a lower water application rate extended the irrigation time, but increased the ultimate infiltration depth. At a constant water application rate, the infiltration depth increased with the increase in slope gradient. As the results showed in the infiltration model validation experiments, the infiltration depths measured were 38.8 cm and 41.1 cm. The relative errors between measured infiltration depth and expected value were 3.1% and 2.7%, respectively, which met the requirement of the soil moisture range consistent with the root boundary. Therefore, this model could be used to determine the optimal irrigation time for developing a sprinkler irrigation control strategy for tea fields in the Yangtze River region.

Keywords: water application rate; slope gradient; infiltration depth; optimal irrigation time

---

1. Introduction

Tea (*Camellia sinensis*) is a subtropical plant, which grows well in a warm and wet climate under an optimal growth temperature around 20 °C with annual rainfall of 1500 mm. The Yangtze River region is the most famous tea production area. Climate abnormality in recent decades caused an uneven spatial distribution of rainfall in this area. Inadequate rainfall cannot provide enough water for tea growth, which seriously affects both the yield and quality of the tea plants [1,2]. Under water stress, the photosynthetic and respiration rates of tea leaves decrease [3], as well as their chlorophyll content, the water content of shoots, the root activity, and the root weight per unit volume [4]. Meanwhile, quality components such as amino acids, caffeine, and water extracts are also reduced, resulting in the deterioration of the tea quality [5–8].

The sprinkler irrigation method is widely applied to save water and counter drought with great improvement with regard to water use efficiency, irrigation uniformity, labor saving, and crop yields [9]. The traditional control strategy for sprinkler irrigation is usually based on the upper limit and lower limit of soil moisture required for the growth of certain plants [10–12]. This kind of control strategy partly provides the required water for tea plant growth and saves some applied water compared to manual irrigation. However, it leads to the ultimate infiltration depth exceeding the boundary of the tea plant root system, resulting in a waste of water resources. Thus, it is necessary to improve the water use efficiency (WUE) of sprinkler irrigation systems, especially for tea fields in the Yangtze River region.

Infiltration is the process of water entering the soil. The main goal of operating an irrigation system is to apply the required infiltration depth for specific plants with high WUE [3,7,9,13,14]. Normally, infiltration rate is determined by measuring, modeling, and predicting the surface runoff [15–17]. To avoid low-WUE problems, it is necessary to have a good understanding of soil infiltration characteristics. Infiltration rate variation results from many causes such as soil property, water application rate, and terrain [18–20]. Infiltration theories and models were developed by several researchers, including the Green–Ampt model, Kostiakov model, modified Kostiakov model, and Smith and Parlange model [21–23]. The suitability of an infiltration model for a particular region is subject to soil type and field conditions. Different infiltration models were applied to certain soil types and certain site conditions [24]. Over the years, several comparative analyses of various infiltration models were conducted to assess the suitability of various models for different soil types under varying field conditions to estimate the infiltration rates and infiltration potentials of soil [25]. Feng, Deng, Zhang, and Guo [26] pointed out that the cumulative infiltration depth firstly decreased then increased with the increase in slope gradient, and the turning point was the threshold of the slope gradient. Recent researches focused on the effects of slope gradient and water application rate on soil water infiltration and redistribution without any plants; however, they failed to reveal the situation with plants [27,28]. To some extent, the morphological, quantitative, depth, type, and distribution characteristics of plant roots affect soil water moisture and its redistribution [29–32]. The infiltration characteristics of soil change with the root length density and root surface area density [33,34]. With the increase in root volume and dry weight of root, the infiltration rate shows an increasing trend [35]. Soil moisture content is lower in soil layers with denser root density, while water content in soil layers without roots is significantly increased [36]. If the moisture content range is not consistent with the root range of the tea plant, it will result in inefficient irrigation or overirrigation. Therefore, an optimal irrigation time and infiltration depth are vital for tea plants in the context of a water-saving irrigation control strategy.

Tea farms in the middle and lower Yangtze River regions are located in a hilly area, with a slope gradient generally less than 15° [27]. Based on the topographic features of tea farms, the specific goals of this study were to (1) investigate the effects of slope gradient and water application rate on soil water infiltration and redistribution, and (2) provide a new infiltration model for determining the optimal stopping time for a tea plantation sprinkler irrigation control system. The model is based on tea root length, water application rate, and slope gradient, and results in the infiltration depth being consistent with tea plant's root system, which validates the precision control for tea plant irrigation.

2. Materials and Methods

2.1. Materials

This study was conducted at a tea farm located in the middle–lower Yangtze River region, east China, which has a moderate sub-tropical climate with a mean annual precipitation of 1029.1 mm and an average annual temperature of 15.5 °C. The annual reference evapotranspiration (ET_0) was 892.24 mm, which was observed over the last 55 years (1961–2015) [37]. The topography of the experimental site is a hilly ground with an average altitude of 18.5 m (latitude 32°01′35″ north (N), longitude 119°40′21″ east (E)).

As shown in Figure 1, the box frame sprinkler irrigation system was composed of a box frame, pump station, main pipe, distribution pipe, lateral pipe, standpipe, and sprinkler. The distance between two sprinklers was 9.0 m. In order to reduce the pressure difference between sprinklers and improve the uniformity of spraying water, lateral pipes with a length of 4.5 m were laid along the slope. Four sprinklers were set up around the box frame with an effective radius of 7.0 m. The box frame was 4.5 m in length, 0.8 m in width, and 0.8 m in depth, and the adjustable range of the gradient is 0–15°. To ensure the box frame was well drained, drain holes (diameter 1.0 cm) were set up at the bottom of the box. The exterior of the box was made of an acrylic plate (80.0 cm × 0.5 cm, transparency 99.0%) to facilitate the measurement of infiltration depth. The sampled tea cultivar was Anji white tea, which was five years old. Six tea plants were planted in the frame box with distances of 0.35, 1.05, 1.75, 2.75, 3.45, and 4.15 m (Figure 2). Three kinds of plastic impact-driven sprinklers were selected. Technical parameters of the sprinklers are listed in Table 1. Three kinds of water application rates were set up using the different types of sprinklers with flow rates of 0.4, 0.6, and 0.8 $m^3 \cdot h^{-1}$. A plastic shade was set up above the experiment area on rainy days to avoid the influence of rainfall.

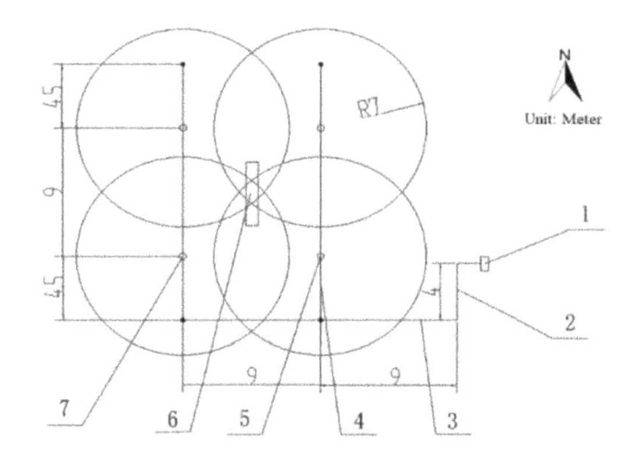

Figure 1. Box frame sprinkler irrigation system: 1—pump station; 2—main pipe; 3—distribution pipe; 4—lateral pipe; 5—standpipe; 6—box frame; 7—sprinkler.

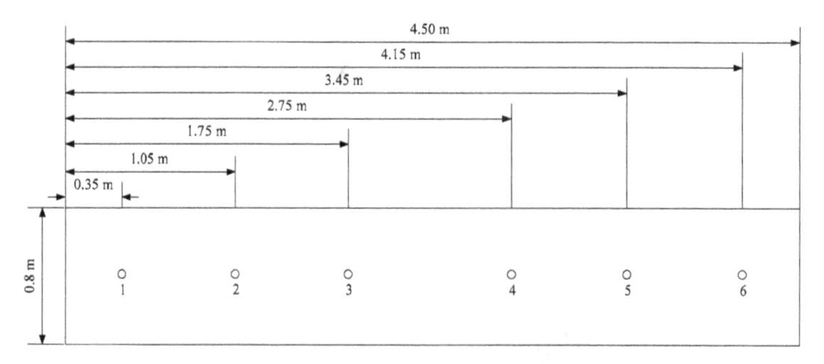

Figure 2. Tea plant distribution in the box frame.

Table 1. Technical parameters of the sprinkler.

No.	Operating Pressure (MPa)	Nozzle Diameter (mm)	Flow Rate ($m^3 \cdot h^{-1}$)	Pattern Radius (m)	Rotation Cycle (s)
1	0.3	3.0	0.4	7.0	18.0
2	0.3	3.5	0.6	7.0	18.0
3	0.3	4.0	0.8	7.0	18.0

The classification of soil texture was based on the World Reference Base (WRB) soil classification system [38]. The soil was collected from the experimental site at depths of 0–10.0 cm, 10.0–20.0 cm, 20.0–30.0 cm, 30.0–40.0 cm, 40.0–50.0 cm, and 50.0–80.0 cm. Bulk density and saturated VWC were measured using the oven dry method. A laser diffraction particle size analyzer (Mastersizer 3000, Malvern Panalytical, UK) was used to measure the particle composition. The soil texture at a depth of 0–40 cm was sandy loam, and that under 40 cm was loam (Table 2). A soil moisture sensor (TRIME PICO 64, IMKO, Ettlingen, Germany) was used to measure VWC, and the measuring accuracy was ± 1%.

Table 2. Physical properties and the particle composition of the soil for experiment. VWC—volumetric water content.

Sampling Depth (cm)	Size Composition (%)			Bulk Density ($g \cdot cm^{-3}$)	Saturated VWC (%)	Soil Texture
	<0.002 mm	0.002–0.02 mm	0.02–2 mm			
0–10.0	0	37.6	62.4	1.2	47.0	Sandy loam
10.0–20.0	0	29.3	70.7	1.4	47.0	
20.0–30.0	0	39.7	60.3	1.4	49.0	
30.0–40.0	0	26.8	73.2	1.5	48.0	
40.0–50.0	4.4	42.8	52.8	1.7	44.0	Loam
50.0–80.0	8.6	40.6	50.8	1.6	43.0	

2.2. Methods

2.2.1. Sensor Layout

The soil in the experimental box frame was taken from the sprinkler irrigation area in the experimental tea farm. The sampling depths were 0–10.0 cm, 10.0–20.0 cm, 20.0–30.0 cm, 30.0–40.0 cm, 40.0–50.0 cm, and 50.0–80.0 cm (Figure 3). The sampled soil was air-dried, ground, and screened before layering it into the experimental box frame. The surface of the filled soil was hacked to reduce the influence of artificial compaction on soil water infiltration.

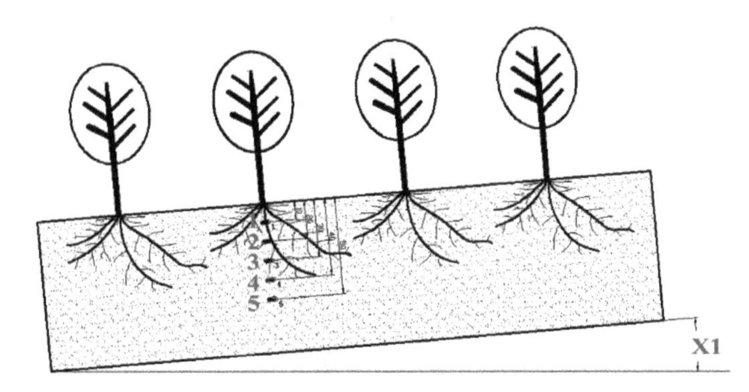

Figure 3. Sensor layout. TRIME PICO 64 soil moisture sensors were arranged along the roots of tea plants. Numbers 1–5 represent the sensors located at depths of 10.0 cm, 20.0 cm, 30.0 cm, 40.0 cm, and 50.0 cm, respectively. X1 represents the slope gradient, which could be adjusted from 0–15°.

Infiltration depth and VWC were measured in this experiment. All soil moisture sensors were calibrated using the oven dry method before set-up. The time interval of data acquisition was 1.0 min.

Soil type in the experimental site was homogeneous sandy loam. The location of the color gradient of the soil was marked, and the depth of position was measured as the infiltration depth. The water infiltration process started with the irrigation and stopped when the irrigation stopped, and then the water continued to infiltrate. When the VWC of each layer no longer increased and showed a decreasing trend, the process of water redistribution stopped.

2.2.2. Characteristics of Soil Water Infiltration

Characteristics of the soil water infiltration experiments were determined in the frame box from 1–26 August 2016 (Table 3). Nine treatments were selected with three typical kinds of slope gradients and water application rates. The irrigation was stopped when the infiltration depth was 20.0 cm due to the average length of the tea plant roots. SPSS 17.0 was used to conduct the multivariate regression analysis of the relationship between slope gradient, water application rate, and the ratio of stopping irrigation depth to infiltration depth. The statistical tools root-mean-squared error (*RMSE*) and coefficient of determination (R^2) were employed to validate the accuracy of the models built in this study. The calculations are presented below.

Table 3. Sprinkler irrigation schedule.

Treatment	Slope Gradient (°)	Water Application Rate (mm·h^{-1})
T1	0	4.0
T2	0	6.0
T3	0	8.0
T4	5.0	4.0
T5	5.0	6.0
T6	5.0	8.0
T7	15.0	4.0
T8	15.0	6.0
T9	15.0	8.0

2.2.3. Infiltration Model Validation Experiments

The soil water infiltration model was established based on tea root depth, water application rate, and slope gradient. The reliability of the model was evaluated in the box frame experiments based on two cases which represented the most common terrains of tea farms in the Yangtze River region: (1) gradient 0° and 8.0 mm·h^{-1} for water application rate; (2) gradient 8° and 4.0 mm·h^{-1} for water application rate. The required irrigation time and expected infiltration depth were calculated. The VWC at depths of 10.0 cm, 20.0 cm, 30.0 cm, and 40.0 cm was measured. The ultimate infiltration depth was compared with the expected value to obtain the relative error between the measured and expected value.

3. Results and Discussion

3.1. VWC at Different Slope Gradients and Water Application Rates

As the sprinkler irrigation stopped, under the action of gravity potential and matric potential, water redistribution began. After 24 h, the VWC no longer increased and showed a downward trend, signifying that the process of water redistribution stopped (Figure 4).

Compared with the VWC before irrigation, VWC at 20.0 cm increased at the time of stopping irrigation, and the increment of VWC at 10 cm under different treatments was different. When the slope gradient was 0°, the increments of VWC at 10.0 cm were 11.1%, 10.0%, and 4.5%, respectively. When the slope was 5°, the increments were 6.1%, 9.1%, and 8.0%, respectively. When the slope was 15°, the increments were 9.0%, 5.9%, and 4.0%, respectively (Figure 4b).

Twelve hours after irrigation, soil water at the soil layer depth of 0–20.0 cm started to redistribute (Figure 4c). Compared with VWC at the time of stopping irrigation, VWC at 10.0 cm decreased by 1.1% for the T2 treatment. For the other eight treatments, the VWC at 10.0 cm increased by 2.3%, 5.0%, 9.7%, 4.1%, 4.3%, 1.3%, 4.3%, and 6.5%, respectively. VWC at 20.0 cm increased for all nine treatments, whereas VWC at 30 cm, 40 cm, and 50 cm showed no change.

Infiltration depth showed an increasing trend under the processing of water redistribution (Figure 4d). Compared with VWC at 12 h after the irrigation, VWC at 10.0 cm showed decreasing trends for all nine treatments, while VWC at 20.0 cm continued increasing. This was probably because the amount of water drawn from the upper soil at 10.0 cm was less than the water absorbed by the lower soil layer, which resulted in a continuous decrease in VWC at 10.0 cm and a sustained increase in VWC below 20.0 cm.

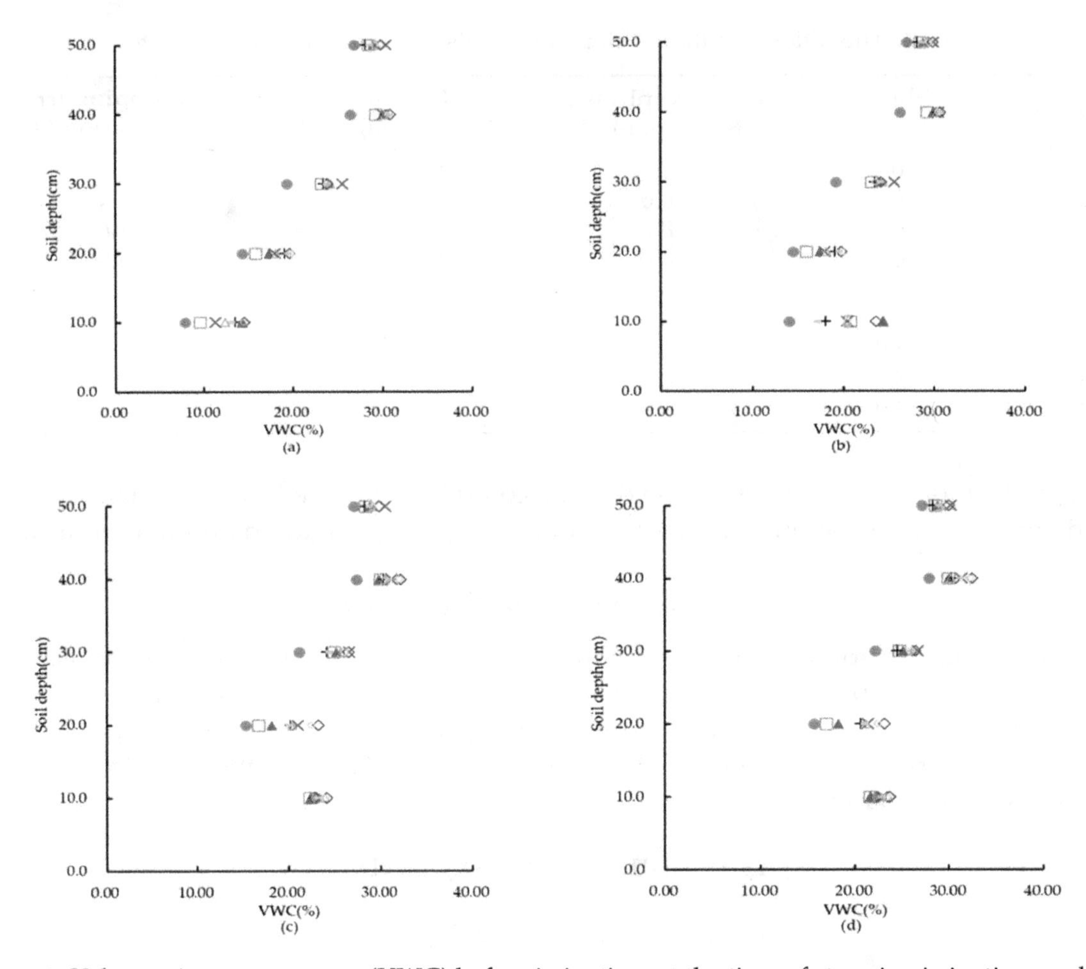

Figure 4. Volumetric water content (VWC) before irrigation, at the time of stopping irrigation, and 12 h and 24 h after sprinkler irrigation with different treatments, T1(□), T2(▲), T3(+), T4(●), T5(×), T6(△), T7(◇), T8(○), and T9(–): (**a**) before irrigation; (**b**) at the time of stopping irrigation; (**c**) 12 h after irrigation; (**d**) 24 h after irrigation.

3.2. *Effect of Water Application Rate and Slope Gradient on the Infiltration Depth and Rate*

When the water application rate was constant, the infiltration depth increased as the slope gradient increased (Figure 5). This is because the pressure of the water perpendicular to the direction of the slope decreased, which increased the infiltration depth (Table 4).

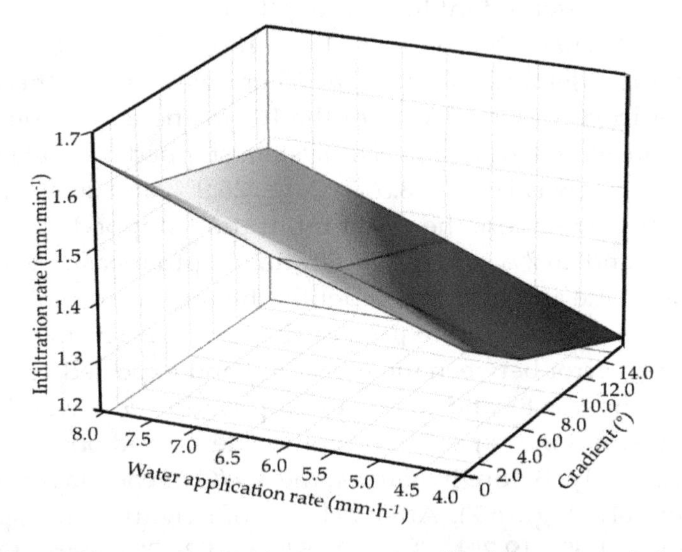

Figure 5. Infiltration rate under various water application rates and slope gradients.

Table 4. The ratios of different slope gradients and water application rates.

Treatment	Slope Gradient (°)	Water Application Rate (mm·h^{-1})	Infiltration Depth (cm)	Ratio of Stopping Irrigation Depth to Infiltration Depth (%)
T1	0	4.0	40.9	48.9
T2	0	6.0	34.9	57.4
T3	0	8.0	32.5	61.6
T4	5.0	4.0	42.6	46.9
T5	5.0	6.0	41.3	48.4
T6	5.0	8.0	35.8	55.9
T7	15.0	4.0	43.2	46.3
T8	15.0	6.0	42.5	47.0
T9	15.0	8.0	37.3	53.6

Through the multivariate regression analysis, it could be concluded that the relationship between slope gradient, water application rate, and the ratio of stopping irrigation depth to infiltration depth was as follows:

$$Y_1 = -0.004X_1 + 0.024X_2 + 0.401, \quad (1)$$

where Y_1 is the ratio of stopping irrigation depth to infiltration depth (mm), X_1 is the slope gradient (°), and X_2 is the water application rate (mm·h^{-1}). The R^2 was 0.83 and the *RMSE* was 0.02 mm.

Using multivariate regression analysis, the linear regression equations for infiltration rate, slope gradient, and water application rate were obtained. The R^2 was 0.92 and the *RMSE* was 0.02 mm·h^{-1}. The linear regression equations were as follows:

$$Y_2 = -0.012X_1 + 0.058X_2 + 1.181, \quad (2)$$

$$L_1 = L \times (-0.04X_1 + 0.24X_2 + 4.01), \quad (3)$$

$$T = \frac{L_1}{Y_2} = \frac{(-0.04X_1 + 0.24X_2 + 4.01)L}{-0.012X_1 + 0.058X_2 + 1.181}, \quad (4)$$

where Y_2 is the infiltration rate (mm·min^{-1}), L_1 is the required infiltration depth (cm), L is the root depth of the tested tea plant (cm), and T is the required sprinkler irrigation time (min).

3.3. The Reliability of Infiltration Model Testing

The average root length of the tested tea plants was 40.0 cm. When the slope gradient was 0° and the water application rate was 8.0 mm·h^{-1}, the required irrigation time was 144 min. The expected stopping irrigation depth and observed infiltration depth were 23.7 cm and 40.0 cm, respectively.

Before irrigation, the VWC of each layer was 11.4%, 18.1%, 22.1%, 28.9%, and 26.7%, respectively (Figure 6). After 63 min of irrigation, the VWC at 10 cm was 11.6%; then, it increased gradually. Additionally, 125 min after irrigation, the VWC for the 10-cm and 20-cm soil layers was 15.8%, 18.1%, respectively, after which it increased gradually. The system stopped after 144 min of irrigation. At this time, the VWC of each soil layer was 19.2%, 18.3%, 22.0%, 28.9%, and 26.6%, and the infiltration depth was 23.2 cm. Then, 24 h after irrigation, the water infiltration stopped. The VWC of each layer was 20.8%, 19.0%, 23.2%, 29.1%, and 26.7%, and the infiltration depth measured was 38.75 cm. Compared with the required irrigation time and infiltration depth, the measured infiltration depth was 23.2 cm, giving an error between the measured and required value of 2.0%. The measured ultimate infiltration depth was 38.8 cm, giving an error between the measured and expected value of 3.1%.

When the slope gradient was 8° and the water application rate was 4.0 mm·h^{-1}, the required irrigation time was 141 mins, and the expected infiltration depth and observed infiltration depth were 18.6 cm and 40.0 cm respectively. Before irrigation, the VWC of each layer was 15.0%, 19.1%, 23.1%, 28.5%, and 26.1%, respectively (Figure 7). After 141 min of irrigation, the sprinkler system stopped. The VWC of each layer was 21.5%, 19.2%, 23.1%, 28.5%, and 26.2%, respectively, and the infiltration depth was 18.7 cm. Then, 24 h after irrigation, the water infiltration stopped. The VWC of each layer was 20.0%, 21.1%, 24.7%, 28.7%, and 26.2%, and the ultimate infiltration depth measured was 41.1 cm. The measured infiltration depth was 18.9 cm, and the relative error between the measured and expected value was 1.4%. The measured ultimate infiltration depth was 41.1 cm, and the relative error between the measured and expected value was 2.7%.

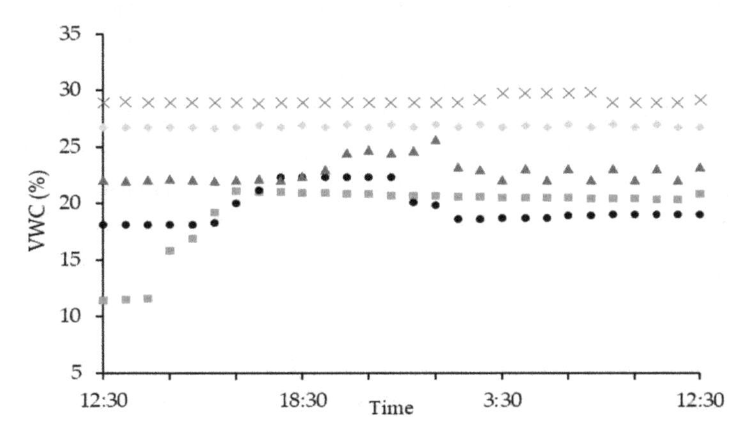

Figure 6. VWC at the slope gradient of 0° and the water application rate of 8.0 $mm \cdot h^{-1}$, with soil depths of 10 cm (■), 20 cm (●) 30 cm (▲), 40 cm (×), and 50 cm (♦).

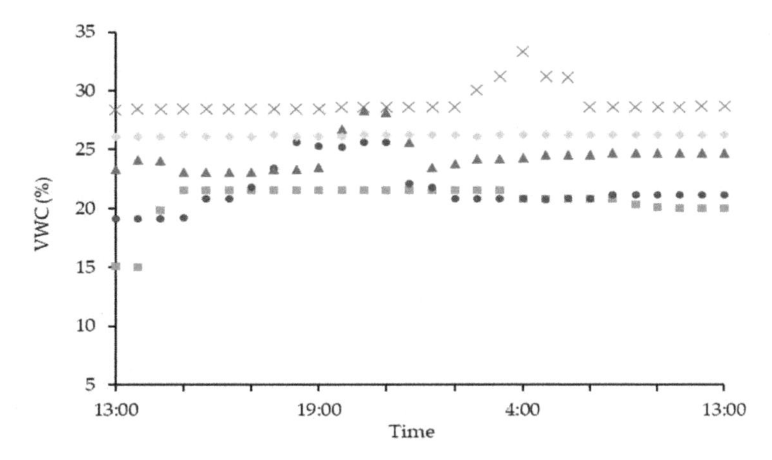

Figure 7. VWC at the slope gradient of 8° and the water application rate of 4.0 $mm \cdot h^{-1}$, with soil depths of 10 cm (■), 20 cm (●) 30 cm (▲), 40 cm (×), and 50 cm (♦).

4. Discussion

With the increase in slope gradient, the required water application rate decreased (Figure 4). When the water application rate was close to the required water application rate, the soil pores were gradually filled with water. This reduced the infiltration capacity and led to a decrease in the increment of VWC.

For a constant water application rate, the infiltration rate decreased as the slope gradient increased. This is because a larger water application rate caused a larger kinetic energy of the sprayed water droplets [39]. Therefore, as the infiltration rate and water application rate increased, the water pressure of the soil surface and soil layers all increased over time. As the kinetic energy of water droplets increased, more pressure was applied on the infiltration water, which resulted in an acceleration of the infiltration rate. The infiltration rate also showed a decreasing trend with the increase in water application rate. The reason was that the component force of the same thickness of the aquifer along the slope direction increased with an increase in slope, and the pressure perpendicular to the slope direction was reduced. Therefore, the pressure of infiltration water reduced, as did the infiltration rate [20,40–42].

At a constant slope gradient, the larger water application rate led to a lower infiltration depth. This was caused by the water application rate being lower than its own infiltration capacity, resulting in the water continuing to infiltrate over time [41]. With the increase in water application rate, the kinetic energy of the water droplets sprayed from the nozzle increased correspondingly, resulting in a reduction in infiltration capacity. A lower water application rate extended the irrigation time, but it was conducive to the vertical movement of water, which deepened the infiltration depth.

As we all know, the infiltration characteristic has a strong relationship with the kinetic energy of water drops and the change in physical properties of the soil surface, such as soil type, vegetation type, and terrain [43–45]. We know our research is limited; however, the results were obvious in the two selected experiments, where the ultimate infiltration was consistent with the boundary of the tea plant root system. Based on the characteristics of irrigation water infiltration and redistribution, the new infiltration model can be used to determine the required irrigation time for developing a sprinkler irrigation control strategy.

5. Conclusions

The sprinkler irrigation method is widely used for tea plants in the Yangtze River region. The traditional control strategy for sprinkler irrigation is based on the upper limit and lower limit of the required soil moisture. However, this strategy always causes the ultimate infiltration depth to exceed the boundary of the tea plant root system, leading to a waste of water. In this study, a new soil water infiltration model was provided by investigating the characteristics of soil water infiltration and redistribution under different water application rates and gradient slopes.

The infiltration characteristics showed that the infiltration rate changed with the water application rate and slope gradient. Water redistribution processes showed that the increment of VWC at 10.0 cm was different for various combinations of water application rate and slope gradient. Those differences affected the ultimate infiltration depth of the soil. When the slope gradient was kept constant, a lower water application rate led to a longer irrigation time, but it increased the ultimate infiltration depth. When the water application rate was kept constant, the infiltration depth increased with the increase in slope gradient. Based on the new soil water infiltration model, the ultimate infiltration depth was basically consistent with the boundary of the tea plant roots. Therefore, the model established in this paper can be generally applied to automatic sprinkler irrigation systems for tea fields in the Yangtze River region.

Author Contributions: Y.-g.H. conceptualized and designed the experiments. Y.-z.L. and P.-f.L. performed the experiments. Y.-z.L. analyzed the data and wrote the manuscript. K.-T.P.U. and A.M. contributed significant comments to improve the quality and language of the manuscript.

Acknowledgments: The authors are grateful for the financial support from the Jiangsu Agriculture Science and Technology Innovation Fund (CX(16)1045), the Key R&D Programs of Jiangsu Province and Zhenjiang (BE2016354, NY20160120037), the Project of Postgraduate Innovation of Jiangsu Province (KYCX17-1788), the China and Jiangsu Postdoctoral Science Foundation (2016M600376, 1601032C), the Six Talent Peaks Program in Jiangsu Province (2015-ZBZZ-021), the Priority Academic Program Development of Jiangsu Higher Education Institutions (2014-37), and the China Scholarship Council (201708320220).

References

1. Ding, Y.; Wang, W.; Song, R.; Shao, Q.; Jiao, X.; Xing, W. Modeling spatial and temporal variability of the impact of climate change on rice irrigation water requirements in the middle and lower reaches of the Yangtze River, China. *Agric. Water Manag.* **2017**, *193*, 89–101. [CrossRef]
2. Song, S.; Xu, Y.P.; Zhang, J.X.; Li, G.; Wang, Y.F. The long-term water level dynamics during urbanization in plain catchment in Yangtze River Delta. *Agric. Water Manag.* **2016**, *174*, 93–102. [CrossRef]
3. Maritim, T.K.; Kamunya, S.M.; Mireji, P.; Mwendia, C.; Muoki, R.C.; Cheruiyot, E.K.; Wachira, F.N.

Physiological and biochemical response of tea [*Camellia sinensis* (L.) O. Kuntze] to water-deficit stress. *J. Hortic. Sci. Biotechnol.* **2015**, *90*, 395–400. [CrossRef]
4. Netto, L.A.; Jayaram, K.M.; Puthur, J.T. Clonal variation of tea [*Camellia sinensis* (L.) O. Kuntze] in countering water deficiency. *Physiol. Mol. Biol. Plants Int. J. Funct. Plant Biol.* **2010**, *16*, 359–367. [CrossRef] [PubMed]
5. Chakraborty, U.; Dutta, S.; Chakraborty, B. Drought induced biochemical changes in young tea leaves. *Indian J. Plant Physiol. India* **2001**, *6*, 103–106.
6. Kigalu, J.M.; Kimambo, E.I.; Msite, I.; Gembe, M. Drip irrigation of tea (*Camellia sinensis* L.): 1. Yield and crop water productivity responses to irrigation. *Agric. Water Manag.* **2008**, *95*, 1253–1260. [CrossRef]
7. Kumar, R.; Bisen, J.S.; Choubey, M.; Singh, M.; Bera, B. Influence of Changes Weather Conditions on Physiological and Biochemical Characteristics of Darjeeling Tea (*Camellia sinensis* L.). *Glob. J. Biol. Agric. Health. Sci.* **2016**, *5*, 55–60.
8. Lin, S.K.; Lin, J.; Liu, Q.L.; Ai, Y.F.; Ke, Y.Q.; Chen, C.; Zhang, Z.Y.; He, H. Time-course of photosynthesis and non-structural carbon compounds in the leaves of tea plants (*Camellia sinensis* L.) in response to deficit irrigation. *Agric. Water Manag.* **2014**, *144*, 98–106. [CrossRef]
9. Saretta, E.; de Camargo, A.P.; Botrel, T.A.; Frizzone, J.A.; Koech, R.; Molle, B. Test methods for characterising the water distribution from irrigation sprinklers: Design, evaluation and uncertainty analysis of an automated system. *Biosyst. Eng.* **2018**, *169*, 42–56. [CrossRef]
10. Ratliff, L.F.; Ritchie, J.T.; Cassel, D.K. Field-Measured Limits of Soil Water Availability as Related to Laboratory-Measured Properties 1. *Soil Sci. Soc. Am. J.* **1983**, *47*, 770–775. [CrossRef]
11. Thompson, R.B.; Gallardo, M.; Valdez, L.C.; Fernández, M.D. Using plant water status to define threshold values for irrigation management of vegetable crops using soil moisture sensors. *Agric. Water Manag.* **2007**, *88*, 147–158. [CrossRef]
12. Wei, Y.; Wang, Z.; Wang, T.; Liu, K. Design of real time soil moisture monitoring and precision irrigation systems. *Trans. Chin. Soc. Agric. Eng.* **2013**, *29*, 80–86.
13. Liu, Y.-Y.; Wang, A.-Y.; An, Y.-N.; Lian, P.-Y.; Wu, D.-D.; Zhu, J.-J.; Meinzer, F.C.; Hao, G.-Y. Hydraulics play an important role in causing low growth rate and dieback of aging *Pinus sylvestris* var. mongolica trees in plantations of Northeast China. *Plant Cell Environ.* **2018**, *41*, 1500–1511. [CrossRef] [PubMed]
14. Vaz, C.M.P.; Jones, S.; Meding, M.; Tuller, M. Evaluation of Standard Calibration Functions for Eight Electromagnetic Soil Moisture Sensors. *Vadose Zone J.* **2013**, *12*. [CrossRef]
15. Al-Ghobari, H.M.; El-Marazky, M.S.; Dewidar, A.Z.; Mattar, M.A. Prediction of wind drift and evaporation losses from sprinkler irrigation using neural network and multiple regression techniques. *Agric. Water Manag.* **2018**, *195*, 211–221. [CrossRef]
16. AL-Kayssi, A.W.; Mustafa, S.H. Modeling gypsifereous soil infiltration rate under different sprinkler application rates and successive irrigation events. *Agric. Water Manag.* **2016**, *163*, 66–74. [CrossRef]
17. Diamond, J.; Shanley, T. Infiltration rate assessment of some major soils. *Ir. Geogr.* **2003**, *36*, 32–46. [CrossRef]
18. Fu, B.; Wang, Y.; Zhu, B.; Wang, D.; Wang, X.; Wang, Y.; Ren, Y. Experimental study on rainfall infiltration in sloping farmland of purple soil. *Trans. Chin. Soc. Agric. Eng.* **2008**, *2008*. [CrossRef]
19. Liu, Z.; Li, P.; Hu, Y.; Wang, J. Wetting patterns and water distributions in cultivation media under drip irrigation. *Comput. Electron. Agric.* **2015**, *112*, 200–208. [CrossRef]
20. Mu, W.; Yu, F.; Li, C.; Xie, Y.; Tian, J.; Liu, J.; Zhao, N.; Mu, W.; Yu, F.; Li, C.; et al. Effects of Rainfall Intensity and Slope Gradient on Runoff and Soil Moisture Content on Different Growing Stages of Spring Maize. *Water* **2015**, *7*, 2990–3008. [CrossRef]
21. Zakwan, M.; Muzzammil, M.; Alam, J. Application of spreadsheet to estimate infiltration parameters. *Perspect. Sci.* **2016**, *8*, 702–704. [CrossRef]
22. Smith, R.E.; Parlange, J.-Y. A parameter-efficient hydrologic infiltration model. *Water Resour. Res.* **1978**, *14*, 533–538. [CrossRef]
23. Parhi, P.K.; Mishra, S.K.; Singh, R. A Modification to Kostiakov and Modified Kostiakov Infiltration Models. *Water Resour. Manag.* **2007**, *21*, 1973–1989. [CrossRef]
24. Mazloom, H.; Foladmand, H. Evaluation and determination of the coefficients of infiltration models in Marvdasht region, Fars province. *Int. J. Adv. Biol. Biomed. Res.* **2013**, *1*, 822–829.
25. Wilson, R.L. Comparing Infiltration Models to Estimate Infiltration Potential at Henry V Events. Bachelor's Thesis, Portland State University, Portland, OR, USA, 2017.

26. Feng, H.; Deng, L.S.; Zhang, C.L.; Guo, Y.B. Effect of Ground Slope on Water Infiltration of Drip Irrigation. *J. Irrig. Drain.* **2010**, *29*, 14–15.
27. Pengfei, L.; Yongguang, H.; Feng, J.; Sheng, W. Influence of sloping tea fields on soil moisture migration. *IFAC Pap.* **2018**, *51*, 565–569. [CrossRef]
28. Zhao, P.; Shao, M.; Melegy, A.A. Soil Water Distribution and Movement in Layered Soils of a Dam Farmland. *Water Resour. Manag.* **2010**, *24*, 3871–3883. [CrossRef]
29. Jha, S.K.; Gao, Y.; Liu, H.; Huang, Z.; Wang, G.; Liang, Y.; Duan, A. Root development and water uptake in winter wheat under different irrigation methods and scheduling for North China. *Agric. Water Manag.* **2017**, *182*, 139–150. [CrossRef]
30. Yan, Y.; Dai, Q.; Yuan, Y.; Peng, X.; Zhao, L.; Yang, J. Effects of rainfall intensity on runoff and sediment yields on bare slopes in a karst area, SW China. *Geoderma* **2018**, *330*, 30–40. [CrossRef]
31. Wang, Y.; Zhang, X.; Chen, J.; Chen, A.; Wang, L.; Guo, X.; Niu, Y.; Liu, S.; Mi, G.; Gao, Q. Reducing basal nitrogen rate to improve maize seedling growth, water and nitrogen use efficiencies under drought stress by optimizing root morphology and distribution. *Agric. Water Manag.* **2019**, *212*, 328–337. [CrossRef]
32. Wang, X.; Zhou, Y.; Wang, Y.; Wei, X.; Guo, X.; Zhu, D. Soil water characteristic of a dense jujube plantation in the semi-arid hilly Regions of the Loess Plateau in China. *J. Hydraul. Eng.* **2015**, *46*, 263–270.
33. Li, N.; Kang, Y.; Li, X.; Wan, S.; Xu, J. Effect of the micro-sprinkler irrigation method with treated effluent on soil physical and chemical properties in sea reclamation land. *Agric. Water Manag.* **2019**, *213*, 222–230. [CrossRef]
34. Li, Z.; Xu, X.; Pan, G.; Smith, P.; Cheng, K. Irrigation regime affected SOC content rather than plow layer thickness of rice paddies: A county level survey from a river basin in lower Yangtze valley, China. *Agric. Water Manag.* **2016**, *172*, 31–39. [CrossRef]
35. Zhang, F.; Niu, X.; Zhang, Y.; Xie, R.; Liu, X.; Li, S.; Gao, S. Studies on the Root Characteristics of Maize Varieties of Different Eras. *J. Integr. Agric.* **2013**, *12*, 426–435. [CrossRef]
36. Liu, X.; Wang, Y.; Ma, L.; Liang, Y. Relationship between Deep Soil Water Vertical Variation and Root Distribution in Dense Jujube Plantation. *Trans. Chin. Soc. Agric. Mach.* **2013**, *44*, 90–97.
37. Chu, R.; Li, M.; Shen, S.; Islam, A.R.M.d.T.; Cao, W.; Tao, S.; Gao, P. Changes in Reference Evapotranspiration and Its Contributing Factors in Jiangsu, a Major Economic and Agricultural Province of Eastern China. *Water* **2017**, *9*, 486. [CrossRef]
38. FAO. *World Reference Base for Soil Resources 2014: International Soil Classification System for Naming Soils and Creating Legends for Soil Maps*; FAO: Rome, Italy, 2014.
39. Cui, S.F.; Pan, Y.H.; Wu, Q.Y.; Zhang, Z.H.; Zhang, B.X. Simulation of Runoff for Varying Mulch Coverage on a Sloped Surface. *Appl. Mech. Mater.* **2013**, *409*, 339–343. [CrossRef]
40. Marshall, S.J. Hydrology. In *Reference Module in Earth Systems and Environmental Sciences*; Elsevier: Amsterdam, The Netherlands, 2013.
41. Ge, S.; Gorelick, S.M. Hydrology, Floods and Droughts|Groundwater and Surface Water. In *Encyclopedia of Atmospheric Sciences*; Elsevier: Amsterdam, The Netherlands, 2015.
42. Lehrsch, G.A.; Kincaid, D.C. Sprinkler Irrigation Effects on Infiltration and Near-Surface Unsaturated Hydraulic Conductivity. *Trans. ASABE* **2010**, *53*, 397–404. [CrossRef]
43. Thompson, A.L.; James, L.G. Water droplet impact and its effect on infiltration. *Trans. Am. Soc. Agric. Eng.* **1985**, *28*, 1506–1510. [CrossRef]
44. Zhu, X.; Yuan, S.; Liu, J. Effect of Sprinkler Head Geometrical Parameters on Hydraulic Performance of Fluidic Sprinkler. *J. Irrig. Drain. Eng.* **2012**, *138*, 1019–1026. [CrossRef]
45. Mohammed, D.; Kohl, R.A. Infiltration response to kinetic energy. *Trans. Am. Soc. Agric. Eng.* **1987**, *30*, 108–111. [CrossRef]

9

Adoption of Water-Conserving Irrigation Practices among Row-Crop Growers in Mississippi, USA

Nicolas Quintana-Ashwell [1,*], Drew M. Gholson [1], L. Jason Krutz [2], Christopher G. Henry [3] and Trey Cooke [4]

[1] National Center for Alluvial Aquifer Research, Mississippi State University, 4006 Old Leland Rd, Leland, MS 38756, USA; drew.gholson@msstate.edu
[2] Mississippi Water Resources Research Institute, 885 Stone Blvd, Ballew Hall, Mississippi State University, Leland, MS 38756, USA; j.krutz@msstate.edu
[3] Rice Research and Extension Center, University of Arkansas Cooperative Extension Service, Stuttgart, AR 72160, USA; cghenry@uark.edu
[4] Delta Farmers Advocating Resource Management, Delta F.A.R.M., Stoneville, MS 38776, USA; trey@deltawildlife.org
* Correspondence: n.quintana@msstate.edu.

Abstract: This article identifies irrigated row-crop farmer factors associated with the adoption of water-conserving practices. The analysis is performed on data from a survey of irrigators in Mississippi. Regression results show that the amount of irrigated area, years of education, perception of a groundwater problem, and participation in conservation programs are positively associated with practice adoption; while number of years farming, growing rice, and pumping cost are negatively associated with adoption. However, not all factors are statistically significant for all practices. Survey results indicate that only a third of growers are aware of groundwater problems at the farm or state level; and this lack of awareness is related to whether farmers noticed a change in the depth to water distance in their irrigation wells. This evidence is consistent with a report to Congress from the Government Accountability Office (GAO) that recommends policies promoting the use of: (1) more efficient irrigation technology and practices and (2) precision agriculture technologies, such as soil moisture sensors and irrigation automation.

Keywords: irrigation; groundwater; alluvial aquifer; water conservation adoption; row crops; Mississippi Delta; precision agriculture; Lower Mississippi River Valley

---

1. Introduction

The Mississippi River Valley Alluvial Aquifer (MRVAA) sustains irrigated agriculture in the Mississippi Delta. Almost 22,000 permitted wells [1] withdrawing more than 370 million m^3 of water per year [2] continue to reduce the stock of groundwater available in the MRVAA at an unsustainable rate [3]. A shortage of irrigation water would be a critical challenge to agricultural production in the region [4]. To address this threat, researchers, regulators, and conservation agencies promote the adoption of water-conserving practices in irrigated agriculture. However, little is known about what drives growers in Mississippi to adopt water conservation practices that improve irrigation efficiency.

Profitability is a primary concern in any sustainable enterprise. Hence, farmers would adopt new practices that result in higher profits or reduced risks. However, profitable practices are not universally adopted; which suggests there are other factors related to the farmers and their ecosystem that influence their choice of agricultural practices. This implies that the combination of practices adopted and the factors that influence their adoption vary by practice and location [4]. In some cases, factors such as farmer age or the practicality of the technology are more important than monetary factors [5]. Recent literature identifies factors likely to be associated with the adoption of certain water management and conservation practices at the state (for example, Nian et al. [4] for Arkansas) and national level [6]. However, no recent study examines the factors driving conservation practice adoption in Mississippi.

This article describes water conservation practices that have the potential to profitably reduce the rate of depletion of the MRVAA and identifies social, economic, and environmental factors associated with the adoption of those practices among irrigators in Mississippi. The adoption of conservation practices and farmer characteristics are identified from a comprehensive survey of irrigators in Mississippi that achieved 148 valid responses. A choice model estimated using probit regression on the dataset identifies which factors have a statistically significant association with each of the practices considered.

The predominant irrigation method in the Delta region of Mississippi is continuous flow furrow irrigation [7] on row-crops. This is a modified gravity irrigation system that employs pipes with holes aligned to deliver the flow of water on the furrows. The system is better suited to the relatively flat Delta area than it would be for other regions. Elevation goes from 62.5 m above sea level in the northern tip just south of Memphis, TN, to 24.4 m above sea level at the southern tip in Vicksburg, MS, while the center of the Delta averages an elevation of 38 m between Greenville and Greenwood. Furthermore, the fields are often precision leveled, which results in less "pooling of water" on the fields and more uniform irrigation. Consequently, the baseline case is a relatively inefficient gravity system in terms of costs and irrigation performance. The conservation practices assessed in this article are modifications, "add-ons", or substitutes to this baseline prevalent system.

There is compelling evidence that even minor modifications to existing irrigation and agronomic practices in the Mid-South USA region can result in noticeable water savings while achieving similar yields at harvest [7–10]. As anticipated, the practices evaluated in those studies and considered here are not universally adopted in the area. Despite the expected profitability of adoption, the costs of these practices occur at the time of adoption while their benefits accrue over time. Consequently, producers may require generous incentives or returns to the cost of investing in conservation practices to adopt them [11] or the assurance of witnessing several years of neighboring farmers employing them.

The adoption of conservation practices in the Delta area of Mississippi is partially driven by regulatory mandate. All wells drilled in the area with a casing diameter of 15 cm or greater are required to have a permit. The permits must be renewed every five years and require crop producers to file an Acceptable Agricultural Water Efficiency Practices (AAWEP) form. Irrigators must claim to employ a high efficiency irrigation system, such as a sprinkler irrigation system; or claim the use or proposed use of three water conservation practices; see [12] for the permit application and list of acceptable practices.

1.1. Conservation Practices

This article considers practices that show potential to profitably conserve irrigation water, are accepted in the AAWEP form, and have been adopted by multiple respondents in the 2016 Survey of Mississippi Irrigators. Although many irrigators may decide on adopting several practices simultaneously, this article analyzes adoption practice by practice in order to maximize the number of valid observations for each case.

Soil Moisture Sensors (SMS) are used in irrigation event decision scheduling to prevent yield-limiting water deficit stress on a given crop. SMS gives producers the knowledge of the moisture within the soil profile to make informed and confident irrigation initiation and termination decisions [8] that typically result in increased irrigation efficiency [7,13].

The simplest and most inexpensive (free) upgrade to the baseline irrigation system is Computerized Hole Selection (CHS). Instead of punching uniform holes in layflat poly-tubing, CHS calculates relatively larger holes for parts of the field with long irrigation runs, while shorter parts of the field receive smaller holes to allow water to uniformly reach the end of the field and minimize water runoff [8]. On-farm studies in the Delta area found that CHS achieves water savings of 20 to 25 percent in most situations [14].

"If it can be measured, it can be managed" goes the adage. Pumping flow-meters (meters) are are an important irrigation water management tool compatible with all irrigation systems. Although they do not provide an intrinsic ability to conserve water, they are crucial components, for example, to calculate the optimal size of the holes with CHS [15]. They are also required for cooperator farmers participating in NRCS conservation contracts.

Surge irrigation (surge) allows fields to be divided in two in order to deliver a higher flow rate of water to each half. Water surges down one half of the field until the surge valve flips to deliver water to the other side of the field [8]. The wetting and soaking cycles reduce surface runoff and deep percolation loss while improving water application efficiency by up to 25 percent on the baseline systems with improved infiltration rates documented in sealing silt loam soils [16].

On-Farm Water Storage systems (OFWS) are irrigation water storage structures that are typically designed by NRCS in the Mississippi Delta with the capacity to apply 77 mm of water per hectare per season and meet irrigation requirements for eight out of 10 years [17]. Depending on seasonal conditions, storage capacity, and farmed area, these systems can completely substitute groundwater pumping in some years. NRCS provides technical and financial assistance to producers interested in building OFWS, but many producers face a high opportunity cost to retire productive land to be used for water storage.

The Tailwater Recovery System (TWS) collects irrigation and storm water runoff on the farm in a reservoir (OFWS). TWS increases the amount of water available to irrigation compared to an OFWS filled only by precipitation. This allows OFWS to occupy a smaller surface area and more hectares to be farmed. NRCS estimates that TWS by itself can reduce groundwater pumping by 25 percent.

Micro-irrigation (micro) is a low pressure, low volume, frequent application of water directly to the plant's root zone [18]. It can increase yields and decrease water use by drastically reducing non-beneficial evaporation and virtually eliminating irrigation water runoff. Micro-irrigation is rarely found in the Delta where the soil types and the water quality make the emitters (i.e., nozzles) prone to clogging.

Center pivot-irrigation (pivot) is a type of irrigation that delivers water through sprinklers that create artificial precipitation and are attached to a wheel-driven frame that rotates radially (the arm pivoting on the center). They are highly configurable for a variety of field and crop requirements. Most center pivot systems in the Delta were installed in the 1980s and designed for cotton [19]. However, the original designs are inappropriate to meet the maximum water demands of corn and soybeans. Consequently, there is both adoption of and migration away from center pivot irrigation in the Delta. Similar to micro-irrigation, the soil types and water quality in the Delta present challenges in the form of nozzle clogging and wheels getting stuck in mud.

Pump timers (timer) are a mechanism to program the time or amount of water at which a pump is shut-off. As it helps to automatically or remotely turn the pumps off, it conserves the excess water that might otherwise be pumped, particularly at night [2]. Timers can be employed across irrigation systems for a variety of crops.

Cover crops (cover) are plants that are typically planted during the off season to cover the soil rather than for the purpose of being harvested. They can help sustain and improve soil health [20], microbial populations, and water infiltration, as well as provide benefits in terms of weed control [21]. This is the only conservation practice being analyzed that is not part of AAWEP.

1.2. Factors Affecting Conservation Practices

This article analyzes data from a regional Crop Irrigation Survey that collected 148 valid observations and include data on farmer practices, perceptions and attitudes, and socio-economic status. The factors selected are irrigated area in the operation, Groundwater (GW) use in irrigation, crop choice (rice), number of years farming, years of formal education, whether the farmer perceives a

GW problem at the farm or state level, average pumping cost in the county of residence, participation in a conservation program, and annual income levels. These factors obtained a sufficient number of valid responses in the sample and were mentioned in two recent comprehensive reviews of the literature [6,22] that identified factors associated with the adoption of agricultural conservation practices or in a recent study of conservation practice adoption in Arkansas [4].

The empirical analysis consists of testing how these factors correlate with adoption of the identified practices. The hypotheses with respect to this association are drawn from the existing literature. Specifically, we draw from a study using similar data in an adjacent area by Nian et al. [4], a comprehensive review of 102 papers in the agricultural conservation practice adoption literature by Prokopy et al. [22], and a 129 page Government Accountability Office report to Congress on Irrigated Agriculture by Pearsons and Morris [6]. In terms of the signs of the regression coefficients, the hypotheses are as follows:

(a) Irrigated area: positive;
(b) (GW) use: positive;
(c) rice farming: negative (most conservation practices are geared towards row-cropping, so their adoption appears less likely for rice farmers.);
(d) years farming: negative;
(e) education: positive;
(f) GW problem: positive;
(g) pumping cost: positive;
(h) conservation program: positive;
(i) cracking soils: negative for surge irrigation, undetermined for other practices; and
(j) income: positive.

2. Materials and Methods

2.1. Data

The data were from a survey of irrigators in Mississippi conducted by the Survey Research Laboratory at the Social Science Research Center at Mississippi State University [23]. A telephone-based survey secured a total of 148 completed interviews in Mississippi from a total 2216 telephone numbers acquired (861 were disconnected or inaccessible after 10 attempts) with an overall cooperation rate of 27.6 percent. The sample is representative of the Delta area of Mississippi with 131 respondents residing in the area and 3 more in neighboring counties. The survey instrument contained questions on growers' characteristics, cultural practices, irrigation management practices, and perceptions and attitudes regarding groundwater availability.

Except for irrigated area, years of education, and pumping cost, the variables included in the analysis were coded as categorical or indicator (dummy) variables. The pumping cost is calculated as [24]:

$$p = \theta_e p_e d, \tag{1}$$

where p_e is the price of the energy source for the power unit, d is depth to water used as a proxy for pumping lift, and θ_e is the amount of energy from source e needed to lift a cubic meter of water a distance of one meter. The distance to water was obtained from the U.S. Geological Survey [25] based on the respondent's claimed county of residence. Average energy prices were obtained from the U.S. Energy Information Administration (EIA).

Cracking soils is the percentage of soils with clay content dominated by smectite. Such soils crack on the surface when a moist soil shrinks due to drying. The data on soil composition is accessible through the USDA-NRCS Websoil survey [26]. This is an important control variable. For example, surge irrigation programs are more difficult to manage in this type of soil because the programming of the alternating cycles is more complicated. The typical program relies on visual cues to switch from one side of the field to the next based on the time water takes to reach the tail of the section. In cracking soils, water infiltrates through the cracks, and actual wetting occurs by 3 or more meters ahead of the wetting on the surface. The program is still applicable and carries the same water savings potential, but becomes harder for the farmer to realize.

Number of years of education was calculated based on a question that was originally categorical. The assigned values were as follows: 10 for less than completed high school, 12 for completed H.S., 13 for some college or vocational program, 14 for completed Associate's degree, 16 for completed Bachelor's, 18 for completed Master's, and 20 for more than Master's. This transformation is helpful in the estimation and interpretation of regression results with a sample that is small relative to the number of variables considered.

The variable GW problem is a dummy variable based on the combination of categorical responses to two different questions in the survey: "In your opinion, do you have a groundwater shortage problem on your farm?" and "In your opinion, do you have a groundwater shortage problem in your state?" Lastly, conservation program is a dummy variable based on the combination of responses to four different questions that would have otherwise yield 19 response categories (see Appendix A).

2.1.1. Choice Model

There are several ways to motivate the empirical strategy. Due to the nature of the survey and structure of the data, a scenario that allows for irrigators to adopt a single practice or a number of them simultaneously is needed. Hence, a model of individual practice adoption is adequate. An irrigator i adopts a water conservation practice w if the grower expects to receive a greater utility from using the practice (U_{iw}) than they would not using it (U_{iNg}). The probability of adopting practice w is the probability that $y^*_{iw} = U_{iw} - U_{iNg} > 0$ and depends on a vector of n identified factors X_i. Following Maddala [27], y^*_i is a latent, unobservable variable defined by the regression relationship:

$$y^*_{iw} = \beta' X_i + u_i \tag{2}$$

where u_i is the error term. The variable that is actually observed is whether a practice is adopted ($y = 1$) or not ($y = 0$).

From these relationships, the probability that any given practice w is adopted can be estimated using probit with the assumption that u_i follows a normal distribution:

$$Pr(y_w = 1) = Pr\left(\sum_j \beta_{jw} X_j > 0\right) = \Phi\left(\sum_j \beta_{jw} X_j\right), \tag{3}$$

where $\Phi(\cdot)$ is the cumulative normal distribution function.

To predict the effect a change in the value of a variable would have on the probability of adopting a given practice, the marginal effects are calculated as:

$$\frac{\partial}{\partial x_{ik}} \Phi\left(X_i'\beta\right) = \phi\left(X_i'\beta\right) \beta_k, \tag{4}$$

where $\phi(\cdot)$ is the normal probability density function. This marginal effect is denoted as dy/dx in the results.

3. Results and Discussion

3.1. The Sample

The survey instrument contained questions on growers' characteristics, cultural practices, irrigation management practices, and perceptions and attitudes regarding groundwater availability. Table 1 summarizes the information gathered on growers' land tenure, education, and income.

Responses were considered to be representative of irrigators operating in the Delta because 88.5 percent of respondents resided in that area and additional respondents lived in neighboring counties. Cracking soils were present in 83.1 percent of the counties where irrigators claimed residence with an average of 27.6 percent of soils classified as cracking.

Table 1. Summary statistics of farmer characteristics from an irrigation survey conducted in the Mississippi Delta in 2016.

Farmer Characteristics	N	%
Delta	131	88.5
Cracking soil	123	83.1
Avg. percentage		27.6
Operator	31	20.9
Landowner and operator	117	79.1
Education:		
Less than high school	5	3.4
Completed high school or GED	23	15.5
Some college	22	14.9
Completed Associate's	18	12.2
Completed Bachelor's	66	44.6
Completed Master's	11	7.4
Beyond Master's	2	1.4
Agriculture-related	63	42.6
Income per year:		
Less than USD50,000	13	8.7
USD50,000 to USD100,000	41	27.7
USD100,000 to USD150,000	17	11.5
USD150,000 to USD200,000	9	6.1
USD200,000 to USD250,000	6	4.1
USD250,000 to USD300,000	5	3.4
More than USD300,000	10	6.8
Unsure or no response	47	31.7

Almost 80 percent of respondents were land-owner operators, and the remaining growers were operators only. Nearly two-thirds of the farmers completed a post-secondary degree (65.6 percent), and 42.6 percent of respondents indicated that part of their formal education was related to agriculture.

Growers also identified the range of income they had achieved the previous year. A total of 101 valid responses were recorded with 31.7 percent of respondents refusing to provide an answer or being unsure with respect to which income bracket they belonged. Amongst the valid responses, fifty-three-point-five percent of farmers claimed an annual income of less than USD100,000. The median income in the sample was between USD75,000 and USD100,000 per year.

Income level is expected to be positively correlated with the adoption of conservation practices [4,6]. Also, the level of farmer education positively influences the adoption of irrigation-related precision agriculture practices [6].

3.1.1. Farming Practices

Data on growers' agricultural experience and practices are summarized in Table 2. In terms of farming experience, respondents represented a wide range of experience, from as little as three years to as much as 80 years of farming experience. Every measure indicated that these were seasoned farmers. On average, growers had 28 years of farming experience with a median and mode of 29 and 30 years of experience, respectively. More than 84 percent of farmers had 10 years or more of farming experience. Approximately two-thirds of the sample were farmers with more than 20 years of experience. The number of years farming was expected to be negatively associated with the adoption of agricultural innovations.

Table 2. Summary statistics of cultural practices.

	N	%	Min	Max	Mean	Std. Deviation
Years farming	148	100	3	80	28.03	14.761
			(Min, Max, and Mean in ha)			
Irrigated area (ha, all)	148	100.0	0	6070	896	1007
Irrigated crops						
Corn	106	71.6	2.02	1821	305	337.7
Cotton	49	33.1	18.1	2833	490	546.7
Rice	41	27.7	32.4	1558	373	407.2
Soybeans	131	88.5	27.1	3804	671	700.2
Cover crop	45	30.4	4.05	1335	-	-
Land leveling	124	83.8				
Zero grade	26	17.6	2.02	971	177	239.6
Precision grade	116	78.4	3.24	4654	733	848.5
Warped or OptiSurface	33	22.3	10.12	1619	266	385.9
Not leveled	76	51.4	4.05	1714	202	240.3
Conservation programs	111	75				
CRP	58	39.2	-	-	-	-
EQIP	87	58.8	-	-	-	-
RCPP	14	9.5	-	-	-	-
CSP	31	20.9	-	-	-	-
NRCS	8	5.4	-	-	-	-
Other	8	5.4	-	-	-	-

Note: CRP is NRCS Conservation Reserve Program; EQIP is NRCS Environmental Quality Incentives Program; RCPP is NRCS Regional Conservation Partnership Program; CSP is NRCS Conservation Stewardship Program; NRCS is USDA Natural Resources Conservation Service unspecified program.

The size of the farming operation is an important factor in deciding the adoption of agricultural practices in general. The average operation involved 896 ha of irrigated farmland with a median of 567 ha and as much as 6070 irrigated ha. More than three-fourths of the responding growers operated 1133 ha or less. Hence, the number of irrigated hectares was expected to be positively correlated with the adoption of conservation practices.

Crop choice is oftentimes associated with the choice of irrigation technology [28,29]. The largest number of growers reported producing irrigated soybeans (n = 131), which occupied the largest cultivated area among the irrigated crops reported: 671 ha on average and as much as 3804 ha. Irrigated corn was the second most popular crop choice with 106 farmers reporting an average of 305 ha and as much as 1821 ha of irrigated farmland dedicated to that crop. Cotton is a traditional crop in the Delta region of Mississippi. About a third of the respondents grew cotton with irrigation dedicating an average of 490 ha and as much as 2833 ha to its production. These are typically row-crops that employ the same or similar irrigation setups when the fields are prepared for furrow irrigation.

Kebede et al. [2] reported that irrigated rice consumes more water than any other crop in the region. Growers in this sample reported an average of 373 ha of irrigated rice farmland with as much as 1558 ha of rice under irrigation. Rice production was expected to be negatively correlated with the conservation practices considered in this article, which were better suited for row-crop irrigation.

Cover crops are typically not harvested. Consequently, these crops were not considered as part of crop choices, but rather as a conservation practice in this article. Nearly a third (30.1 percent) of growers in the sample claimed to plant cover crops. Responses to the survey varied widely in terms of crop and area matching. Wheat and radishes received the highest reported area of 1335 ha, while the least was reported for Asian mustard greens (4 ha). The adoption of this practice was tested against the identified explanatory factors.

Land leveling is a relatively common practice in the area with 84 percent of growers reporting having at least a part of their fields land-leveled in some way. It is also no surprise that 51.4 percent of the participating growers reported that some fields were not leveled because the Delta in Mississippi is unusually flat. Precision grade was the most common land leveling method with 78 percent of growers employing it on an average of 733 ha and in up to 4654 ha of their operation.

Awareness and participation in conservation programs were expected to positively influence the adoption of irrigation water conservation practices [4,6]. Three-fourths of the growers claimed participation in a conservation program. The program most commonly cited was the NRCS Environmental Quality Incentives Program (EQIP) with 59 percent participation.

3.1.2. Irrigation and Water Conservation Practices

Grower irrigation practices are summarized in Table 3. Groundwater was the principal source of water for irrigation with 93 percent of respondents identifying it as a source. On average, eight-hundred eighty-nine hectares were irrigated with groundwater with a maximum of 4856 ha relying on that source for irrigation. Surface water was also employed for irrigation including streams and bayous, which are the source for 178 ha on average. The surface sources can also be complemented with OFWS, 11.5 percent of responses, and TWS in 14.2 percent of responses. Some growers built OFWS and TWS capable of fully supplying the irrigation water needs for some of their fields. Producers relying on groundwater from a depletable aquifer were expected to be more inclined to adopt water conservation practices.

The predominant irrigation practice was furrow irrigation for row-crops, which was employed by 86 percent of respondents on an average of 823 ha. Practices that improve the performance of furrow irrigation are deep tilling, employed by 71 percent of the respondents, computerized hole selection (CHS), adopted by 59 percent of growers, and surge irrigation, adopted by 24 percent of farmers in the sample. The last two are considered water-conserving practices for which we sought to find determining adoption factors. Irrigation systems with higher application efficiency are also considered water conservation practices. Micro irrigation was very rare with only 3.4 percent of respondents employing it on an average of 65 ha; while center pivot sprinkler use was more widespread with 60 percent of respondents having used it on an average of 370 ha.

Irrigation scheduling is a crucial component of irrigation water management. The use of Soil Moisture Sensors (SMS) for scheduling has the potential to save as much as 50 percent of total water applied [30]. Agronomic studies in the area showed that SMS could help improve water use efficiency in furrow irrigated soybeans [7] and corn [10] by reducing water use without reductions in yields when compared to conventional farmer-managed scheduling.

Flow-meters at the irrigation wells are another important management tool. Nearly 70 percent of participant growers owned them. A voluntary metering program in Mississippi encourages their use, and participation in NRCS incentive programs makes participation in that program mandatory for their cooperators. Another pump accessory is the pump timer, which allows the irrigation events to be started or stopped automatically. Almost 44 percent of respondents employed pump timers on an average of 11 pumps.

Finally, the energy source for the pump power units varied with most growers having more than one type of energy source. Electricity was the most common energy source with 85 percent of farmers claiming it. Diesel was the second most common source with almost 80 percent of respondents using it. The energy source mix is important in calculating irrigation pumping costs. The average cost of pumping was estimated at USD 0.0538 per megaliter.

Table 3. Summary statistics of irrigation practices.

	N	%	Min (ha)	Max (ha)	Mean (ha)	Std. Dev.
Irrigation by source of water						
Groundwater	137	92.6	2.83	4856	889	909
Stream or bayou	39	26.4	7.69	599	178	142
Stream or bayou and OFWS	17	11.5	2.02	977	182	253
Stream or bayou and TWS	21	14.2	16.2	707	178	185
No outside source w/OFWS/TWS	16	10.8	4.45	304	81	84
Irrigation by practice						
Flood (row-crops)	69	46.6	6.07	4047	737	939
Furrow (row-crops)	127	85.8	1.62	4452	823	942
Deep till (furrow)	104	70.7	1.62	2428	512	522
Computerized hole selection	87	58.8	1.62	3462	734	706
Surge	35	23.6	12.1	607	140	146
Border (row-crops)	26	17.6	10.1	769	158	274
Micro (row-crops)	5	3.4	2.02	223	65	106
Pivot (row-crops)	88	59.5	2.43	2428	370	391
Irrigation scheduling						
Soil moisture sensors	72	48.6	0.4	6070	554	1027
Visual crop stress	103	69.6	-	-	-	-
Computerized scheduling	5	3.4	80.9	243	146	102
Routine	29	19.6	-	-	-	-
Probe/feel	27	18.2	-	-	-	-
ET or Atmometer	4	2.7	0	202	99	84
Watch other farmer	6	4.1	-	-	-	-
			Min (units)	**Max (units)**	**Mean (units)**	**Std. Dev.**
Irrigation pumps	146	98.6	1	120	21	24.0
Pump timers	65	43.9	1	90	11	16.6
Flow-meters	103	69.6	1	46	8	7.9
Power unit						
Electric	126	85.1	1	80	11	14.8
Diesel	117	79.1	1	85	13	14.0
Propane	24	16.2	1	45	7	9.3
Natural gas	3	2.0	1	5	3	2.0
Pumping cost (USD/ML)	94	-	0.02	0.088	0.054	0.022

Note: OFWS is On-Farm Water Storage; TWS is Tailwater Recovery System.

3.1.3. Grower Perceptions and Attitudes

The data collected on farmer perceptions and attitudes towards groundwater availability issues are summarized in Table 4. Less than one-third of growers observed a change in their wells' depth to water while over two-thirds of respondents indicated they perceived there was not a problem with the groundwater supply at their farm or in the state. A test of independence in these responses indicated that they had a statistically significant dependence: those who perceived there had been a change in their wells depth to water were more likely to believe there was a groundwater problem in their farm or at the state level.

The U.S. Geological Survey (USGS) has recently published a map of the Potentiometric Surface of the Mississippi River Valley Alluvial Aquifer for the Spring 2016 [25] that shows the location and gradient of the aquifer's cone of depression. A cross-tabulation of farmer perceptions and their location in the center of the cone of depression is presented in Table 5. A Pearson test of independence of the responses showed evidence that farmers located in the cone of depression were more likely to observe a change in their well levels and think there was a groundwater problem at the farm or state level. Half of those located in the center of the cone of depression believed there was a groundwater problem as opposed to 29 percent amongst those located outside that area. Similarly, forty-six percent of those in the cone of depression area observed a change in the depth-to-water in their wells while only 26 percent of those outside the area noticed such a change.

Table 4. Summary statistics of farmer perceptions and attitudes.

	Thinks There Is a GW Problem		
Frequency	**No**	**Yes**	**Total**
Well Depth to Water:			
No change	60	24	**84**
Increased	9	11	**20**
Decreased	12	13	**25**
Do not know	16	2	**18**
Refused	1	0	**1**
Total	**98**	**50**	**148**
Percentage	**No**	**Yes**	**Total**
Change in Depth to Water:			
No/cannot tell	51	17	**68.9**
Changed	14	16	**30.4**
Refused	1	0	**0.7**
Total	**66.2**	**33.8**	**100**
Pearson χ_4^2 = 13.4 with Pr = 0.009.			

Note: GW is Groundwater.

Table 5. Aquifer "cone of depression" and farmer perceptions and attitudes.

	Cone of Depression		
Percentages	**No**	**Yes**	**Total**
Depth to Water:			
No change	56	14	69
Changed	20	11	31
Total	76	24	100
Pearson χ_1^2 = 4.3 with Pr = 0.038			
Groundwater Problem:			
No	54	12	**66**
Yes	22	12	**34**
Total	**76**	**24**	**100**
Pearson χ_1^2 = 5.6 with Pr = 0.018.			

3.2. Probit Regression Analysis

The estimated models of practice adoption fit the data relatively well with pseudo-R^2 (McFadden's) ranging between 0.156 to 0.545. Except for micro-irrigation (only five adopters), the conservation practices being analyzed had at least one factor with a statistically significant coefficient. The probit regression coefficients are detailed in Table 6 for all factors except cracking soils and income, which are detailed in Table 7.

Tailwater Recovery System (TWS) adoption was positively and significantly influenced by the farmers perception that a groundwater problem existed (GW prob.). The marginal effect (dy/dx) indicated that a producer who becomes aware of the groundwater problems in the Delta area would be associated with a 25 percent higher likelihood of adopting TWS. The data indicated that farmers who do not use groundwater for irrigation have not adopted TWS.

For OFWS, the number of irrigated hectares under operation (Irr.area) was positively and significantly associated with the adoption of OFWS. The calculated marginal effect indicated that for an additional 40 hectares of farmed land, the probability of a farmer adopting OFWS was 0.8

percent higher. The data indicated that farmers who did not use groundwater for irrigation have not adopted OFWS.

Table 6. Results from probit regressions (coefficients by income level in Table 7). Irr, Irrigation.

	Irr. Area	GW Use	Rice	Years Farm	Educ.	GW Prob.	P. Cost	Cons.pr.
TWS	0.0001	(a)	−0.398	0.009	0.177	0.857 **	0.464	0.691
s.e.	(0.0001)		(0.435)	(0.015)	(0.132)	(0.448)	(1.052)	(0.588)
dy/dx	0.00003		−0.115	0.003	0.051	0.25 **	0.134	0.199
McFadden's R^2 = 0.228								
OFWS	0.0003 **	(a)	−0.007	0.027	0.214	0.779	1.45	0.186
s.e.	(0.005)		(0.507)	(0.018)	(0.159)	(0.54)	(1.18)	(0.67)
dy/dx	0.00008 ***		−0.002	0.006	0.052	0.189	0.353	0.045
McFadden's R^2 = 0.324								
CHS	0.0002 *	0.59	−0.176	−0.01	0.153	1.13 **	−0.58	−0.24
s.e.	(0.0001)	(1.11)	(0.49)	(0.015)	(0.124)	(0.51)	(1.03)	(0.495)
dy/dx	0.00005 *	0.16	−0.048	−0.003	0.041	0.30 **	−0.157	−0.065
McFadden's R^2 = 0.271								
Surge	0.00007	(a)	−0.516	−0.019	0.258	1.172 *	−2.462 **	0.735
s.e.	(0.00011)		(0.508)	(0.017)	(0.151)	(0.601)	(1.23)	(0.76)
dy/dx	0.00002		−0.116	−0.004	0.058 **	0.26 **	−0.55 **	0.165
McFadden's R^2 = 0.362								
SMS	0.0009 **	(a)	−2.72 **	−0.016	0.03	-0.81	−1.1	2.45 **
s.e.	(0.0003)		(0.94)	(0.019)	(0.168)	(0.9)	(1.31)	(1.06)
dy/dx	0.0002 ***		−0.5 ***	−0.003	0.005	−0.15	−0.204	0.45 **
McFadden's R^2 = 0.545								
Micro	−0.00003	(a)	(b)	0.02	0.47	(b)	−4.42	(a)
s.e.	(0.0004)			(0.05)	(0.63)		(10.6)	
dy/dx				0.0003	0.009		−0.074	
McFadden's R^2 = 0.464								
Pivot	0.0001	(a)	−0.622	0.0005	0.24 *	−0.028	1.28	−0.29
s.e.	(0.0001)		(0.471)	(0.015)	(0.13)	(0.48)	(1.08)	(0.58)
dy/dx	0.00003		−0.18	0.001	0.068 **	−0.0087	0.363	−0.083
McFadden's R^2 = 0.156								
Timer	0.0003 **	(a)	−0.84	0.003	−0.14	1.67 **	−0.43	(b)
s.e.	(0.0001)		(0.52)	(0.016)	(0.13)	(0.54)	(1.23)	
dy/dx	0.00006 **		−0.2 *	0.001	−0.034	0.41 ***	−0.1	
McFadden's R^2 = 0.355								
Flow-meter	0.0005 **	0.181	−0.43	0.017	0.167	0.932	−0.56	0.33
s.e.	(0.0003)	(1.15)	(0.61)	(0.016)	(0.143)	(0.67)	(1.26)	(0.56)
dy/dx	0.0001 **	0.037	−0.09	0.004	0.034	0.19	−0.12	0.068
McFadden's R^2 = 0.34								
Cover	0.0001	−0.76	0.43	−0.04 **	0.068	−0.44	−0.015	0.78
s.e.	(0.0001)	(1.55)	(0.47)	(0.018)	(0.125)	(0.50)	(0.96)	(0.59)
dy/dx	0.00003	−0.21	0.12	−0.011 **	0.02	−0.12	−0.004	0.22
McFadden's R^2 = 0.21								

Note: Educ. is years of formal education; GW Prob. is an indicator variable for perception of a groundwater problem at the farm or state level; P. Cost is the cost of pumping; and Cons. pr. is participation in a conservation program. Standard errors are in parentheses. *,**,***: significant at $p < 0.1$, $p < 0.05$, $p < 0.01$, respectively. (a) negative cases predict failure; (b) positive cases predict failure.

Computerized Hole Selection (CHS) was positively and significantly associated with the number of irrigated hectares, the perception of the existence of a groundwater problem, and having an income between $100,000 and $150,000. The marginal effects indicated a 0.5 percent higher probability of CHS adoption for an additional 40 ha of land irrigated, and the probability of adoption increased by 30 percent when a farmer realized there was a groundwater problem at the farm or state level.

Table 7. Results from probit regressions (continued) on "cracking" soils and income levels.

	Cracking	50 k to 100 k	100 k to 150 k	150 k to 200 k	200 k to 250 k	250 k to 300 k	≥300 k
TWS	0.18	−0.49	−0.5	−1.195	0.302	(dropped)	0.604
s.e.	(0.02)	(0.6)	(0.637)	(1.02)	(1.063)		(1.185)
dy/dx	0.005	-0.14	−0.144	−0.345	0.087		0.174
OFWS	−0.008	−1.021	−0.7	(dropped)	0.615	(dropped)	0.186
s.e.	(0.03)	(0.684)	(0.676)		(1.33)		(0.671)
dy/dx	−0.002	−0.248	−0.17		0.15		0.045
CHS	0.035	0.611	1.325 *	0.59	0.273	0.613	1.33
s.e.	(0.02)	(0.625)	(0.694)	(0.878)	(1.05)	(1.25)	(1.24)
dy/dx	0.009	0.165	0.358 **	0.159	0.074	0.166	0.359
Surge	0.028	−1.26 *	−0.241	0.66	(dropped)	(dropped)	1.44
s.e.	(0.03)	(0.766)	(0.669)	(0.963)			(1.165)
dy/dx	0.006	−0.28 *	−0.05	0.148			0.324
SMS	0.014	0.99	2.48 **	−0.496	0.258	−0.019	−1.38
s.e.	(0.03)	(0.95)	(1.06)	(0.982)	(1.091)	(1.26)	(2.0)
dy/dx	0.003	0.183	0.458 **	−0.092	0.048	−0.004	−0.255
Micro	−0.052	(dropped)	(dropped)	(dropped)	(dropped)	(dropped)	(dropped)
s.e.	(0.06)						
dy/dx	-0.001						
Pivot	0.021	−0.45	−0.49	−0.365	(b)	0.548	(a)
s.e.	(0.02)	(0.64)	(0.64)	(0.927)		(1.183)	
dy/dx	0.06	−0.13	−0.14	−0.1		0.156	
Timer	0.046	−0.486	−0.129	0.421	−0.028	(c)	(c)
s.e.	(0.029)	(0.711)	(0.78)	(0.9)	(1.03)		
dy/dx	0.01	−0.12	−0.03	0.1	−0.01		
Flow meter	0.03	−0.88	0.09	−1.15	(c)	(c)	(c)
s.e.	(0.03)	(0.78)	(0.87)	(1.07)			
dy/dx	0.005	−0.18	0.02	−0.24			
Cover	−0.003	0.15	0.55	0.36	1.97 *	0.57	1.23
s.e.	(0.019)	(0.68)	(0.68)	(0.94)	(1.06)	(1.15)	(1.16)
dy/dx	−0.001	0.041	0.155	0.1	0.554 **	0.16	0.34

Note: base income is $50,000 or less; k represents thousands of dollars. Standard errors are in parentheses. *,**: significant at $p < 0.1$, $p < 0.05$, respectively. (a) negative cases predict failure; (b) positive cases predict failure; (c) positive cases predict success.

The adoption of surge irrigation (surge) was positively and significantly associated with the perception of the existence of a groundwater problem and negatively by the pumping cost. The negative influence of the pumping cost variable was a departure from the hypothesized relations in the GAO report. Because this variable is a combination of various data with a fundamental rooting in the county of residence claimed by the farmer, there may be confounding of factors, the identification of which is beyond the scope of this study. However, the result was driven in part by the fact that nobody claiming to reside in the cone of depression (highest pumping cost) used surge irrigation. Surge irrigation is harder to manage in the cracking clays that are a common soil type in that area , but this effect did not appear statistically significant in this regression. The marginal effect calculations suggested that the probability of adoption of surge increased by 0.2 percent for an additional 40 hectares of irrigated land added to the operation, but an increase of one percent in the cost of pumping would decrease the adoption probability by 0.55 percent. The data indicated that farmers who do not use groundwater for irrigation have not adopted surge.

The use of SMS was significantly associated with irrigated area (positive at five percent), rice production (negative at five percent), participation in a conservation program (positive at five percent), and increasing income (from the baseline to the $100 k to $150 k income bracket, positive at five percent). The estimates suggested that an increase of 40 hectares in irrigated land would result in a two percent higher probability of adopting SMS; the choice of growing rice would reduce that probability by 50 percent, and the participation in a conservation program would add 45 percent to the SMS adoption probability. The data indicated that farmers who do not use groundwater for irrigation have not adopted SMS.

With respect to micro-irrigation, the probit regressions did not find statistically significant effects. This may be due in part to the relatively few respondents who claimed to practice it. For center pivot irrigation, however, there was enough variability to show a statistically significant and positive effect of the number of years of farmer formal education. For every additional year of formal education completed, the farmer was 0.68 percent more likely to adopt center pivot irrigation. The data indicated that farmers who do not use groundwater for irrigation have not adopted center pivot.

The adoption of a pump timer was positively and significantly associated with the number of irrigated hectares and the perception that a groundwater problem existed at the farm or state level. An additional 40 hectares or irrigated land was associated with a 0.6 percent higher probability of adoption, and the realization that a problem with groundwater stock existed in the state implied a 41 percent higher probability of employing a pump timer. All farmers in the sample with incomes above $250,000 used pump timers.

Flow meter adoption was also positively and significantly associated with the number of irrigated hectares. An additional 40 hectares or irrigated land were associated with a one percent higher probability of adoption. All farmers in the sample with incomes above $200,000 used flow meters.

Growing cover crops was negatively and significantly associated with farmer experience as represented by the number of years farming and positively associated with the $200,000 to $250,000 income bracket. An additional year of farming experience was associated with a 1.1 percent lower probability of growing cover crops. Since the agronomic benefits of cover crops payoff over a longer time-horizon, a farmer getting closer to retirement may be less eager to invest in a practice for which she/he will not see most of the benefits.

In terms of identifying factors that are associated with the adoption of conservation practices, the results indicated the following: irrigated area (positive), GW use (positive), rice (negative), years farming (negative), education years (positive), perception of GW problem (positive), pumping costs (negative), conservation program participation (positive), and income (positive). These results confirmed the stated hypotheses with respect to factor association except for the effect of pumping cost, which was expected to have a positive association with the adoption of conservation practices. It is not clear from the data what drives this result, which is limited to the adoption of surge irrigation only and does not appear statistically significant for other practices.

4. Discussion

Promoting the adoption of water conservation practices in irrigated agriculture has been a principal initiative to slow down the decline of the Mississippi River Valley Alluvial Aquifer (MRVAA). Ongoing agronomic research from scientists at Mississippi State University's Delta Research and Extension Center (DREC) and the USDA Agricultural Research Service in Stoneville, MS, continue to show the potential for these practices to reduce water use while maintaining farm yield and revenue levels. The Mississippi Department of Environmental Quality and the Yazoo Mississippi Delta Joint Water Management District require the use of a minimum number of these practices to issue groundwater well drilling and use permits in the area. Grower associations such as the Delta Farmers Advocating Resource Management (F.A.R.M.) sponsor and promote the use of these practices among their members. Financial and technical assistance from USDA Natural Resource Conservation Service (NRCS) is geared towards minimizing farmer risk exposure associated with the implementation of these conservation practices. However, little is known about the farmer factors that drive their decision to adopt any given conservation practice. This article provides insights that help identify and understand the determinants of conservation practice adoption in the Delta region in Mississippi.

The regression analyses indicated that no single factor consistently predicted the adoption of every conservation practice, but many factors influence a farmer's decision to adopt a given practice. The size of the farming operation is an important factor in deciding the adoption of agricultural practices in general. Indeed, it is the factor with the most statistically significant coefficients (positive) in the regression analyses. From an economic perspective, this may be attributed to the fact that practices that are marginally beneficial on a per hectare basis may not be worth the managerial cost to small operations, but could add-up to significant economies of scale for larger operations in which fixed and overhead costs associated with a practice can be spread over more hectares.

Results from the 2016 Mississippi Irrigation Survey indicated that groundwater is a source for irrigation for almost 93 percent of growers in the sample. Yet, only a third of the growers think there is a groundwater problem at the farm or state level. The analysis presented suggests that this lack of awareness is significantly related to whether the growers observe a change in the depth to water in their wells or not.

Perception of the existence of a GW problem at the farm or state level was the second most important factor identified. This is an encouraging finding because the results show a strong influence of this variable on the probability of the adoption of several adoption practices and because this is, essentially, an awareness issue. Perception of water quantity issues in regions of high rainfall can be difficult to overcome. This suggests that additional incentives are necessary to bring those who do not perceive a problem to the realization that it actually exists.

Increasing producer awareness is a task that fits, for example, the mission of university extension services, which can aid the communications efforts of federal and state research and regulatory entities in that regard. Regulatory agencies and universities can work together to have a consistent message regarding the projection of groundwater in the state and to increase grower awareness of the issue with a focus on county-based expected changes in wells' depth-to-water distances. However, such efforts have been carried out by these organizations, which suggests there is a need for an additional signal to help convince farmers of the seriousness of the situation.

At nearly 70 percent of respondents, this sample far surpasses the national average of 30 percent flow-meter use [31]. However, only 10 to 15 percent of permittees report individual water-use every year. There is promising evidence from areas with mandatory water use reporting that producers become better informed and more concerned with the health of the aquifer.

For example, growers near Sheridan county in Kansas widely supported a self-imposed Local Enhancement Management Area to create a five year allocation of groundwater that resulted in a 26 percent reduction in water use [32]. A similar case where the threat of regulation from the state level induced irrigators to form the Groundwater Subdistrict No. 1 in the San Luis Valley of Colorado and "formulate a homegrown governance response" that reduced water use by 33 percent in the district [33].

The survey results in our study indicated that farmers may require stronger signals that the aquifer problem is real and important. Furthermore, the experience documented in the aforementioned studies suggested that aquifer problem awareness, resulting largely from individual water use monitoring, and the threat of top-down regulation can induce more active farmer participation in water conservation.

Although participation in conservation programs was a statistically significant factor for only one of the practices (SMS), Figure 1 shows how NRCS expenditures [34] and practice adoption in the region track similar time trends. In particular, the rate of adoption of CHS, SMS, and surge starting in 2009 are noteworthy. This suggests there is an effect of NRCS technical and financial assistance in terms of adoption. However, perhaps due to insufficient data, this association cannot be validated empirically in this study.

These results can inform policy makers, regulatory entities, university extension services, and producers about the salient aspects of conservation practice adoption in Mississippi. Conservation agencies can use the insights in this study to better target their incentive programs, for instance focusing on incentivizing relatively young farmers to adopt practices with long-term benefits such as cover crops. Similarly, further research, extension, and incentives are necessary to devise incentives and training to facilitate practice adoption by smaller farms. Lastly, periodic surveys (2–5 year intervals) may be necessary to track trends and assess the effectiveness of conservation practice adoption programs.

Our results are largely consistent with the most recent literature [4,6,22]; see Table 8 for a comparison. The GAO report in particular describes policy options at the federal level including their benefits and challenges. With respect to irrigation technology, it recommends that policy makers promote: (1) the use of more efficient irrigation technology and practices, in conjunction with appropriate agreements to use the technology and practices to conserve water; and (2) the use of precision agriculture technologies, in conjunction with appropriate agreements to use the precision agriculture technologies to conserve water. These recommendations are consistent with the ongoing efforts mentioned above, and this study lends empirical validity to the recommendations for the case of Mississippi.

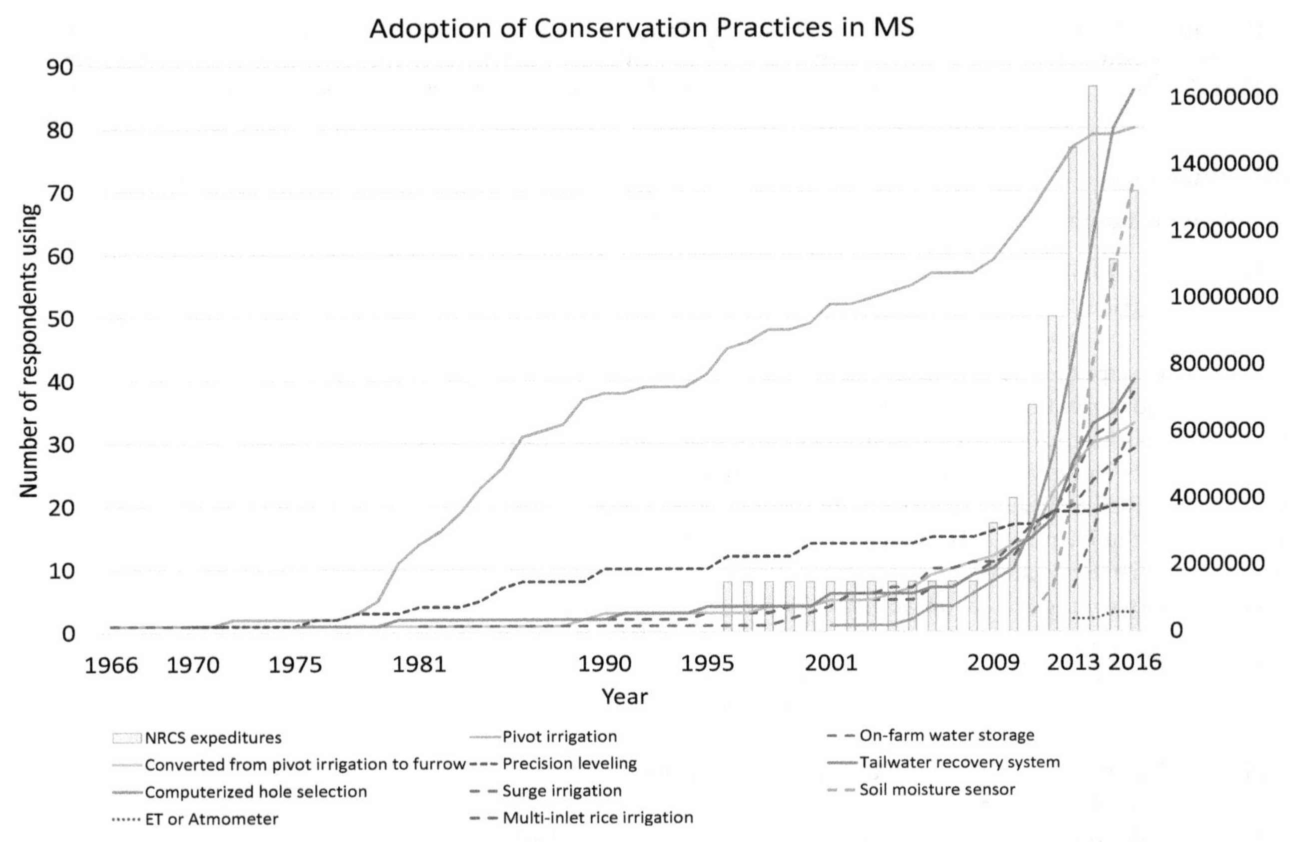

Figure 1. Timeline of the adoption of conservation practices and USDA-NRCS EQIP expenditures in the MS Delta area.

Table 8. Impact of conservation practice adoption factors compared to existing literature.

Factor	Results	Nian et al. Morris (2020)	Prokopy et al. (2019)	Pearsons and (2019)
Irrigated area	Pos	Pos	Pos	Pos
GW use	Pos	Pos	Pos	
Rice farming	Neg		Mix	
Years farming	Neg	Neg	Neg	Neg
Education	Pos		Mix	Pos
Perceives GW problem	Pos	Pos	Pos	Pos
Pumping cost	Neg		Pos	Neg
Participation in conservation programs	Pos	Pos	Pos	Pos
Cracking soils	Mix			Neg
Farmer income	Mix	Pos	Pos	Pos

Note: Pos denotes a Positive influence of the factor on adoption; Neg denotes a Negative influence of the factor on adoption; Mix denotes Mixed or undetermined influence of the factor on adoption.

The main limitation of this study is the small sample relative to the number of practices and factors being considered. This limitation determined the choice of probit regression models rather than multivariate and sequential modeling. There is evidence that practices, especially add-ons to furrow irrigation, are adopted in bundles. This feature is not incorporated in this study and constitutes a promising avenue for future research.

Author Contributions: Conceptualization, N.Q.-A. and D.M.G.; methodology, N.Q.-A.; software, N.Q.-A.; validation, D.M.G. and L.J.K.; formal analysis, N.Q.-A.; investigation, L.J.K. and C.G.H.; resources, L.J.K., C.G.H., and T.C.; data curation, N.Q.-A.; writing, original draft preparation, N.Q.-A.; writing, review and editing, N.Q.-A., D.M.G., and L.J.K.; funding acquisition, L.J.K., C.G.H., and T.C. All authors read and agreed to the published version of the manuscript.

Acknowledgments: The authors acknowledge and thank Paul Rodrigue, USDA-NRCS Supervisory Engineer Area 4, for helpful advice and regional NRCS expenditure data. Additionally, the authors are grateful for financial support for the survey from the Mississippi Soybean Board, Mid-South Soybean Board, and United Soybean Board.

Abbreviations

The following abbreviations are used in this manuscript:

AAWEP	Acceptable Agricultural Water Efficiency Practices
AWEP	Agricultural Water Enhancement Program
ARS	USDA Agricultural Research Service
CHS	Computerized hole selection
CRP	Conservation Reserve Program
DREC	Mississippi State University Delta Research and Extension Center
EIA	U.S. Energy Information Administration
EQIP	NRCS Environmental Quality Incentives Program
F.A.R.M.	Delta Farmers Advocating Resource Management
FSA	Farm Service Agency
GAO	Government Accountability Office
GW	Groundwater
MRVAA	Mississippi River Valley Alluvial Aquifer
NRCS	USDA Natural Resources Conservation Service
OFWS	On-Farm Water Storage
RCPP	Regional Conservation Partnership Program
SMS	Soil Moisture Sensors
TWS	Tailwater Recovery System
USACE	U.S. Army Corps of Engineers
USDA	U.S. Department of Agriculture
WRP	Wetlands Reserve Program

Appendix A. Conservation Programs in Mississippi

The Delta region in Mississippi has a diversity of habitats in which urban, agricultural, and wildlife habitat landscapes coexist. Growers and landowners in the area actively seek guidance in attaining land and water resource stewardship. Hunting and fishing clubs are among the highest priced memberships in the areas. Agricultural land and water projects often include wildlife habitat enhancing features that further increase the value of such clubs. The National Wildlife Federation work in the area focuses on protecting and restoring healthy rivers and estuaries, conserving wetlands, springs, and aquifers, and protecting wildlife habitats. These goals overlap with producer interests and are prominent features in existing conservation programs in the area. Table A1 contains a list and brief description of the conservation programs in which survey respondents claimed to participate.

Table A1. List and brief description of the conservation programs in which survey respondents claimed to participate.

Program	Sponsor	Description
AWEP	NRCS	Agricultural Water Enhancement Program is a conservation initiative that provides financial and technical assistance to agricultural producers to implement agricultural water enhancement activities on agricultural land for the purposes of conserving surface and groundwater and improving water quality.
CRP	FSA	Conservation Reserve Program is a land conservation program to remove environmentally sensitive land from agricultural production and plant species that will improve environmental health and quality.
CSP	NRCS	Conservation Stewardship Program participants earn performance-based CSP payments: higher payment to higher performance.
Delta F.A.R.M.	Public-private partnership	Farmers Advocating Resource Management is an association of growers and landowners that strive to implement recognized agricultural practices, which will conserve, restore, and enhance the environment.
EQIP	NRCS	Environmental Quality Incentives Program provides incentive payments and cost-sharing for conservation practice adoption.
RCPP	NRCS	Regional Conservation Partnership Program promotes coordination of NRCS conservation activities with partners that offer value-added contributions to expand their collective ability to address on-farm, watershed, and regional natural resource concerns.
Rice stewardship	Public-private partnerships	USA Rice-Ducks Unlimited Rice Stewardship Partnership provides financial assistance for conserving water and wildlife in ricelands.
Soil erosion	Unspecified	Unspecified
Unspecified	USACE	The U.S. Army Corps of Engineers may enroll farmers adjacent to their projects as part of environmental or habitat enhancement features.
WRP	NRCS	Wetlands Reserve Program offers landowners the opportunity to protect, restore, and enhance wetlands on their property.

References

1. Christy, D. (Yazoo Mississippi Delta Joint Water Management District, Mississippi State University, Starkville, MS, USA). Personal communication, 2014.
2. Kebede, H.; Fisher, D.K.; Sui, R.; Reddy, K.N. Irrigation methods and scheduling in the Delta region of Mississippi: Current status and strategies to improve irrigation efficiency. *Am. J. Plant Sci.* **2014**, *5*, 2917. [CrossRef]
3. Barlow, J.R.; Clark, B.R. *Simulation of Water-Use Conservation Scenarios for the Mississippi Delta Using an Existing Regional Groundwater Flow Model*; U.S. Geological Survey Scientific Investigations Report 2011–5019; US Department of the Interior, US Geological Survey: Washington, DC, USA, 2011; 14p.
4. Nian, Y.; Huang, Q.; Kovacs, K.F.; Henry, C.; Krutz, J. Water Management Practices: Use Patterns, Related Factors and Correlations with Irrigated Acres. *Water Resour. Res.* **2020**, *56*, e2019WR025360. [CrossRef]
5. Balogh, P.; Bujdos, Á.; Czibere, I.; Fodor, L.; Gabnai, Z.; Kovách, I.; Nagy, J.; Bai, A. Main Motivational Factors of Farmers Adopting Precision Farming in Hungary. *Agronomy* **2020**, *10*, 610. [CrossRef]
6. Persons, T.M.; Morris, S.D. Irrigated Agriculture: Technologies, Practices, and Implications for Water Scarcity. In *Report to Congressional Requesters, Technology Assessment*; GAO-20-128SP; U.S. Government Accountability Office: Washington, DC, USA, 2019.
7. Bryant, C.; Krutz, L.; Falconer, L.; Irby, J.; Henry, C.; Pringle, H.; Henry, M.; Roach, D.; Pickelmann, D.; Atwill, R.; et al. Irrigation water management practices that reduce water requirements for Mid-South furrow-irrigated soybean. *Crop Forage Turfgrass Manag.* **2017**, *3*, 1–7. [CrossRef]

8. Henry, W.B.; Krutz, L.J. Water in agriculture: Improving corn production practices to minimize climate risk and optimize profitability. *Curr. Clim. Chang. Rep.* **2016**, *2*, 49–54. [CrossRef]
9. Wood, C.; Krutz, L.; Falconer, L.; Pringle, H.; Henry, B.; Irby, T.; Orlowski, J.; Bryant, C.; Boykin, D.; Atwill, R.; et al. Surge irrigation reduces irrigation requirements for soybean on smectitic clay-textured soils. *Crop Forage Turfgrass Manag.* **2017**, *3*, 1–6. [CrossRef]
10. Spencer, G.; Krutz, L.; Falconer, L.; Henry, W.; Henry, C.; Larson, E.; Pringle, H.; Bryant, C.; Atwill, R. Irrigation Water Management Technologies for Furrow-Irrigated Corn that Decrease Water Use and Improve Yield and On-Farm Profitability. *Crop Forage Turfgrass Manag.* **2019**, *5*, 1–8. [CrossRef]
11. Carey, J.M.; Zilberman, D. A model of investment under uncertainty: modern irrigation technology and emerging markets in water. *Am. J. Agric. Econ.* **2002**, *84*, 171–183. [CrossRef]
12. Yazoo Mississippi Delta Joint Water Management District, Permit Application. Available online: https://www.ymd.org/permitting (accessed on 15 June 2020).
13. Krutz, L. *Utilizing Moisture Sensors to Increase Irrigation Efficiency*; Mississippi State University Extension Service: Starkville, MS, USA, 2016.
14. Krutz, L. *Pipe Planner: The Foundation Water Management Practice for Furrow Irrigated Systems*; Mississippi State University Extension Service: Starkville, MS, USA, 2016.
15. Roach, D. *Flow Meters Available at County Extension Offices*; Mississippi State University Extension Service: Starkville, MS, USA, 2018.
16. Krutz, L. *Surge Valves Increase Application Efficiency*; Mississippi State University Extension Service: Starkville, MS, USA, 2016.
17. Tagert, M.L.; Paz, J.; Reginelli, D. *On-Farm Water Storage Systems and Surface Water for Irrigation*; Mississippi State University Extension Service: Starkville, MS, USA, 2018.
18. Zotarelli, L.; Fraisse, C.; Dourte, D. *Agricultural Management Options for Climate Variability and Change: Microirrigation*; EDIS UF/IFAS (HS1203); University of Florida: Gainesville, FL, USA, 2015.
19. Coblentz, B.A. *Pivot Irrigation, Not Furrows, Is Most Economical for Delta*; Mississippi State University Extension Service: Starkville, MS, USA, 2014.
20. Burdine, B. *Cover Crops: Benefits and Limitations*; Mississippi State University Extension Service: Starkville, MS, USA, 2019.
21. Coblentz, B.A. *Cover Crops Present Challenges, Advantages*; Mississippi State University Extension Service: Starkville, MS, USA, 2018.
22. Prokopy, L.S.; Floress, K.; Arbuckle, J.G.; Church, S.P.; Eanes, F.; Gao, Y.; Gramig, B.M.; Ranjan, P.; Singh, A.S. Adoption of agricultural conservation practices in the United States: Evidence from 35 years of quantitative literature. *J. Soil Water Conserv.* **2019**, *74*, 520–534. [CrossRef]
23. Edwards, J.F. *Crop Irrigation Survey*; Final Report; Mississippi State University, Social Science Research Center, Survey Research Laboratory: Starkville, MS, USA, 2016; unpublished.
24. Rogers, D.H.; Alam, M. *Comparing Irrigation Energy Costs*; Agricultural Experiment Station and Cooperative Extension Service, Kansas State University: Manhattan, KS, USA, 2006; MF-2360.
25. McGuire, V.L.; Seanor, R.C.; Asquith, W.H.; Kress, W.H.; Strauch, K.R. *Potentiometric surface of the Mississippi River Valley Alluvial Aquifer, Spring 2016: U.S. Geological Survey Scientific Investigations Map 3439*; Technical Report; US Geological Survey: Washington, DC, USA, 2019; 14p, 5 sheets. [CrossRef]
26. USDA-NRCS. Soil Survey Staff, Natural Resources Conservation Service, United States Department of Agriculture. In *Soil Survey Geographic (SSURGO) Database for Sunflower County, Mississippi*; USDA-NRCS: Washington, DC, USA, 2010.
27. Maddala, G.S. *Limited-Dependent and Qualitative Variables in Econometrics*; Cambridge University Press: Cambridge, UK, 1986; Volume 3.
28. Pfeiffer, L.; Lin, C.Y.C. Does efficient irrigation technology lead to reduced groundwater extraction? Empirical evidence. *J. Environ. Econ. Manag.* **2014**, *67*, 189–208.
29. Fenichel, E.P.; Abbott, J.K.; Bayham, J.; Boone, W.; Haacker, E.M.; Pfeiffer, L. Measuring the value of groundwater and other forms of natural capital. *Proc. Nat. Acad. Sci. USA* **2016**, *113*, 2382–2387. [CrossRef] [PubMed]
30. Hassanli, A.M.; Ebrahimizadeh, M.A.; Beecham, S. The effects of irrigation methods with effluent and irrigation scheduling on water use efficiency and corn yields in an arid region. *Agric. Water Manag.* **2009**, *96*, 93–99. [CrossRef]

31. USDA. *Irrigation: Results from the 2013 Farm and Ranch Irrigation Survey;* Census Agricultyre Highlights ACH12-16; USDA: Washington, DC, USA, 2014.
32. Drysdale, K.M.; Hendricks, N.P. Adaptation to an irrigation water restriction imposed through local governance. *J. Environ. Econ. Manag.* **2018**, *91*, 150–165.
33. Smith, S.M.; Andersson, K.; Cody, K.C.; Cox, M.; Ficklin, D. Responding to a groundwater crisis: The effects of self-imposed economic incentives. *J. Assoc. Environ. Resour. Econ.* **2017**, *4*, 985–1023. [CrossRef]
34. USDA-NRCS. *The PROTRACTS Database, Mississippi;* USDA-NRCS: Washington, DC, USA, 2019.

10

Assessing Heat Management Practices in High Tunnels to Improve the Production of Romaine Lettuce

Muzi Zheng [1,*], Brian Leib [1], David Butler [2], Wesley Wright [1], Paul Ayers [1], Douglas Hayes [1] and Amir Haghverdi [3]

[1] Department of Biosystems Engineering and Soil Science, University of Tennessee, 2506 E.J. Chapman Drive, Knoxville, TN 37996-4531, USA; bleib@utk.edu (B.L.); wright1@utk.edu (W.W. & P.A.); dhayes1@utk.edu (D.H.)
[2] Department of Plant Sciences, University of Tennessee, 2431 Joe Johnson Dr., Knoxville, TN 37996-4531, USA; dbutler@utk.edu
[3] Department of Environmental Sciences, University of California, Riverside, 900 University Avenue, Riverside, CA 92521, USA; amirh@ucr.edu
* Correspondence: mzheng3@vols.utk.edu

Abstract: A three-year experiment evaluated the beneficial effects of independent and combined practices on thermal conditions inside high tunnels (HTs), and further investigated the temperature impacts on lettuce production. Specific practices included mulching (polyethylene and biodegradable plastic films, and vegetative), row covers, cover crops, and irrigation with collected rainwater or city water. The study conducted in eastern Tennessee was a randomized complete block split-split plot design (RCBD) with three HTs used as replicates to determine fall lettuce weight (g/plant) and lettuce survival (#/plot), and the changes in soil and air temperature. The black and clear plastic mulches worked best for increasing plant weight, but when compared to the bare ground, the higher soil temperature from the plastics may have caused a significant reduction in lettuce plants per plot. Moreover, the biodegradable mulch did not generate as much soil warming as black polyethylene, yet total lettuce marketable yield was statistically similar to that for the latter mulch treatment; while the white spunbond reduced plant weight when compared with black plastic. Also, row covers provided an increased nighttime air temperature that increased soil temperature, hence significantly increased lettuce production. Cover crops reduced lettuce yield, but increased soil temperatures. Additionally, irrigation using city water warmed the soil and provided more nutrients for increased lettuce production over rainwater irrigation.

Keywords: cover crop; lettuce production; irrigation; mulch; row cover; temperature variations

1. Introduction

High tunnels (HTs) are simple, plastic covered, greenhouse-like structures that do not utilize heaters or ventilation fans. Even without heaters or fans, HTs allow producers to lengthen the growing season and protect plants from extreme weather conditions (e.g., hail, frost, or strong wind), hence increasing the profitability and sustainability of organic farms [1]. Compared to crop production in open fields, other advantages of HTs highlighted by Lamont et al. [2] also include: (1) preventing rainfall from wetting and splashing soil onto fruit and foliage, thus creating cleaner products with less disease; (2) increasing water-use efficiency; (3) improving crop environmental conditions; and (4) reducing soil compaction and insect invasion. Additionally, the benefits of less electricity consumption, lower startup costs, and fewer maintenance efforts make HT systems more competitive than standard greenhouses [3].

High-valued crops, such as lettuce, tomatoes, and other leafy greens, produced inside HT systems can also compensate for the increased cost of HT systems when compared to open field production. Studies also show that the microclimates inside HTs can be modified so that the growing season can be lengthen from 1 to 4 weeks in spring and 2 to 8 weeks in fall [4]. Therefore, HTs are currently being adopted by small- and mid-scale producers in order to take advantages of market seasonality with higher profits [5,6].

However, there are several limitations for the sustainability of HT systems. First, since HTs block rainfall, ground/well or treated water needs to be used for crop irrigation. As well, lettuce is considered moderately sensitive to irrigation salinity, since the salts in soil and water might cause the retardation of crop growth, reduction of lettuce head formation, or even the necrotic lesions on the leaf margins [7]. To combat this sensitivity, rainwater can be collected and reserved in black polyethylene tanks, and then used inside HTs through drip irrigation. A rain harvesting system (RHS) is considered beneficial in regions where treated water is inadequate and costly, and helps in reducing local flooding in some low-lying areas. This study investigates additional benefits of RHS when black polyethylene tanks are used; black tanks can capture solar energy and warm up enclosed rainwater, and this warmer water can warm the soil during drip irrigation.

Secondly, the poor insulation properties of the plastic-covered HT structure do not provide significant heat retention at night under cold climates. Without a favorable growth microclimate, crop performances inside HTs might be inhibited or even terminated. Accordingly, surface mulch and row covers are practices that could help retain heat while providing other advantages, such as reducing weed competition, alleviating soil crusting, modifying the radiation budget of the soil surface around the plants, and conserving water by inhibiting evaporation from the soil surface [8]. A two-year lettuce experiment in the open field found that colored polyethylene mulches (clear, white, and black) significantly increased the overall lettuce production by 33%, 40%, and 39%, respectively, compared to bare ground [9]. Additionally, for other high-value crops, a study showed that black polyethylene mulch combined with an HT system significantly increased overall pepper production in regard to crop height and leaf area, while in the same system, clear plastic mulch significantly raised ground temperature, thus decreasing the number of days to the first flower when compared with black polyethylene mulch and bare ground [10]. As well, biodegradable mulch can eliminate plastic disposal issues, including the labor cost of disposal [11,12]. There is limited research on how biodegradable mulches within HT systems affect soil temperature and crop productivity. But in open field tests, outdoor use of biodegradable mulches had no significant effect on overall pepper yield and quality in Australia; although, it raised soil temperature slightly compared with paper mulch [13]. A study from Moreno [14] also indicated that biodegradable mulches did not significantly increase tomato production in Central Spain, but Miles et al. [15] observed improved tomato production using biodegradable mulches when compared with bare ground in northwestern Washington, USA. Since lettuce production could be limited from weed competition and seedling establishment problems, appropriate polyethylene and biodegradable mulches might reduce the costs of weed management, while minimizing root damage, also retaining favorable soil temperatures for lettuce growth and development [16]. Similar to surface mulches, floating row covers are considered as additional thermal protection for crops from wind and frost inside a HT. There have been many studies showing that the combination of mulching and row covers under field conditions significantly increased soil and air temperatures for favorable crop growth conditions [17]. However, some studies indicated that although HT air temperature was slightly higher than outside, lettuce development was still slow, since the air temperature around the plant's full height was not optimal [18]. Therefore, more investigation on the impacts of row cover and colored plastic/biodegradable mulches under HTs are needed to maximize heat retention.

The last limitation is related to the decomposition of soil organic matter due to continuous cropping in HTs, which may negatively influence nutrient/water holding capacity in soil. The common solution of adding organic supplements to the soil requires energy at each step as long as the supplements need to be collected and delivered to HTs. Accordingly, an effective practice is to pre-plant a legume cover crop which has the ability to fix nitrogen, supply crop nutrients and improve soil structure in HT systems. This study investigated incorporating cover crop residues into the ground and leaving the residues on the soil surface (vegetative mulch). Teasdale and Daughtry [19] found that cover-crops have positive effects on weed control and Liebman and Davis [20] observed reduced maximum-soil-temperature under vegetative mulches in the daytime, along with the benefit of increased summer production. Even though cover crops provide many benefits, there is limited research on the effects of cover crop residues incorporated into the ground in combination with polyethylene mulches covering the soil surface, and their effects on thermal protection in HT systems.

Overall, this study aims to provide recommendations to small- and mid-scale producers on how to improve thermal protection with low input sustainable practices to improve lettuce production in HTs. Rain water, surface mulch, row covers, and cover crops have many benefits, but this study will focus on the benefits to heat management. Zheng et al. [21]'s study showed how thermal energy conservation benefited spring pepper production, while this study aims to assess the yield performance of fall Romaine lettuce using the same thermal energy conservation practices.

2. Material and Methods

2.1. The Gothic Type High Tunnel

Three experimental high tunnels were used in the experiment, that were orientated N–S, and located at the University of Tennessee's Organic Crop Unit in Knoxville, TN in the United States (latitude 35.88° N, longitude 83.93° W, and elevation 252 m). These HTs have peaked roofs, vertical side walls and the dimension of the structures are 9 m in width and 15 m in length (Figure 1). Each end door covers an opening which is 2.45 m tall and 3.35 m wide, and side curtains run the entire length of the HT. The end doors were opened almost every day to provide natural ventilation while the side-curtains were only opened during warm weather when more air was need for proper ventilation. In cold periods, doors and side curtains were closed at night to preserve thermal energy, and only the end doors were opened in the daytime to reduce accumulated heat and humidity inside HTs. Rainwater was collected and stored in black polyethylene tanks and was delivered using gravity pressure and solar power. Rainwater storage tanks in the first HT were elevated with cinder blocks to provide gravity pressure for irrigation with drip tubing, while the other two HTs utilized a solar powered pump to deliver rainwater via drip tapes with 10 psi of pressure [21].

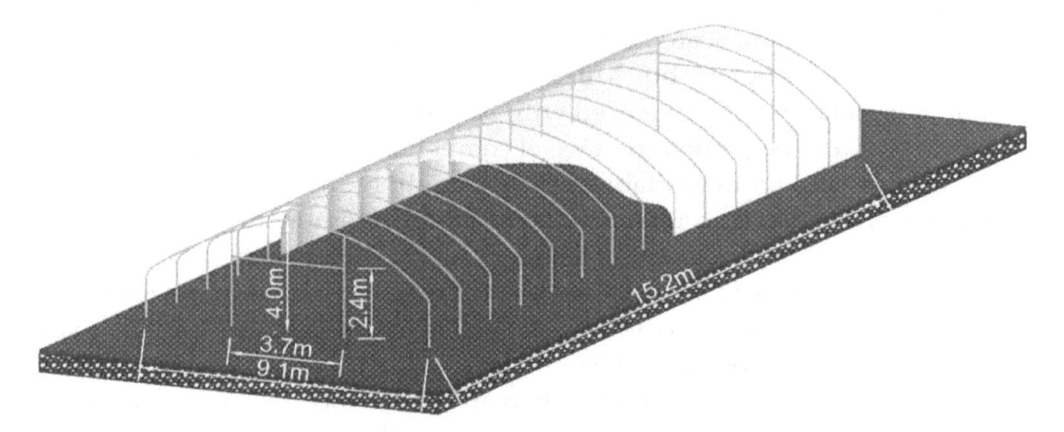

Figure 1. Isometric view of a high tunnel with the covering cut away for clarity) [21].

Romaine Lettuce ('coastal star') was grown during the fall seasons of 2011, 2013, and 2014. Specifically, five different surface mulches with a bare soil control, and two sources of irrigation water were applied in fall 2011. Thus, there were six beds in each HT, and these six different treatments were randomly arranged in different beds. Figure 2a shows that mulch treatment was laid out in six rows, including: (1) white biodegradable spunbond (spun), (2) vegetative mulch from a leguminous cover crop (veg), (3) bare ground (bare), (4) black polyethylene film (black), (5) biodegradable brown paper (paper), and (6) black biodegradable biobag (biobag). The dimension of each bed was 12.19 m in length and 0.91 m in width with double rows of lettuce planted in each bed. Double planting rows were placed 0.3 m apart and 0.46 m apart between lettuce transplants. There were 48 plants per row and rows were divided into four plots. Lettuce was transplanted on 30 September 2011 in half of each house (snapdragon flowers were added to the other half of the HT for a different research project), and then harvested on 7 December 2011. Next, in the fall of 2013, there were five surface mulches and a bare control, with or without a row cover. Figure 2b shows that the mulch distribution was: (1) black polyethylene film (black), (2) vegetative mulch from a cover crop (veg), (3) bare ground (bare), (4) black polyethylene film with a cover crop (blackCC), (5) clear polyethylene film with a cover crop (clearCC), and (6) clear polyethylene film (clear). Once inside temperature fell below approximately 7 °C, row covers were applied to half of the crop during the night. Lettuce was transplanted on 2 October 2013 and harvested on 4 December 2013. Finally, in the fall of 2014, a total of 12 treatments were used, including two water resources (rainwater and city water), two surface mulches (black and blackCC) and a bare ground control, with or without a row cover (Figure 2c). Lettuce was transplanted in each half of the HT on 29 September 2014 and harvested on 5 December 2014.

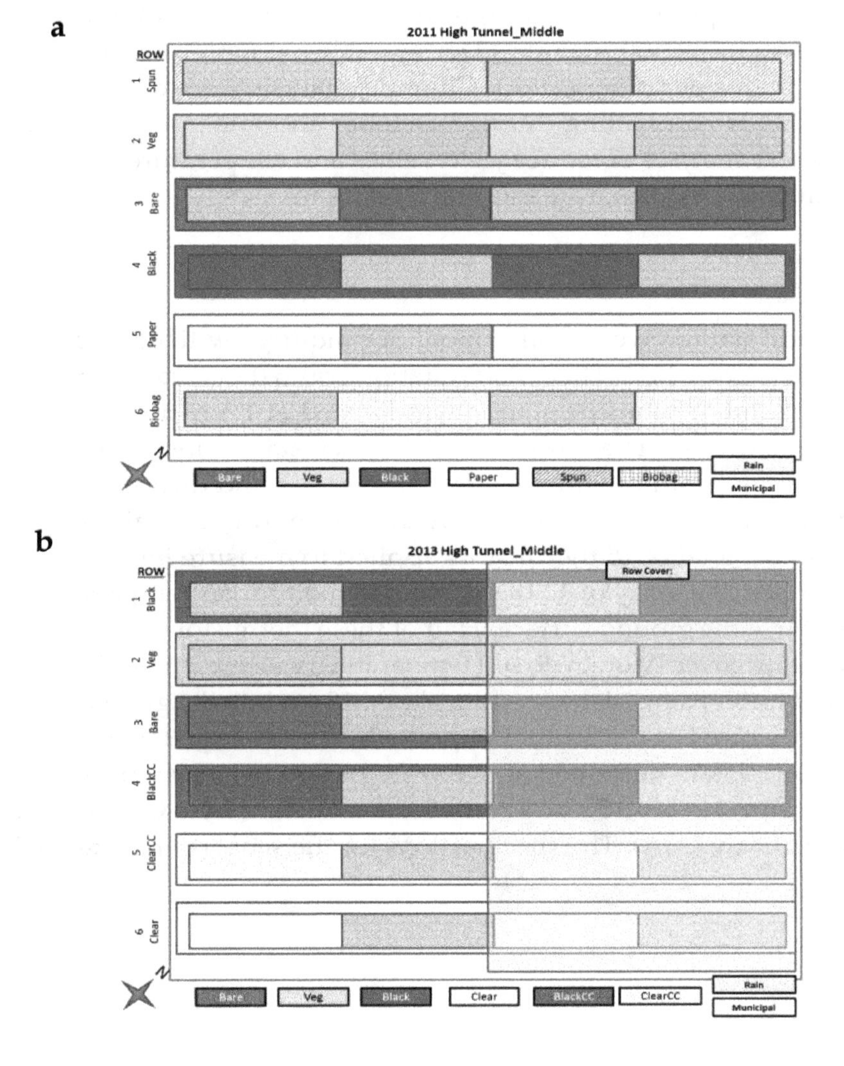

Figure 2. *Cont.*

c

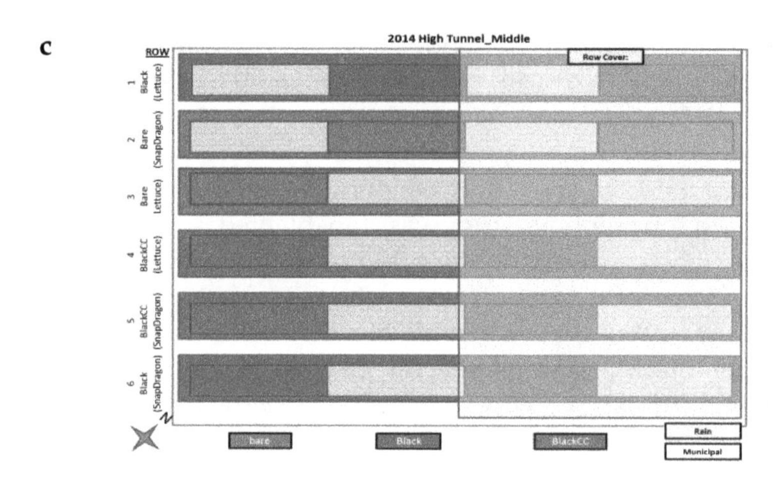

Figure 2. The lettuce production of high tunnels (HTs) in 2011, 2013, and 2014 followed a split-split-plot design.

To determine statistical differences, the experiment was a randomized complete block design (RCBD), based on a split-split-plot sub-design for lettuce production and temperature variations. In addition, soil at the experimental site was a Dewey silt loam. Lettuce yield was determined by plant weight (g/plant) and plants per plot (#/plot). Climatic analysis were divided into two time periods, including early fall (EF) from planting through October and late fall (LF) from November to early December, and statistical analyses of lettuce yield values and temperatures variations were performed by SAS statistical software (SAS Institute Inc., Cary, NC, USA).

2.2. *Climatic Monitoring and Instrumentation*

Two meteorological stations were used to monitor conditions with one placed outside and another placed inside the high tunnels [21]. The outside station was 6.0 m away from the middle HT and set to 4.0 m above the ground. The inside station was located at 1.5 m above the ground at the center of the middle experimental HT. The sensors used for each station included a pyranometer for solar radiation, a cup anemometer and wind vane for wind speed and direction, along with a capacitive chip and platinum resistance thermometer, for relative humidity and air temperature, respectively. Additionally, there was a total of 15 thermistors applied to measure inside air temperature at 1.2 m above the ground surface. An array of 12 thermistors to monitor air temperatures at the canopy level were located 30 cm above the ground surface. Half of these thermistors were under the row cover and half were outside the row cover. Moreover, soil temperature was measured in each combination for the mulch, row cover, and water using 24 thermistors placed 10 cm into the ground. Additionally, the water temperatures were measured at the bottom of polyethylene storage tanks and in the supply line of city water. All the thermistors described above were potted in solar reflecting white epoxy and all the sensors were measured every 15 s and recorded every hour by a CR1000 datalogger (Campbell Scientific, Inc.: Logan, Utah, USA). The specifications for the sensors are listed in Table 1.

Table 1. Environmental parameters measured during the experiments [21].

Measured Parameters	Sensors	Range and Accuracy	Location
Solar radiation Q_{solar} (Wm^{-2})	LI200X Pyranometer (LI-COR.INC)	0 to 3000 Wm^{-2} with ±5%	Inside and outside weather station
Air temperature T_i (K) Relative humidity H_R (%)	HMP60 Probe (Vaisala.INC)	−40 to 60 °C with ±0.6% ±3% RH over 0 to 90%, ±5% RH over 90 to 100%	Inside and outside weather station
Wind speed U (ms^{-1}) Wind direction (°)	R.M.Young Wind Sentry (Campbell Sci. Inc)	0 to 50ms^{-1} with ±1% 0 to 360° with ±5°	Outside weather station
Air temperature (°C)	Thermistor (PS104J2) †	±0.1 °C with 0.026 °C	An array of 15 thermistors all over the inside of middle HT located 1.5 m above the ground
Canopy temperature (°C)	Thermistor (PS104J2) †	±0.1 °C with 0.026 °C	12 thermistors located 30 cm above the ground between crop rows
Soil temperature (°C)	Thermistor (PS104J2) †	±0.1 °C with 0.026 °C	24 thermistors located 10.16 cm below the ground
Water temperature (°C)	Thermistor (PS104J2) †	±0.1 °C with 0.026 °C	Polyethylene storage tank bottom for rainwater and supply line of city water

† means the thermistors are made in our lab, which are non-linear sensors following a polynomial response to temperature that is defined by the Steinhart–Hart equation.

3. Results and Discussion

The HT affected inside microclimate in terms of solar radiation, air and soil temperature, and relative humidity and wind speed. There were 1380 observations of weather data measured over three falls of 2011, 2013, and 2014. Generally, 9% of total solar radiation was reflected by the HT's polyethylene covering (dataset in 2011 was disregarded due to sensor failure). During the EF (early fall) with the end-doors and side-curtains fully opened, air temperatures inside the HT were consistently higher by 0.5–6.0 °C compared to outside; while in LF (late fall) with the side curtains rolled down, the inside maximum air temperature during the day was from 1 to 11 °C higher when compared to the outdoor conditions. Moreover, outside and inside thermistor readings indicated that night time air temperatures inside the HT were 0.6–2.0 °C higher than outdoor temperatures, and the average nighttime soil temperature inside the HT was 1–3 °C higher than outside over the span of the three falls [21].

Statistically, mulch treatments were considered the main contributing factor to the lettuce productivity in the falls of 2011, 2013, and 2014 ($p < 0.001$) (Figure 3). In 2013 and 2014, black consistently produced a higher yield than bare, with an increase of 71% (2013) and 37% (2014) in g/plant. However, no significant difference was found in lettuce yield between using black and bare plots in 2011. Clear mulch provided the maximum production in g/plant in 2013 but with no significant difference compared to the black mulch treatment. In addition, Table 2 shows that the increased crop yield using black was related to its soil-warming ability when compared with bare. Although some temperature values were missing due to sensor failures in LF 2011, black had a positive impact on soil temperature, which was constantly higher than the bare treatment by 0.8–1.3 °C (2011), 0.6–0.9 °C (2013), and 0.9–1.3 °C (2014). The soil temperature rise under black mulch can promote faster lettuce growth and development, although the lettuce survival (#/plot) with black was reduced approximately 20% when compared to bare in 2013. The soil-warming ability underneath clear was even greater than that under black, by as much as 1.0 °C, but the overall lettuce yield did not improve significantly in spring 2013. After transplanting in EF, soil temperatures were warmer by 1.4 °C (day) and 2.1 °C (night) for clear, and 0.6 °C and 0.9 °C for black when compared to the bare. The higher measured soil temperatures under the mulches do not adequately reflect, the extremely hot air emitted from the transplant holes or the high temperature of the mulch itself before the crop canopy can shade the plastic mulches. These higher temperatures seem beneficial to lettuce growth (lettuce weights per plant) but they may not be beneficial to lettuce survival at transplant, since the plants per plot were significantly reduced in clear (10%) and black (20%) plots compared with the bare.

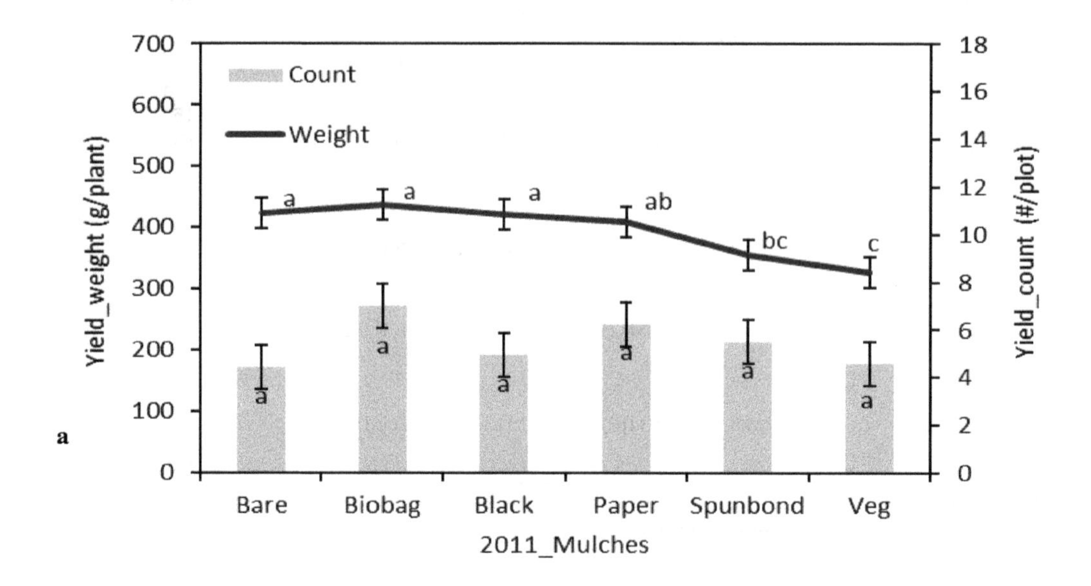

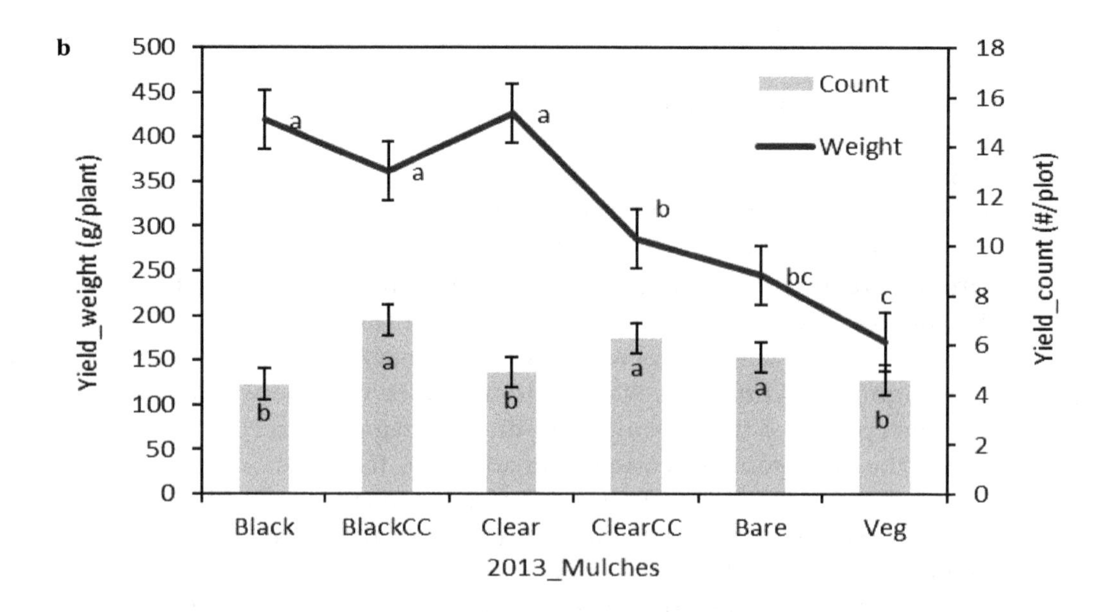

Figure 3. *Cont.*

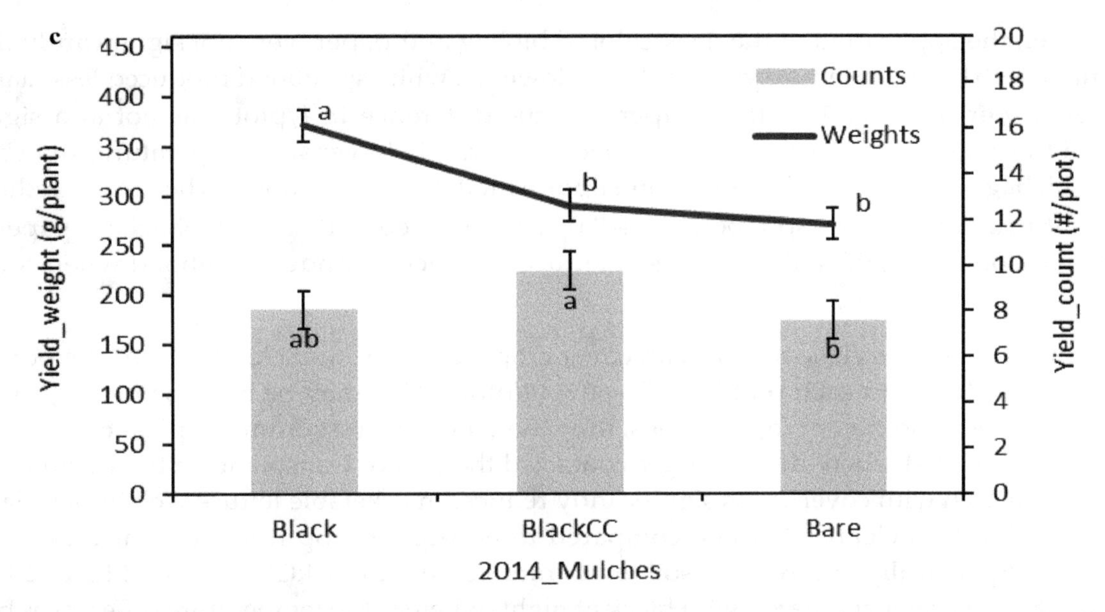

Figure 3. Marketable lettuce weight per plant and lettuce numbers per plot under the effects of different mulch over the falls of the years 2011 (**a**), 2013 (**b**), and 2014 (**c**) in HTs. Values followed by the same letter are not significant different according to Fisher's LSD (p = 0.1).

Table 2. Marketable lettuce weight per plant and lettuce numbers per plot, and average soil temperatures at 10 cm depth under different mulch treatments during the day and night in the early and late falls of 2011, 2013, and 2014.

Year	Mulch	Lettuce Yield		Soil Temp_Day		Soil Temp_Night	
		Weight	Count	EF†	LF†	EF†	LF†
		(g/plant)	(#/plot)	(°C)		(°C)	
2011	Black	421.6 a	7.7 a	18.22 a	-	14.42 a	15.39 a
	Biobag	436.9 a	8.3 a	17.12 b	-	13.26 c	14.17 c
	paper	409.1 ab	8.7 a	16.68 bc	-	-	-
	Spunbond	355.7 bc	9.3 a	16.21 c	-	13.01 d	13.99 d
	Bare	424.2 a	9.2 a	17.06 b	-	13.59 b	14.13 cd
	Veg	327.5 c	9.2 a	17.06 b	-	12.96 d	13.52 d
	Sig.level	*0.0311*	*0.73*	*0.0115*	-	*<0.0001*	*<0.0001*
2013	Black	419.4 a	4.4 b	19.97 c	14.12 b	19.74 c	14.31 e
	BlackCC	361.4 a	7.0 a	20.08 c	14.18 b	20.52 b	14.66 b
	Clear	425.8 a	4.9 b	20.80 b	14.20 b	20.86 b	14.66 b
	ClearCC	285.6 b	6.3 a	21.47 a	14.69 a	21.81 a	15.02 a
	Bare	244.9 bc	5.5 a	19.41 d	13.26 d	18.82 d	13.43 e
	Veg	170.4 c	4.6 b	19.22 e	13.01 d	18.36 d	13.28 e
	Sig.level	*<0.0001*	*<0.0001*	*<0.0001*	*<0.0001*	*<0.0001*	*<0.0001*
2014	Black	371.1 a	8.1 ab	18.38 a	13.07 a	19.48 a	13.86 a
	BlackCC	290.0 b	9.8 a	18.24 b	12.97 a	19.32 a	13.76 a
	Bare	271.2 b	7.6 b	17.30 c	12.12 b	18.16 b	12.86 b
	Sig.level	*<0.0001*	*0.1765*	*<0.0001*	*<0.0001*	*<0.0001*	*<0.0001*

* EF and LF were characterized by diurnal soil temperature in specific early and late spring. Temperature values are daily means from 10:00 a.m. to 16:00 p.m. and nightly means from 21:00 p.m. to 8:00 a.m. Values followed by the same letter are not significantly different according to Fisher's LSD (p = 0.1).

Moreover, the application of the dark colored biobag and paper were not significantly different when compared to black for lettuce yield in 2011. However, white spunbond produced less marketable lettuce weight per plant by 13% than paper, but the difference in #/plot was not at a significant level. All of the degradable mulch plots provided significantly lower soil temperatures than black by 1.0–1.2 °C (biobag), 1.4–2.0 °C (spunbond), and approximately 1.6 °C (paper). The greater reduction of soil temperatures produced by spunbond (1.4–2.0 °C) compared to those of biobag and paper mulch may be the reason for significantly decreased lettuce production under spunbond when compared to black.

In additional, polyethylene mulch with cover crops (blackCC and clearCC) had the potential to save more lettuce heads in each plot by 21%–60% (#/plot). This may be because more air was able to penetrate through the cover crop residues, thus the thermal stress from the plastic mulch surface was able to be mitigated when the film edges contacted the lettuce transplants in the warm EF period. But the plastic mulch with cover crops significantly reduced marketable lettuce weight per plant (18% in blackCC and 33% in clearCC), when compared to non-cover crop treatments as an average for 2013 and 2014. Specifically, the average soil temperatures under blackCC in EF and LF of 2013 were around 0.3–0.8 °C warmer compared with black at night, whereas the temperature differences between clearCC and clear were 0.5–0.7 °C and 0.3–0.9 °C warmer during the day and night, respectively. Thus, although the cover crops had a higher soil-warming ability and saved more lettuce in each plot, limited nutrient availability caused by immobilization of nitrogen during the decomposition of cover crop residues may have inhibited the growth of each lettuce head, as the incorporated cover crop and residual soil nutrients served as the only fertilizer source. In comparison, veg always significantly provided the lowest lettuce production. It appears that the vegetative mulch shielded the transfer of solar energy to the soil and cooled the soil to a point where lettuce development was detrimentally delayed. Results confirmed that veg had a significant decrease of soil temperature (0.6 °C in 2011 and 0.3 °C in 2013) than that in bare. Therefore, veg treatment with the coolest soil temperature was working against a goal of making better use of solar energy to extend the growing season.

For the LF, additional thermal protection was provided by row covers for lettuce growth (Figure 4). Row cover played a significant role in increasing g/plant for 2014 ($p < 0.0001$). The lettuce weight per plant utilizing a row cover was improved approximately 36% as an average of all treatments (black, blackCC, and bare) when compared to all treatments without row covers. Moreover, positive interactions between the applications of mulches and row covers were found on yield ($p < 0.1$) in fall 2013. The black mulch with row cover provided the largest lettuce production and the row covers increased lettuce weight by 28% over black mulch alone. While some of the lower yielding plots had greater lettuce weight when a row cover was applied, bare and veg yielded significantly more lettuce weight, approximately 45% and 73% more lettuce weight per plant respectively, under row cover. Moreover, blackCC, clear, and clearCC with a row cover had the potential to improve the total marketable production (g/plant) compared with those without using row covers in fall 2013, but the improvement was not to a significant level. In addition, the row cover can significantly increase canopy and soil temperature by 0.8–2.3 °C and 0.1–0.9 °C at night, respectively. Although soil temperature increases from row covers interacted differently with the mulch treatments—0.4 °C in black, 0.6 °C in blackCC, 0.1 °C in clear, 0.8 °C in clearCC, 0.7 °C in bare, and 0.2 °C in veg—the row covers had a positive effect on lettuce production in all plots.

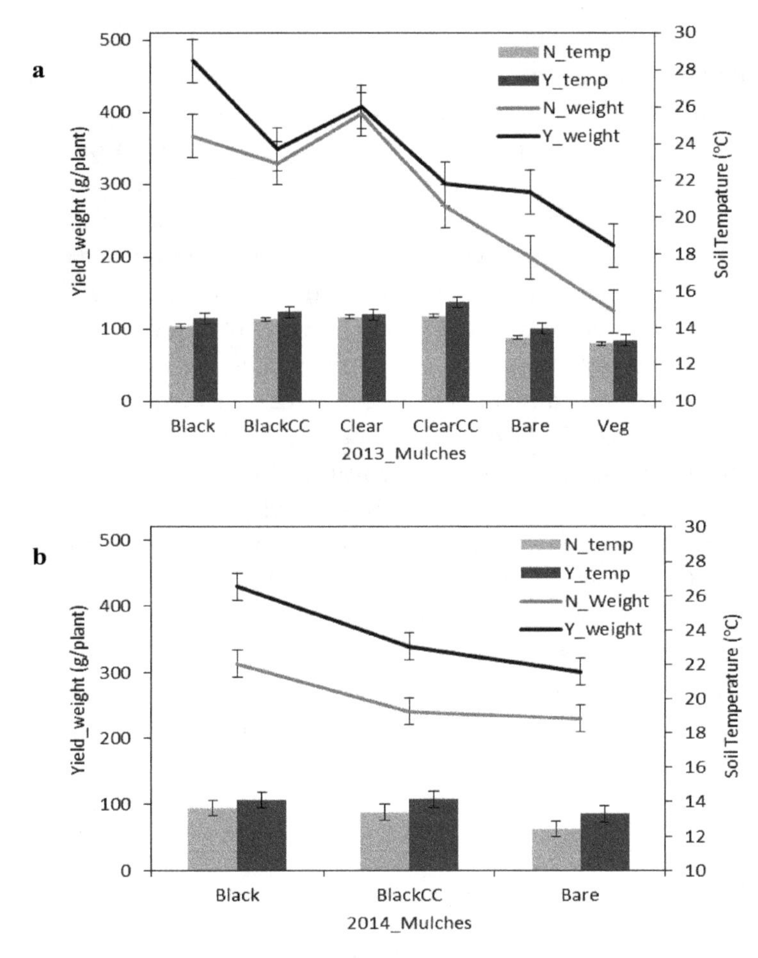

Figure 4. Interactions of mulch and row cover on lettuce yield (weight/plant), and soil temperature (°C) inside high tunnels during the late fall (LF) of 2013 and 2014. *'Y'* represents the mulch with a row cover and *'N'* means the mulch without a row cover.

Finally, lettuce yield was improved by using city water compared to rainwater applied with drip irrigation (Figure 5). As an average of all mulches, city water potentially increased the total lettuce weight per plant by 16% in 2011, 12% in 2013, and 8% in 2014, and also increased total lettuce numbers per plot by 13% in 2013 (Table 3). Although the numeric values of lettuce number per plot were potentially higher by using city water than rainwater, the differences were not at a significant level. In 2011, the biobag mulch and bare irrigated using city water significantly improved lettuce weights by 23% and 36%, respectively, compared to rainwater ($p < 0.0001$), although in subsequent years, city water did not significantly increase yield in the bare soil treatment. In 2013, black and veg mulch yield was significantly increased when city water was used for irrigation by 36% and 20% in lettuce weight per plant, respectively; and in 2014, black mulch with city water also significantly improved the lettuce production approximately by 17% (g/plant) over rainwater. The differences of lettuce yield using two water types were assumed to be explained by the impacts of water temperature on soil temperature. Figure 6 shows that the rainwater temperature from the solar tank was consistently cooler than that in gravity tanks. Additionally, in the EF of 2011, 2013, and 2014, the average temperature of rainwater stored in the solar tank was significantly higher than the city water in all years by 2.4 °C; but during the LF periods, the rainwater from solar tanks was significantly cooler than the city water by 2.2 °C in 2011, 1.3 °C in 2013 and 0.8 °C in 2014. Although water temperatures changed inconsistently between the EF and LF, soil temperatures responded in a similar manner over different mulches. Table 3 shows that city water generally produced warmer soil temperatures than rainwater during the entire fall seasons, with several exceptions in 2011 (biobag, spunbond, and bare), but all these variations cannot change the fact of greater lettuce production in city water plots (23% for biobag, 13% for spunbond, and 35% for bare). In summary, the temperature of city water was cooler than rainwater in EF and warmer during LF. However, the overall effect was for city water to warm the soil on average during each time period. As early fall progresses into late fall, rain water would be expected to cool in relation to city water, because the above ground tanks are exposed to less daily solar radiation and longer cooler nights, while soil temperature around the city water pipes lags behind the above ground heat loss of the rain water tank; i.e., soil temperature decline lags air temperature decline in the fall. The comparison of water temperature data does follow this expected pattern. However, the contradiction between EF water temperature and soil temperature differences is less easily explained. Perhaps the rainwater temperature was not representative of the water delivered for irrigation due to the sensor's location in the tank and temperature gradients within the tank caused by tank surface heating and cooling. In addition, the temperature transition from warmer rainwater to warmer city water would not occur precisely at the interface of EF to LF. There was most likely an intermixing of more and less heat transfer from the irrigation sources to the soil, in combination with air temperature's effects on soil temperature. Overall. the response of different mulches to irrigation water type shows that lettuce yield was improved by applying city water when compared to rainwater, and warmer soil temperatures appear to be an important factor where city water was applied. Another reason for increased lettuce production using city water may be related to water quality. Periodic water samples were analyzed and revealed the pH of city water was consistently around 6.4, while the rain water changed over time, but mainly was around at 6.7 to 7. Additionally, city water had more nutrients than rainwater: around 11 to 16 times higher levels of K, Mg, and Ca which may help promote lettuce growth and development.

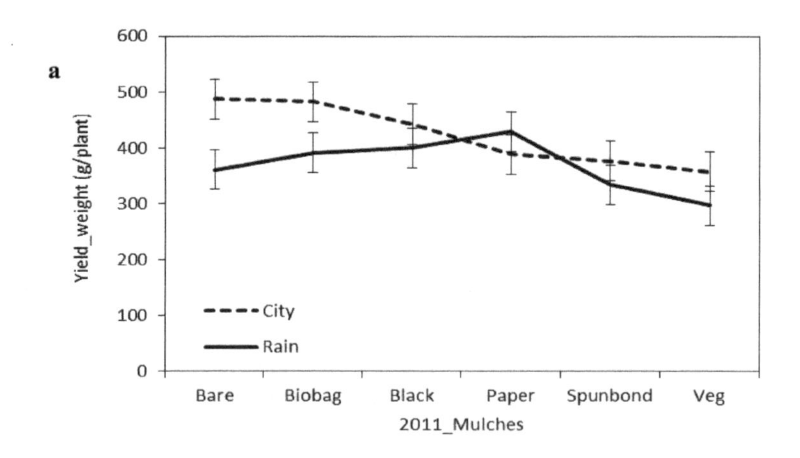

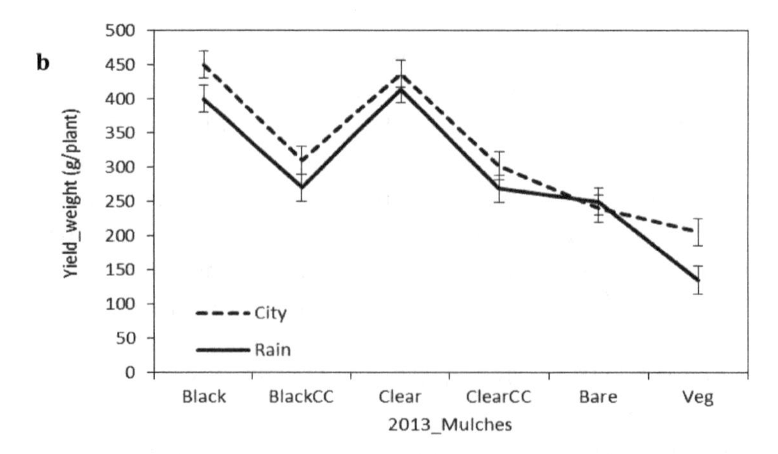

Figure 5. *Cont.*

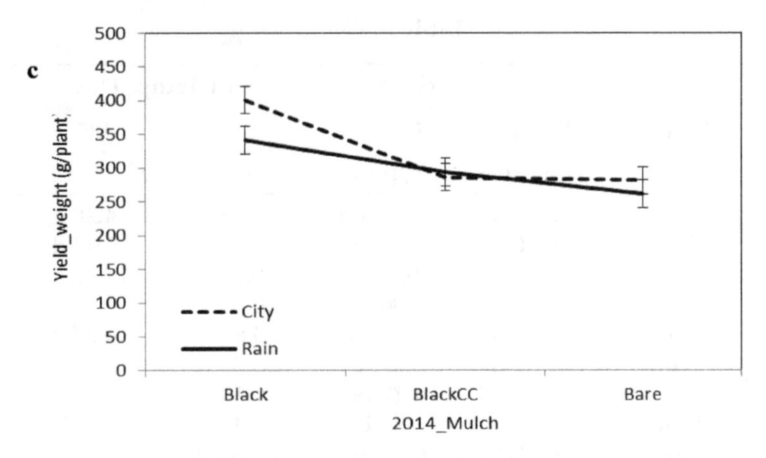

Figure 5. Marketable weight of lettuce (mean ± SE) under the effects of different irrigation water (rain and city) over three fall seasons of the years 2011 (**a**), 2013 (**b**), and 2014 (**c**) in high tunnels.

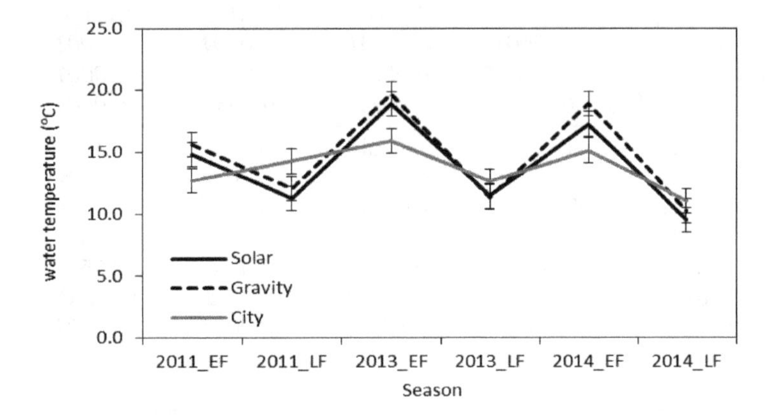

Figure 6. Fluctuations in average water temperature (mean ± SE) for three fall seasons of years 2011, 2013, and 2014. Gravity and Solar represent the rainwater inside the gravity and solar tanks, respectively, and City represents the municipal water from the underground pipes.

Table 3. Interactions of mulch and water on the lettuce yield and soil temperature inside high tunnels during all three falls of years 2011, 2013, and 2014.

Year	Mulch	Water	Lettuce Yield		Soil Temp_Day		Soil Temp_Night	
			Weight	Count	EF†	LF†	EF†	LF†
			(g/plant)	(#/plot)	(°C)		(°C)	
2011	Black	City	442.67 abc	10.00 a	18.37 a	-	14.44 a	15.45 a
		Rain	400.58 bcd	7.33 ab	18.07 ab	-	14.40 a	15.34 a
	Biobag	City	482.52 ab	8.67 ab	16.94 bcd	-	13.10 fg	13.83 e
		Rain	391.33 cd	8.00 ab	17.30 abc	-	13.42 def	14.52 bc
	Paper	City	388.67 cd	9.67 ab	16.80 bcd	-	13.48 cde	14.53 bc
		Rain	429.50 abc	9.00 ab	16.56 c	-	-	-
	Spunbond	City	377.33 cde	8.33 ab	15.90 d	-	12.87 g	13.77 e
		Rain	334.00 de	10.00 a	16.53 c	-	13.15 efg	14.21 d
	Bare	City	487.33 a	8.33 ab	17.11 bc	-	13.19 efg	13.79 e
		Rain	361.04 cde	7.0 b	17.01 bc	-	13.99 b	14.48 c
	Veg	City	357.69 cde	9.00 ab	17.34 abc	-	13.61 cd	14.72 b
		Rain	297.33 e	9.00 ab	16.78 bcd	-	13.78 bc	14.55 bc
	Mulch		*0.0311*	*0.73*	*0.0115*	-	*<0.0001*	*<0.0001*
	Water		*0.0158*	*0.38*	*0.8917*	-	*0.0018*	*<0.0001*
	Mulch × Water		*0.3177*	*0.59*	*0.5621*	-	*0.0678*	*<0.0001*

Table 3. *Cont.*

Year	Mulch	Water	Lettuce Yield		Soil Temp_Day		Soil Temp_Night	
			Weight	Count	EF†	LF†	EF†	LF†
			(g/plant)	(#/plot)	(°C)		(°C)	
2013	Black	City	450.0 a	5.17 bcd	20.15 c	14.28 c	20.04 cd	14.57 d
		Rain	400.0 a	3.67 d	19.79 d	13.97 d	19.43 ef	14.04 e
	BlackCC	City	310.3 abc	8.00 a	20.09 c	14.16 c	20.48 c	14.64 cd
		Rain	270.0 ab	6.00 cd	20.07 c	14.20 c	20.57 c	14.68 cd
	Clear	City	437.0 a	5.00 bcd	20.83 b	14.24 c	21.23 b	14.74 c
		Rain	414.7 a	4.8 bc	20.77 b	14.16 c	20.49 c	14.57 d
	ClearCC	City	302.2 bcd	6.17 bc	21.54 a	14.80 a	21.78 ab	15.10 a
		Rain	269.0 cd	6.33 bc	21.40 a	14.59 b	21.83 a	14.93 b
	Bare	City	239.7 de	6.50 ab	19.46 ef	13.48 e	18.97 fg	13.57 g
		Rain	250.1 cd	6.50 ab	19.35 f	13.03 g	18.68 g	13.29 h
	Veg	City	205.6 de	4.83 cd	19.63 de	13.86 d	19.62 de	13.88 f
		Rain	135.0 e	4.33 d	18.81 g	13.28 f	19.33 ef	13.80 f
	Mulch		*<0.0001*	*<0.0001*	*<0.0001*	*<0.0001*	*<0.0001*	*<0.0001*
	Water		*0.3362*	*0.0576*	*0.0002*	*<0.0001*	*0.1511*	*<0.0001*
	Mulch × Water		*0.0718*	*0.4984*	*0.005*	*<0.0001*	*0.2181*	*0.0023*
2014	Black	City	400.9 a	8.83 ab	18.51 a	13.17 a	19.62 a	13.98 a
		Rain	341.3 b	7.33 ab	18.26 b	12.96 b	19.34 b	13.74 b
	BlackCC	City	286.2 c	10.17 a	18.24 b	12.97 b	19.36 b	13.78 b
		Rain	293.8 c	9.33 ab	18.24 b	12.98 b	19.27 b	13.74 b
	Bare	City	281.0 c	6.00 b	17.49 c	12.25 c	18.18 c	12.89 c
		Rain	261.4 c	5.6 ab	17.11 d	11.98 d	18.13 c	12.82 c
	Mulch		*<0.0001*	*0.1765*	*<0.0001*	*<0.0001*	*<0.0001*	*<0.0001*
	Water		*0.12*	*0.9542*	*0.0021*	*0.0072*	*0.0954*	*0.0192*
	Mulch × Water		*0.1956*	*0.2697*	*0.0587*	*0.0893*	*0.4201*	*0.217*

For values within mulch, water followed by the same letter are not significantly different according to Fisher's LSD ($p = 0.1$). EF† and LF† were characterized by nighttime soil temperature in specific EF and LF.

4. Conclusions

The three-years of data demonstrated that high tunnels combined with the application of surface mulches and the row cover, along with different irrigation water can be expected to have significant impacts on soil temperature, which influenced lettuce growth and development. The mulches had significant impacts on soil temperature, which was related to total lettuce yield over the falls of 2011, 2013, and 2014. Clear and black plastics all have good soil-warming ability, thus produced the highest lettuce weights. However, the thermal stress with the higher soil temperatures produced in clear and black plastics at transplanting may have caused a reduction of plants per plot when compared to the bare. The biodegradable biobag, spunbond and paper mulches did not heat the soil as much as the clear or black mulch, but the marketable lettuce yield produced by biobag and paper mulches were not statistically different compared with the black polyethylene mulch. In addition, the cover crops incorporated into the soil underneath the black and clear mulch had the potential to save more lettuces in each plot, but overall lettuce weight per head was reduced significantly when compared to the non-cover crop treatments. The reduction in yield from the cover crop may be due to less available nitrogen for plant use, as the organic matter is being decomposed. Moreover, vegetative mulch produced the coolest soil temperatures, and consistently generated the lowest lettuce yield, so is not recommended to be applied in HTs. Additionally, during cold nights, row covers can add protection in the LF, thus produced higher lettuce production when compared to groups without row covers. Finally, the temperature of city water was generally higher than rainwater except in the EF, and it could increase the soil temperature by 0.2–0.8 °C compared to the rainwater plots. Although there were several exceptions when soil temperatures were warmer using rainwater than city water, overall lettuce production was still increased by using city water. Additionally, city water generally provided more nutrients than rainwater in terms of K, Mg, and Ca which may promote lettuce growth and development. Therefore, city water irrigation may not have just warmed the soil but also provided more nutrients for increased lettuce production.

Author Contributions: Conceptualization, M.Z. and B.L.; methodology, M.Z., B.L. and D.B.; software, M.Z. and B.L.; validation, M.Z. and B.L.; formal analysis, M.Z.; investigation, M.Z.; resources, B.L.; Data curation, M.Z., B.L., W.W., D.B., A.H.; writing—original draft preparation, M.Z.; writing—review and editing, M.Z., B.L., D.H., P.A., and D.H.; visualization, M.Z.; supervision, B.L.; funding acquisition, B.L.

References

1. Hinrichs, C. The practice and politics of food system localization. *J. Rural. Stud.* **2003**, *19*, 33–45. [CrossRef]
2. Lamont, W.J. Plastics: Modifying the Microclimate for the Production of Vegetable Crops. *HortTechnology* **2005**, *15*, 477–481. [CrossRef]
3. Kittas, C.; Karamanis, M.; Katsoulas, N. Air temperature regime in a forced ventilated greenhouse with rose crop. *Energy Build.* **2005**, *37*, 807–812. [CrossRef]
4. Wells, O.S.; Loy, J.B. Rowcovers and high tunnels enhance crop production in the northeastern United States. *HortTechnology* **1993**, *3*, 92–95. [CrossRef]
5. Blomgren, T.; Frisch, T. *High Tunnels: Using Low Cost Technology to Increase Yields, Improve Quality, and Extend The Growing Season*; University of Vermont Center for Sustainable Agriculture: Burlington, VT, USA, 2007.
6. Carey, E.E.; Jett, L.; Lamont, W.J., Jr.; Nennich, T.T.; Orzolek, M.D.; Williams, K.A. Horticultural Crop Production in High Tunnels in the United States: A Snapshot. *HortTechnology* **2009**, *19*, 37–43. [CrossRef]
7. Shannon, M.; Grieve, C. Tolerance of vegetable crops to salinity. *Sci. Hortic.* **1998**, *78*, 5–38. [CrossRef]
8. Liakatas, A.; Clark, J.A.; Monteith, J.L. Measurements of the heat balance under plastic mulches: Part I. Radiation balance and soil heat flux. *Agric. For. Meteorol.* **1986**, *36*, 227–239. [CrossRef]
9. Siwek, P.; Kalisz, A.; Wojciechowska, R. Effect of mulching with film of different colours made from original and recycled polyethylene on the yield of butterhead lettuce and celery. *Folia Hort. Supl.* **2007**, *19*, 25–35.
10. Iqbal, Q.; Amjad, M.; Asi, M.R.; Ali, M.A.; Ahmad, R. Vegetative and reproductive evaluation of hot peppers under different plastic mulches in poly/plastic tunnel. *Pak. J. Agri. Sci.* **2009**, *46*, 113–118.
11. Kapanen, A.; Schettini, E.; Vox, G.; Itävaara, M. Performance and Environmental Impact of Biodegradable Films in Agriculture: A Field Study on Protected Cultivation. *J. Polym. Environ.* **2008**, *16*, 109–122. [CrossRef]
12. Kasirajan, S.; Ngouajio, M. Polyethylene and biodegradable mulches for agricultural applications: A review. *Agron. Sustain. Dev.* **2012**, *32*, 501–529. [CrossRef]
13. Olsen, J.K.; Gounder, R.K. Alternatives to polyethylene mulch film—A field assessment of transported materials in capsicum (*Capsicum annuum* L.). *Aust. J. Exp. Agric.* **2001**, *41*, 93. [CrossRef]
14. Moreno, M.; Moreno, A. Effect of different biodegradable and polyethylene mulches on soil properties and production in a tomato crop. *Sci. Hortic.* **2008**, *116*, 256–263. [CrossRef]
15. Miles, C.; Wallace, R.; Wszelaki, A.; Martin, J.; Cowan, J.; Walters, T.; Inglis, D. Deterioration of Potentially Biodegradable Alternatives to Black Plastic Mulch in Three Tomato Production Regions. *HortScience* **2012**, *47*, 1270–1277. [CrossRef]
16. Brault, D.; Stewart, K.; Jenni, S. Optical Properties of Paper and Polyethylene Mulches Used for Weed Control in Lettuce. *HortScience* **2002**, *37*, 87–91. [CrossRef]
17. Ibarra-Jiménez, L.; Quezada-Martín, M.R.; De La Rosa-Ibarra, M. The effect of plastic mulch and row covers on the growth and physiology of cucumber. *Aust. J. Exp. Agric.* **2004**, *44*, 91–94. [CrossRef]
18. Gimenez, C.; Otto, R.; Castilla, N. Productivity of leaf and root vegetable crops under direct cover. *Sci. Hortic.* **2002**, *94*, 1–11. [CrossRef]
19. Teasdale, J.R.; Daughtry, C.S.T. Weed Suppression by Live and Desiccated Hairy Vetch (*Vicia villosa*). *Weed Sci.* **1993**, *41*, 207–212. [CrossRef]

20. Liebman, M.; Davis, A.S. Integration of soil, crop and weed management in low-external-input farming systems. *Weed Res.* **2000**, *40*, 27–47. [CrossRef]
21. Zheng, M.; Leib, B.; Butler, D.; Wright, W.; Ayers, P.; Hayes, D.; Haghverdi, A.; Feng, L.; Grant, T.; Vanchiasong, P.; et al. Assessing heat management practices in high tunnels to improve organic production of bell peppers. *Sci. Hortic.* **2019**, *246*, 928–941. [CrossRef]

11

Pearl Millet Forage Water use Efficiency

Bradley Crookston [1,*], Brock Blaser [2,*], Murali Darapuneni [3] and Marty Rhoades [2]

[1] Department of Plants, Soils and Climate, Utah State University, Logan, UT 84322, USA
[2] Department of Agricultural Sciences, West Texas A&M University, Canyon, TX 79016, USA; mrhoades@wtamu.edu
[3] Department of Plant and Environmental Sciences, New Mexico State University, Tucumcari, NM 88401, USA; dmk07@nmsu.edu
[*] Correspondence: bradley.crookston@aggiemail.usu.edu (B.C.); bblaser@wtamu.edu (B.B.)

Abstract: Pearl millet (*Pennisitum glaucum* L.) is a warm season C_4 grass well adapted to semiarid climates where concerns over scarce and depleting water resources continually prompt the search for water efficient crop management to improve water use efficiency (WUE). A two-year study was conducted in the Southern Great Plains, USA, semi-arid region, to determine optimum levels of irrigation, row spacing, and tillage to maximize WUE and maintain forage production in pearl millet. Pearl millet was planted in a strip-split-plot factorial design at two row widths, 76 and 19 cm, in tilled and no-till soil under three irrigation levels (high, moderate, and limited). The results were consistent between production years. Both WUE and forage yield were impacted by tillage; however, irrigation level had the greatest effect on forage production. Row spacing had no effect on either WUE or forage yield. The pearl millet water use-yield production function was $y = 6.68 \times x$ (mm) $- 837$ kg ha^{-1}; however, a low coefficient of determination ($r^2 = 0.31$) suggests that factors other than water use (WU), such as a low leaf area index (LAI), had greater influence on dry matter (DM) production. Highest WUE (6.13 Mg ha^{-1} mm^{-1}) was achieved in tilled soil due to greater LAI and DM production than in no-till.

Keywords: optimum water use; forage

1. Introduction

Crop yield loss occurs under water deficit, however, many studies have found that higher crop water use efficiency (WUE) is often achieved under water stress conditions [1], albeit, with reduced yield. It is therefore, imperative to identify crop management strategies that optimize WUE without sacrificing attainable yield under limited water availability. The ratio of crop fodder (forage) or grain biomass produced per unit water used (transpiration and losses to soil evaporation) is considered crop water use efficiency [2]. Although there are many location characteristics and environmental conditions that influence WUE (e.g., climate regime, soil type) that cannot easily be manipulated by land managers, there are various crop management schemes known to increase water availability and promote greater WUE [1,2]. It is up to researchers to identify useful combinations of practices fit for particular regions based on the best science available from around the globe.

Pearl millet is a C_4 warm season grass predominately in production for grain for human consumption and forage for livestock feed throughout Africa and India and is noted for its tolerance to semiarid conditions where there is low rainfall and limited levels of soil nutrients and organic matter [3,4]. It is typically grown on rainfed (dryland) areas in systems with grain sorghum (*Sorghum bicolor* L.), maize (*Zea mays* L.), or often integrated with legume crops such as cowpea (*Vigna unguiculata* (L.) Walp) in Africa [4,5]. Pearl millet is also gaining recognition in the Southern Great Plains, USA, semi-arid region, as a potential forage crop to be integrated into grain sorghum-winter wheat (*Triticum aestivum*, L.) livestock feed cropping systems by replacing summer fallow [6–8]. The United Nations Food and Agriculture Organization Statistical Database (FAO) reported that worldwide millet production in 2018 was approximately 31 million metric tons, with Africa and Asia accounting for 51 and 46% of total production, respectively; the Americas make up only 1% of global millet production [9]. Pearl and other millets (e.g., foxtail (*Setaria italica*, (L.) P. Beauvois), proso (*Panicum miliaceum*, L.), and Japanese (*Echinochloa esculenta*, A.Braun)) are common in semi-arid regions for livestock feed because of their rich nutrient value and fit in cropping rotations with short growing seasons that have limited water supply [6,10].

Water use efficiency can be improved using soil management, crop husbandry, water management, genetic selection, and crop competition management [1]. Many of these methods to improve WUE have been reported in studies of pearl and other millet species in locations around the world over the past 25 years (Table 1). Husbandry studies have tested various planting arrangement methods and population densities in hill type sowing [11–13]; forage harvest intervals have also been tested [8]. Studies of irrigation technology have investigated sprinkler, surface drip, and subsurface drip methods or irrigation scheduling throughout the season [10,14]. Other researchers explored increasing water stress and nutrient management; many of those studies utilized the factorial combination of the treatments to benefit from the interaction of improved WUE with nutrient management under limited water supply [15–20]. However, throughout all these studies, the range of millet WUE is 1–92 kg ha^{-1} mm^{-1} with a mean and median of 20.6 and 13.4 kg ha^{-1} mm^{-1}, respectively (Table 1). In a global meta-analysis of WUE in various crops, [2], reported many environmental factors impact WUE variability within and among crop species. For example, greater WUE was observed in tropical versus desert climates, and greater organic matter content was strongly correlated with positive WUE, whereas clay content was negatively correlated with WUE [2]. Ultimately, [2] and co-authors admonished researchers to investigate the mechanisms that might further explain WUE variability within a species.

To our knowledge, no studies with pearl millet have investigated the combination of irrigation levels to limit water availability, plant arrangement in narrow versus wide row spacing and soil management with conventional rotary tillage versus no-till. Goals of this research were to determine the optimum levels irrigation, row spacing, and tillage to maximize WUE while maintaining stable forage production [21] in pearl millet in the semi-arid region of the Southern Great Plains, USA. The underlying hypotheses guiding the objectives were (a) WUE will increase with decreased water availability [22]; (b) no-till soil management techniques can promote more efficient WU [23]; (c) pearl millet crop canopy can be modified through row spacing to increase incidence of leaf area index and light interception [24].

Table 1. Location, soil type, millet crop, experimental treatment, soil tillage management, crop row spacing, water source, crop water use (WU) and water use efficiency (WUE) of pearl and other millet species.

Reference	Location	Soil Type [a]	Crop [b]	Treatment [c]	Tillage	Row Spacing (cm)	Water Source [d]	WU (mm)	WUE (kg ha^{-1} mm^{-1})	Comments
[15]	Texas, USA	CL	Pearl, F	Irr levels, cultivars	Disk	100	Subdrip	193–480	87–92	Max WUE in limited Irr only 1 yr.
[8]	Texas, USA	CL	Pearl, F	Harvest intervals	Rotary till	19	Surfdrip	594	16.2	WUE not different across harvest intervals
[12]	Tanzania	Snd	Pearl, G	Cultivation, micro-dosing	various	na	Rf	481	1.0–4.0	Max WUE in tide-ridges with micro-dose
[25]	China	SndL	Foxtail, Japanese, G	Species, sowing dates	Rotary till	42	Rf	356	10.2–22.8	Max WUE in early sowing
[17]	Niger	Snd	Pearl, G	Manure, micro-dosing	na	100	Rf	582	4.9–12.8	Manure and fertilizer increased WUE
[11]	India	SndL	Pearl, G	Planting arrangement	na	100	Rf	263	5.8–7.7	Tide-ridge recommended
[16]	Jordan	CL	Pearl, F	Species, irr levels	Disk	50	Surfdrip	325–516	21.3	Deficit increased millet WUE
[10]	Saudi Arabia	SndL	Pearl, F	Irr type	na	20	Sprinkler, Surfdrip, subdrip	100–1300	8.4–13.4	Max WUE in subdrip
[19]	Australia	CL	Pearl, F	Irr levels	na	na	Sprinkler	496	22.7	Forage species, high WUE
[20]	Iran	CL	Pearl, F	Irr levels, N levels	na	60	Flood	321–755	28.4–41	Max WUE in deficit with N applied
[18]	Niger	na	Pearl, G	Animal integrated cropping	na	na	Rf	372–393	0.9–6.2	Max WUE with animal integrated cropping
[26]	Colorado, USA	SltL	Proso, F	Cropping systems	No-till	na	Rf	100–250	3.0–22.0	WUE improved in forage-based systems
[14]	Nebraska, USA	SltL	Pearl, G	Irr scheduling	No-till	76	Flood	353	13.4–28.5	Sorghum had higher WUE than millet
[27]	Texas, USA	CL	Pearl, G/F	Cropping systems	No-till	25	Rf	268	4–31.0	Forage crops improved system WUE
[13]	Niger	Snd	Pearl, G	Population, varieties	na	100	Rf	366	7.6–9.7	Max WUE in high populations with nutrients applied
[22]	India	SndL	Pearl, F	Irr scheduling, cultivars	na	30	Flood	223–568	13.8–17.9	Max WUE under water stress

Note: na, not available. [a] Cl, clay loam; SltL, silt loam; Snd, sand; SndL; sandy loam. [b] F, forage; G, grain. [c] Irr, irrigation. [d] Rf, rainfed; Subdrip, subsurface drip; Surfdrip, surface drip.

2. Materials and Methods

2.1. Location and Field Preparation

This forage pearl millet study was conducted during the 2016 and 2017 growing seasons at West Texas A&M University Nance Ranch near Canyon, TX (34°58′6″ N, 101°47′16″ W; 1097 m above sea level elevation) located on Olton clay loam (fine, mixed, superactive, thermic, Aridic Paleustoll). Treatments were arranged as strip-split-plots with four replications with irrigation as main plots, row spacing as subplots, and tillage as sub-subplots.

The plot area was prepared for planting in 2016 by first a using a rotary mower to reduce native perennial grasses, which had established in an unused portion of the field, and application of Roundup© PowerMax™ (Monsanto, St. Louis, MO, USA) herbicide at 9.5 L ha on 31 May 2016. The field was further prepared in 2016 by applying 88 mm of preplant irrigation through drip tape (system described below) to increase the ease of mechanical tillage. In 2017, pearl millet was planted following winter wheat sown in October 2016 terminated in March using Roundup© PowerMax™ herbicide at the same 2016 application rate. In both years, tilled plots were cultivated 20 cm depth using a CountyLine rotary tiller (King Kutter, Inc., Winfield, AL 35594, USA) the day of planting. 'Graze King' BMR pearl millet (176,405 seeds kg^{-1}, 85% germination, 98% purity) was planted on 17 June 2016. Seed used in 2017 was procured from Winfield United (Shoreview, MN 55126, USA) (116,280 seeds kg^{-1}, 85% germination, 98% purity). Sowing utilized a Great Plains 3P500 grain drill (Great Plains Mfg., Inc., Salina, KS, USA). The 3.05 × 6.1 m plots were seeded at 125 seeds m^{-2} and 95 seeds m^{-2} in 2016 and 2017, respectively. Plot width being 3.05 m, we accommodated 16 rows and 4 rows in 19 and 76 cm row spacing treatments, respectively. Different seeding rates were the result of different seed sizes and planter limitation to accommodate the small seed size at the same rate for both years.

2.2. Irrigation Management

Irrigation was administered with a metered surface drip line system with two lines 152 cm apart in each plot with emitters spaced 61 cm, each applying 7.5 L $hour^{-1}$ $emitter^{-1}$. In-line control valves were used to regulate water flow to the irrigation treatment levels so that when irrigation was applied to high (H) only, flow was restricted to moderate (M) and limited (L). However, when L was irrigated, the other levels were irrigated as well, and when M was irrigated, H also received water (Figure 1).

Irrigation management was preplanned to simulate grower conditions where a specific amount of irrigation allowable by aquifer pumping would be provided during the season. Approximately 225 mm was selected as the upper quantity for H. The M and L levels were to be approximately 60 and 30% of the high level, respectively. Weekly applications of 25 mm were planned for H, biweekly in M, and only early season irrigation in L. Ultimately, irrigation management decisions were influenced by growing season climatic conditions, therefore, irrigation events took place when necessary.

Crop WU was defined as the sum of growing season precipitation, irrigation, and plant available soil water (PAW) at planting, minus PAW remaining at harvest. Water lost to runoff or drainage was not measured. Water use efficiency is DM divided by total WU.

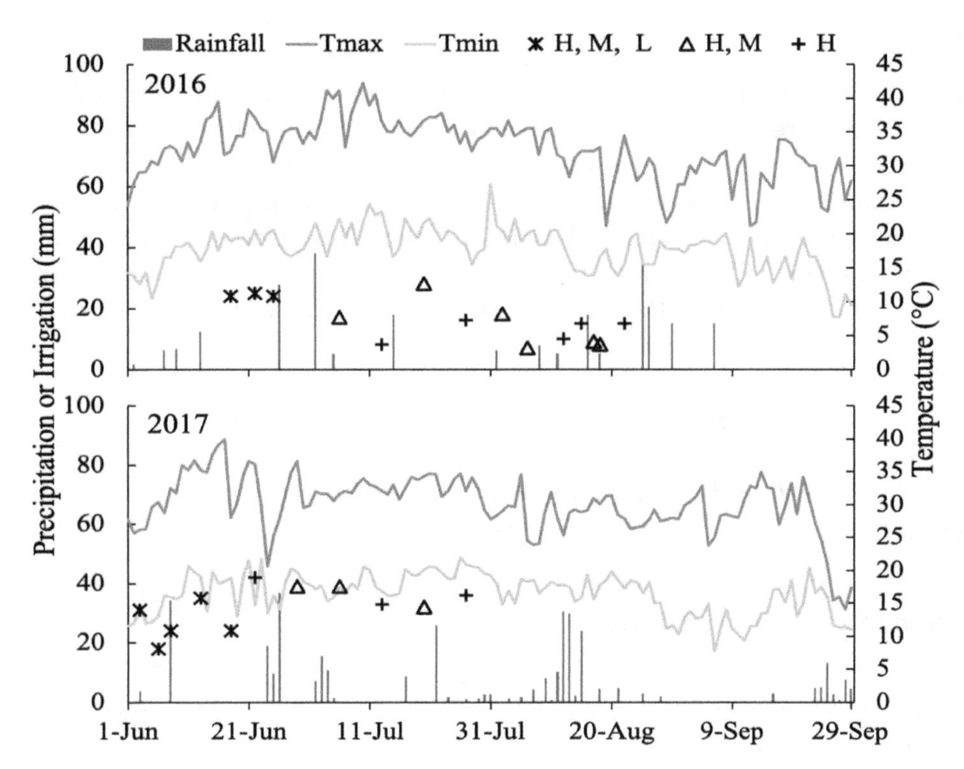

Figure 1. Precipitation, irrigation by water level, and maximum and minimum daily temperature from 1 June to 30 September for 2016 and 2017 near Canyon, TX. Tmax, maximum daily temperature; Tmin, minimum daily temperature; H, high, M, moderate, L, limited irrigation levels.

2.3. Soil Moisture

Soil cores used to calculate average soil volumetric water content at planting were obtained by sampling four random locations throughout the plot area using a tractor mount Giddings hydraulic press (Giddings Machine Company, Inc., Windsor, CO, USA). Soil samples were taken the day prior to planting to a depth of 60 cm in 2016 and to a depth of 75 cm in 2017. Each core sample was divided into three incremental depths: 0–15, 15–30, 30–60 cm in 2016 and 0–15, 15–45, 45–75 cm in 2017. The soil cores were weighed, and oven dried at 104 °C for 72 h until a constant dry weight was attained. Soil volumetric water content at harvest in both years was determined using one core sample from each plot divided into segments of 0–15, 15–45, 45–75 cm.

Soil characteristics and properties utilized data obtained from the United States Department of Agriculture Natural Resources Conservation Service Web Soil Survey [28] (Table 2). Plant available soil water (PAW) was calculated by subtracting volumetric water content at permanent wilting point (−1500 kPa) from volumetric water content at −33 kPa and multiplied by the soil depth.

Table 2. Soil moisture characteristics for Olton clay loam [a].

		Water Content (kPa) [b]			
Depth	**ρ_b** [c]	**−33**	**−1500**	**Plant Available Water**	
cm	g cm^{-3}	% by Volume		cm cm^{-3}	cm
0–15	1.54	34.8	23.5	0.13	
15–30	1.49	35.3	24.5	0.16	
30–60	1.46	35.6	25	0.17	
60–90	1.45	35.2	24.3	0.16	
Mean	1.49	35.23	24.33	0.16	
Profile					13.95

[a] NRCS, Web Soil Survey, Randall County, Texas (TX381), Olton Clay Loam, 0–1% slopes. [b] Water content at field capacity (−33 kPa) and permanent wilting point (−1500 kPa). [c] ρ_b, soil bulk density.

2.4. Weather Data

Precipitation during the 2016 growing season was collected with a graduated plastic rain gauge at the study site and cross referenced with National Weather Service in Amarillo, TX (approximately 32 km from the study location). Temperature values were taken from the National Weather Service in Amarillo. Weather data for 2017 was collected on site utilizing a Campbell Scientific (Logan, UT, USA) weather station approximately 100 m from the plot location. The American Society of Civil Engineers reference evapotranspiration (ET_o) for a well-watered grass crop was calculated using the REF-ET [29] macro software for Microsoft Excel (Microsoft Corp., Redmond, WA, USA). Growing degree-days (GDD) were calculated using Equation (1):

$$\mathrm{GDD} = \mathrm{crop}\ max\left(\frac{T_{max} + T_{min}}{2} - T_{\mathrm{base}},\ 0\right) \tag{1}$$

where crop maximum is 35 °C and base temperature is 10 °C [30]. Growing degree-days were calculated for the time period between planting date and harvest.

2.5. Crop Growth and Forage Yield Measurements

Plant emergence and density were observed and counted at 308 GDD (19 days after planting; DAP) and 255 GDD (18 DAP) in 2016 and 2017, respectively. Crop canopy height was determined using a 0.25 m^2 circular clear plastic disk and the method described by [31] and used by [32]. Harvest DM was the mean of two 1 m^2 quadrats from each plot, arranged in the 76 cm row spacing to include two rows, and cut at 15 cm above the soil surface [33].

Photosynthetically active solar radiation (PAR) intercepted by the crop canopy was measured every 7 to 12 d beginning 7 July 2016 and 19 June 2017 using the AccuPAR Linear PAR ceptometer, model 80 light measuring instrument (Decagon Devices, Pullman, WA, USA). Light interception measurements were collected by placing the instrument diagonally across three 19 cm spaced rows and perpendicular across two 76 cm spaced rows. Percent light intercepted by the canopy was determined by calculating the difference of one above (incident) measurement from the mean of two below canopy measurements, divided by the above measurement.

Leaf area index (LAI) was measured and calculated utilizing the LAI-2200 Plant Canopy Analyzer (Li-Cor, Inc., Lincoln, NE, USA). Measurements were obtained for row crops of narrow and wide row spacing per the Li-Cor instruction manual to calculate a mean from two sequences of one above canopy and four below canopy readings. Measurements were obtained at sunrise or sunset every 10 to 14 d beginning 12 July 2016 and 19 June 2017.

2.6. Statistical Design and Analysis

Year, irrigation, row spacing, and tillage were treated as fixed effects. Analysis of variance of main effects utilized Proc MIXED and GLM of the Statistical Analysis System, Version 9.4 (SAS Institute, Cary, NC, USA). Linear regression and analysis of covariance utilized Proc REG and Proc GLM with contrast statements to determine differences among slopes and intercepts of the regression equations [26]. Significance was determined at $p < 0.05$ for all means separation tests and used the P diff statement with Tukey's adjustment for all main effects.

Interactions for all main effects were initially tested. When no interactions were detected, only main effect results are presented. Years were analyzed and are presented separately for analysis of variance as a consequence of management differences between the seasons: planting in 2016 was into native perennial grass, into wheat stubble in 2017; different seed used in each year. Years were combined for regression analysis after observing similar trends for both years.

3. Results

3.1. Weather, Irrigation, and Soil Moisture

3.1.1. Seasonal Weather

Total growing season precipitation was 217 and 309 mm in 2016 and 2017, respectively (Figure 1; growing seasons: 17 June to 30 September 2016; 1 June to 30 August 2017). The 30-year precipitation average for the Amarillo, TX area for June through September is 275 mm. Precipitation received was 89% and 137% of normal for the respective years (Table 3). Higher temperatures in July 2016 resulted in 10% greater cumulative degree days than in July 2017. Lower temperatures and greater precipitation in August 2017 generated 13% lower GDD accumulation for that month compared to 2016 (Table 3). The local 30-year average GDD accumulation for a 90-d pearl millet growing season is 1676 GDD for a 1 June planting date (Table 3).

Table 3. Average maximum daily temperature, total precipitation, and estimated cumulative thermal time by month for 2016 and 2017 growing seasons near Canyon, TX from planting to harvest. [a]

Month	Avg. [b] Max Daily Temp.			Precipitation			Heat Units		
	°C			mm			GDD		
	2016	2017	Avg.	2016	2017	Avg.	2016	2017	Avg.
June	33	32	30	57	103	80	437	411	405
July	36	32	33	61	79	72	528	478	486
August	31	28	31	113	127	74	440	389	461
September	29		28	15		49	362		324
Total				246	309	275	1767	1279	1676
Percent of normal				89%	137% [c]		105%	94%	

[a] Growing Season, 17 June to 30 September 2016; 1 Jun to 30 Aug 2017. [b] Averages for temperature and precipitation are averages for Amarillo, TX, USA. 1981–2010. [c] 2017 percent of normal for precipitation and heat units is taken from historical avg. of June–August. The percent of normal is indicated for each season's precipitation.

3.1.2. Irrigation

Total irrigation applied in 2016 for each irrigation level was 220, 157, and 72 mm for the H, M, and L levels, respectively (Table 4). In 2017, 356, 242, and 132 mm were applied to the H, M, and L irrigation treatments, respectively (Table 4). The L level received three irrigation events in 2016, and five events in 2017; greater water was applied in 2017 early in the season due lower soil moisture at planting and to prevent soil surface crusting. The M level received nine events in 2016 and eight events in 2017. The H level received 14 events in 2016 and 11 events during the 2017 season. As a result of more frequent rainfall during the second half of July and in August of 2017, there were fewer irrigation events needed than in 2016.

Table 4. Millimeters of irrigation applied to pearl millet by month for the 2016 and 2017 growing seasons [a] near Canyon, TX.

	2016			2017		
	High	Moderate	Limited	High	Moderate	Limited
June	72	72	72	214	171	132
July	66	43		142	71	
August	82	42				
September						
Total	220	157	72	356	242	132

[a] Growing Season, 17 June to 30 September 2016; 1 June to 30 August 2017.

Applied irrigation accounted for approximately 24–53% of seasonal total water (irrigation and precipitation) for the crop at the respective irrigation levels over both growing seasons (Table 5). During the growing seasons, irrigation accounted for an average range of 50–53%, 41–43%, and 24–29% of the total water received by the crop (excluding stored soil moisture) in H, M, and L levels, respectively (Table 5). The total water is compared to ET [29] to evaluate the efficacy of irrigation management to meet climatic demand on the crop (Table 5). The ET_o climatic demand was estimated at 772 and 902 mm in 2016 and 2017, respectively. The range of satisfied climatic demand across water regimes was 48–73% in 2017 vs. 37–56% in 2016 (Table 5). Although 2017 had fewer irrigation events for H and M levels, additional rainfall in 2017 was available to meet the climatic demand.

Table 5. Precipitation and irrigation by month and irrigation level for 2016 and 2017 near Canyon, TX. Percent of total water as irrigation and total as water percent of ET_o are also included.

	2016 [a]				**2017**			
	mm							
	High	**Moderate**	**Limited**	**ET_o** [b]	**High**	**Moderate**	**Limited**	**ET_o**
Total	437	374	289	772	665	551	441	902
Percent of total as irrigation	50%	41%	24%		53%	43%	29%	
Total water as percent of ET_o	56%	48%	37%		73%	61%	48%	

[a] Growing Season, 17 June to 30 September 2016; 1 June to 30 August 2017. [b] ET_o, ASCE reference evapotranspiration (ET) for a well-watered grass crop, calculated using REF-ET [29].

The authors of [14] reported applying an average 353 and 350 mm in two study years on pearl millet and grain sorghum (*Sorghum bicolor* L.), respectively, in Nebraska where annual precipitation was 285 and 480 mm in the study two locations. In studies of rainfed millet production, [27] at Bushland, TX reported pearl millet forage WU range was 236 to 289 mm (precipitation and PAW from the soil). In a study testing the response of sorghum, maize, and pearl millet to four irrigation levels in India, [22], reported total WU ranged from 242 to 568 mm during the field study. In Akron, CO, during the five-year study of opportunity cropping systems [34], millet growing season precipitation ranged from 57 to 298 mm, similar to other semiarid locations.

3.1.3. Soil Moisture

Plant available water at planting was 3.35 cm for the 60 cm profile in 2016 and was 1.9 cm for the 75 cm profile in 2017. Mean PAW (soil depth 75 cm) at harvest across all treatments in 2016 (0.63 cm) was 20% of mean PAW in 2017 (3.09 cm). Harvest PAW in 2016 was influenced by irrigation, there was no influence of row spacing or tillage treatments, averaging 0.64 cm. High and M irrigation levels averaged 0.86 cm, which was different from L at 0.19 cm of PAW. Differences in PAW were found in 2017 among irrigation levels and in an irrigation × tillage interaction. Plant available water was highest in M (3.52 cm) and lowest in L (2.49 cm), a difference of 41%; H (3.1 cm) was not different from the M and L levels. The interaction resulted from L irrigation × no-till, 1.8 cm PAW, being 47% lower than the average of all other treatments (3.42 cm; SE = 0.42 cm), ranging from 2.7 to 3.8 cm. The primary reason harvest PAW in 2016 was 80% less than 2017 is the additional 31 d of growing season in 2016 and the rainfall prior to harvest in 2017 (Figure 1).

3.2. *Treatment Effects: Irrigation, Row Spacing and Tillage*

3.2.1. Plant Growth and Forage Dry Matter

Average pearl millet plant population was 14.6 and 30.9 plants m^{-2} in 2016 and 2017, respectively, a difference of 52%. A year x till interaction was observed, most likely due to low plant population in no-till in 2016 compared to no-till in 2017 (7.9 vs. 45.8 plants m^{-2}, respectively). Additionally, in 2016, average population in no-till was 67% less than till (7.9 vs. 21.2 plants m^{-2}, respectively). However, in 2017, average till plant population was 65% less than no-till (16.1 vs. 45.7 plants m^{-2}). The no-till planting in 2016 was into herbicide killed native perennial grasses which prevented adequate seed to soil contact for proper germination. In 2017, an afternoon rainstorm following a morning irrigation event three days after sowing caused runoff that was observed in the till treatment. In contrast, no-till in 2017 was protected by the winter wheat stubble remaining on the soil surface after planting.

Forage DM was influenced by year, irrigation level, and tillage ($p < 0.05$) as main effects in both years; row spacing had no effect on forage production (Table 6). Average pearl millet DM in 2016 was approximately 38% less than in 2017 (1712 and 2759 kg DM ha^{-1}, respectively). Average WU was 523 and 395 mm in 2017 and 2016, respectively. Average WUE was 5.28 and 4.26 kg ha^{-1} mm^{-1} in 2017 and 2016, respectively (Figure 2). Forage DM in H irrigation was 46% greater than L, ranging from 1219 to 2213 kg ha^{-1} in 2016. The H level was 54% greater than L in 2017, which ranged from 2213 to 3450 kg ha^{-1} (Figure 2). Dry matter in M was not different from H in either year but was different from L ($p < 0.05$; Figure 2). Water use in H was 15% greater than M and 43% greater than L in 2016. In 2017, H received 59% more water than L and 21% more than M (Figure 2). Water use efficiency was not different among irrigation levels ranging from 3.78 to 4.79 kg ha^{-1} mm^{-1}. In contrast, a C_3 bioenergy crop, hybrid poplar (*Populus generosa* Henry × *P. nigra* L.), achieved differing WUE values of 5.6 and 8.9 kg ha^{-1} mm^{-1} grown under 115 mm and 240 mm of applied irrigation, respectively [35]. Row spacing did not influence pearl millet forage DM, WU, or WUE in 2016 or 2017 (Figure 2, Table 6).

Table 6. The *p*-values from ANOVA of pearl millet dry matter (DM), water use (WU), and water use efficiency (WUE).

		2016	
Treatment [a]	**DM**	**WU**	**WUE**
I	0.001	<0.0001	ns
R	ns	ns	ns
I × R	ns	ns	ns
T	0.0001	ns	<0.0001
I × T	ns	ns	ns
R × T	ns	ns	ns
I × R × T	ns	ns	ns
		2017	
	DM	**WU**	**WUE**
I	0.0097	<0.0001	ns
R	ns	ns	ns
I × R	ns	ns	ns
T	0.0064	ns	0.0095
I × T	ns	ns	ns
R × T	ns	ns	ns
I × R × T	ns	ns	ns

[a] I, irrigation; R, row spacing; T, tillage; ns, not significant at $p = 0.05$.

Pearl millet DM and WUE in the tilled plots were greater than in no-till in both years, although, WU was not affected by tillage management in either year (Figure 2). Forage DM in till was 66 and 39% greater than no-till, in 2016 and 2017, respectively. Pearl millet WUE in tilled soil was 68% greater than no-till in 2016 and 39% greater in 2017 (Figure 2).

Forage DM in 2016 till was greater as a consequence of the low plant population in no-till due to poor seed to soil contact. The low forage yields in 2017 are attributed to N deficiency observed in the no-till plants. This deficiency in no-till may be a result of higher plant population, compared to the till treatment, leading to the deficiency observed mid-season and no deficiency observed in the till plants.

Crop canopy height in 2016 was not influenced by irrigation, row spacing, or tillage; the end of season canopy height was 50 cm. For the 2017 season, canopy height was affected by irrigation over time. Height at harvest was 72, 67, and 59 cm for H, M, and L, respectively. In addition, there was an irrigation × tillage interaction in 2017, where L × no-till was 38 cm and H × till was 49 cm at harvest. No difference for crop canopy height was measured in the tillage treatment for either year.

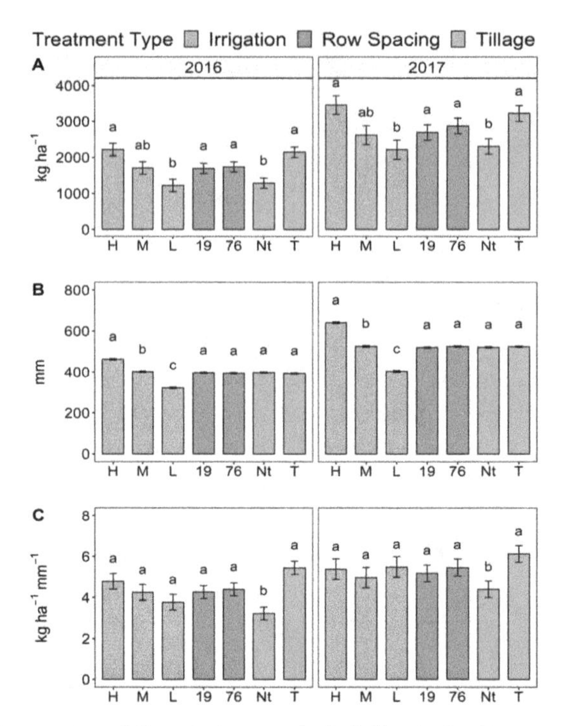

Figure 2. (**A**) pearl millet dry matter; (**B**) water use = (rainfall + irrigation + plant available soil moisture at planting)–plant available soil moisture at harvest; (**C**) water use efficiency = dry matter/water use. Lowercase letters indicate significance at $p < 0.05$ within treatment and year; bars indicate standard error within treatment. H, high irrigation level, M, moderate irrigation level, L, limited irrigation level; 19 cm and 76 cm row spacing; N-t, no-till, T, tilled.

3.2.2. Light Interception and Leaf Area Index

Crop LAI during the 2016 season was impacted by irrigation, which reached a maximum of 2.47 at 1262 GDD (Figure 3). In 2017, LAI was similarly affected by irrigation and reached a maximum of 3.07 at 1203 GDD (Figure 3). Incidence of crop canopy light interception responded to changes in irrigation through time in 2016 only, reaching 74.6% interception at 967 GDD (Figure 3).

Light interception was influenced by row spacing though the 2016 season, which reached 74% in 19 cm and 65% in 76 cm at 1262 GDD (Figure 4). During 2017, LAI reached 2.87 in 19 cm and 2.59 in 76 cm at 1203 GDD (Figure 4). In both years, light interception and LAI were lower in wide row spacing. However, other work with pearl millet and row spacing has shown that in years where rainfall was less than the local average, wide row spacing outperformed narrow rows [36]. Weeds were not controlled either year of this field study and in 2016 rainfall was 11% less than the 30 year normal (Table 3), yet, row spacing had no effect on pearl millet DM.

During 2016, no differences for LAI nor light interception were observed between tillage treatments (Figure 5). Through 2017, mid-season LAI was different for the tillage treatments but there was no difference by harvest time, reaching a maximum value of 2.8 and 2.6 in till and no-till, respectively (Figure 5). Light interception was greater in no-till through the 2017 season, reaching 77 vs. 70% in till at end of season (Figure 5). For the 2017 season, light interception interactions were found for irrigation × tillage and row spacing × tillage. In the till treatment, there was an increasing trend following increasing irrigation and narrow row spacing, however, that trend was interrupted in the no-till where M had 6% less light interception than L, and 76 cm row spacing had 20% less than 19 cm.

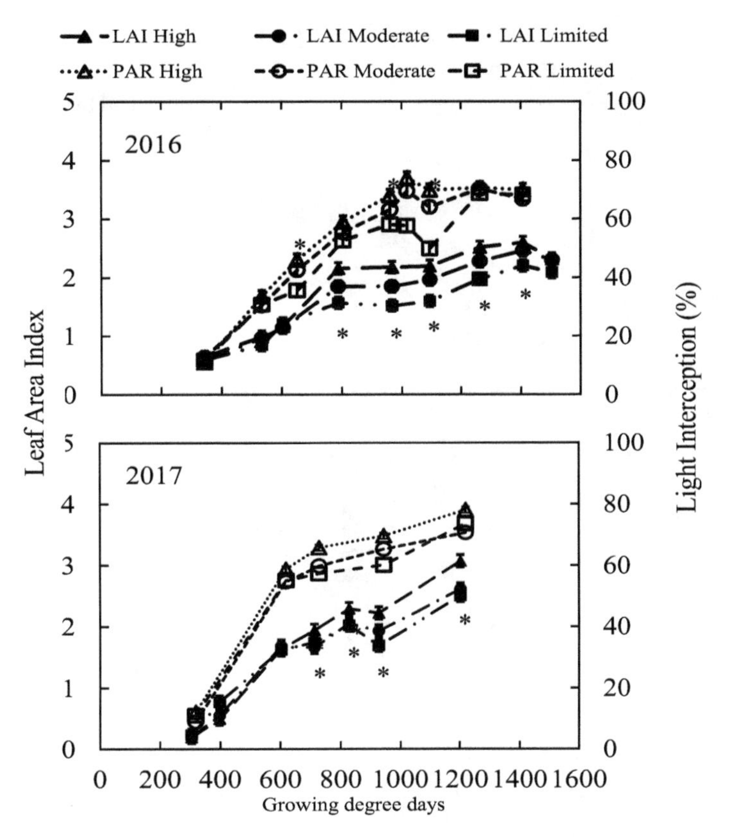

Figure 3. Leaf area index (LAI) and percent light interception (PAR) for irrigation levels in 2016 and 2017 in pearl millet at Canyon, TX. Asterisks indicate treatment $p < 0.05$ on measurement date.

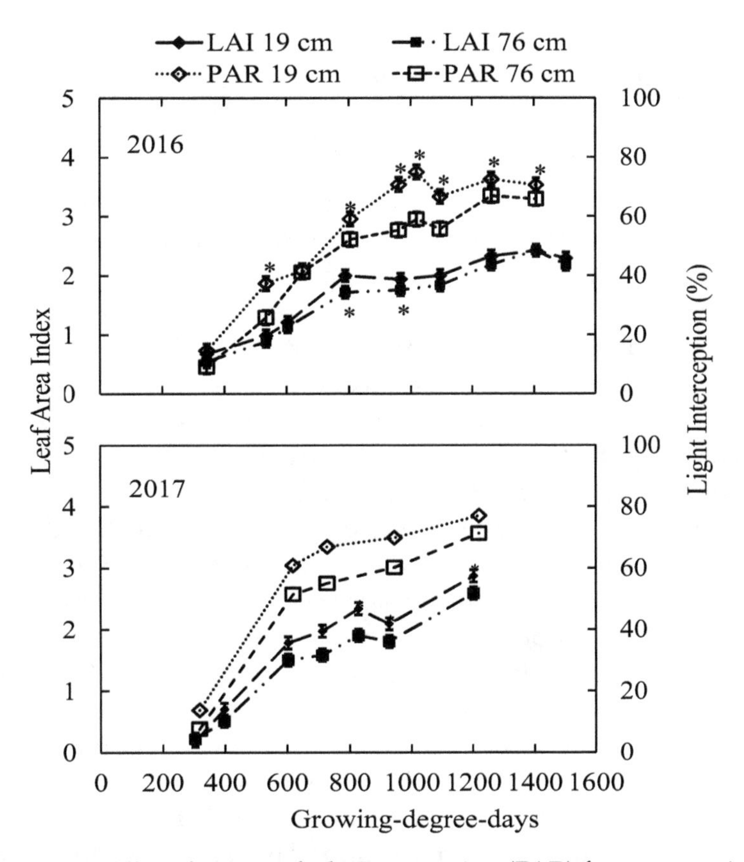

Figure 4. Leaf area index (LAI) and percent light interception (PAR) for row spacing in 2016 and 2017 in pearl millet at Canyon, TX. Asterisk indicate treatment $p < 0.05$ on measurement date.

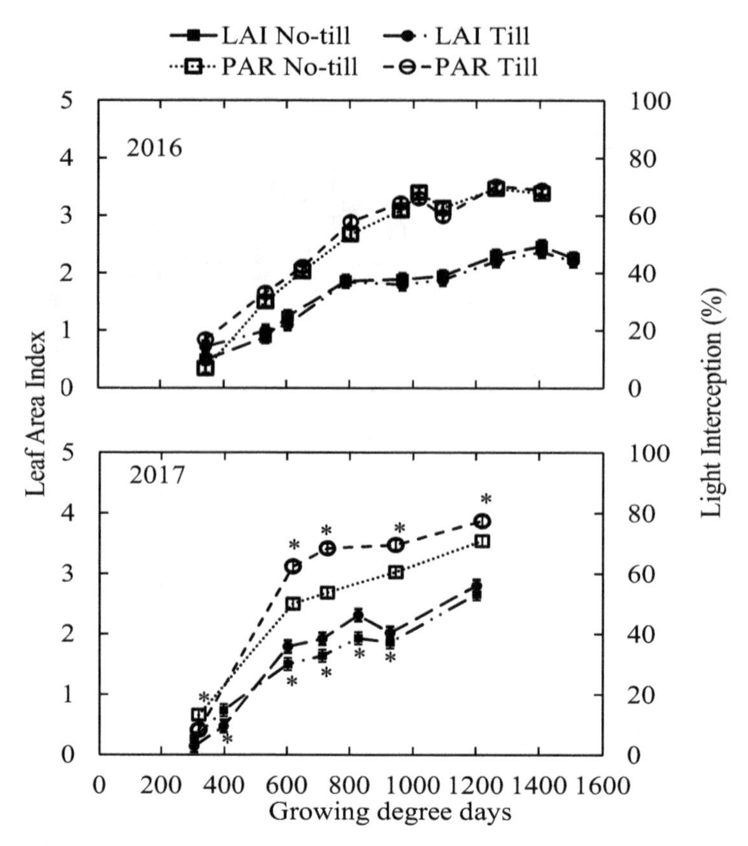

Figure 5. Leaf area index (LAI) and percent light interception (PAR) for tillage levels in 2016 and 2017 in pearl millet at Canyon, TX. Asterisk indicate treatment $p < 0.05$ on measurement date.

4. Discussion

4.1. Explaining Water Use Efficiency

4.1.1. Water Use Efficiency and Soil Nutrient Supply

Although nutrient management was not a treatment in this study, field observations of possible nutrient deficiencies prompted investigation to the relationship between WUE and nutrient management. Proper soil nutrient supply improves plant growth, photosynthetic rate, transpiration, root growth and ultimately yield, all result in increased WUE [23]. For example, pearl millet, being tested under four levels of increasing N and irrigation supply, achieved 41 kg ha^{-1} mm^{-1} WUE in the most water stressed treatment with the highest level of N application, 225 kg N ha^{-1} [20]. The authors of [37] also illustrated the effect that irrigation and N application can have on WUE in an experiment with irrigated and rainfed wheat when "adequate" N was applied with irrigation, mean WUE was 78% greater and grain production was 16% greater in irrigated than in rainfed wheat without N application. The 2016 and 2017 of this pearl millet study, DM was influenced by observed nutrient deficiency, especially in the no-till treatment of 2017. The authors of [8] conducted an experiment with forage pearl millet at the same time just adjacent to the plot where this forage pearl millet experiment took place. Agronomic management similar to the H irrigation, 19 cm row spacing, and tilled treatment was similar in both studies except that in [8] N was applied in both study years at rates of 84 kg ha^{-1} and 78 kg ha^{-1} in 2016 and 2017, respectively. Pearl millet DM reported by [8] used approximately 514 and 602 mm of water in 2016 and 2017, respectively, which produced 6287 and 9874 kg ha^{-1}, with WUE of 12.2 and 16.4 kg ha^{-1} mm^{-1} in those years, respectively. The DM yields from [8] were approximately three times greater than the mean across all treatments of this study in 2016 and 2017 having used a maximum of 462 and 641 mm in those years without N application. The lack of N, demonstrated by comparison to [8], lead to lack of nutrients reduced plant productivity and negatively impacted WUE [38].

4.1.2. Weed Influence on Pearl Millet Water Use Efficiency

Native grass/weed DM may have impacted pearl millet WUE due to no weed control in either year of this study. In 2016 grass/weed DM was not influenced by row spacing, however grass/weed DM was different among irrigation and tillage levels (Table 7). In 2017, only row spacing influenced grass/weed DM. The authors of [39] demonstrated in forage sorghum, when weeds are present, there was no difference in DM yields among wide (76 cm) or narrow (25 cm) row spacing. Although weed control was not a treatment in this study, regression analysis showed that grass/weed DM did influence pearl millet WUE across treatments and years (Figure 6). As weed proliferation was not checked by high leaf area index and resulting light interception correlated with crop competitive ability [40], weed/grass DM accumulated, thus reducing pearl millet WUE.

Table 7. Grass/weed dry matter (DM) at pearl millet harvest by treatment near Canyon, TX in 2016 and 2017.

Treatment	DM	
	kg ha^{-1}	
	2016	2017
Irrigation		
High	710a [a]	324a
Moderate	561b	363a
Limited	484b	297a
SE	37.8	44.7
Row Spacing		
19 cm	576a	218b
76 cm	594a	438a
SE	30.9	36.5
Tillage		
No-till	792a	318a
Till	379b	338a
SE	30.9	36.5

[a] Lowercase letters indicate mean separation within effect, $p < 0.05$.

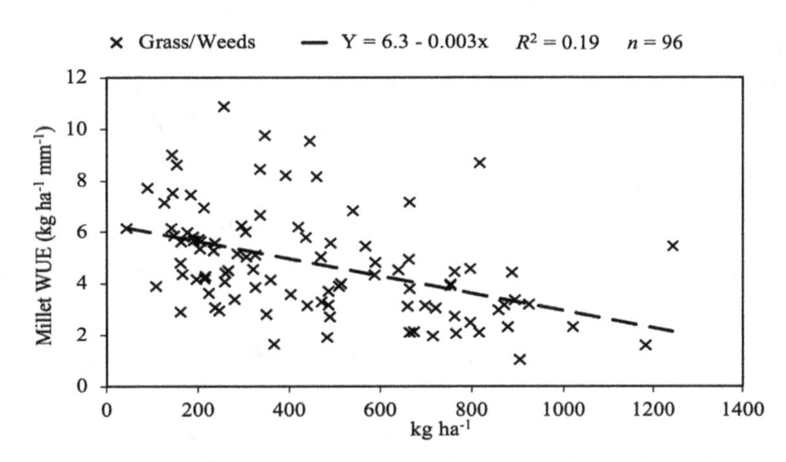

Figure 6. Pearl millet water use efficiency (WUE) as a function of grass/weed dry matter (DM) across treatments and years.

4.1.3. Water Use Efficiency Response to Agronomics

Pearl millet WUE reported in the literature has been consistent yet wide ranging. During a five-year rotation study by [27], mean pearl millet production from two seasons was 3670 kg ha^{-1}. Although weather conditions were not considerably different between the two years, [27] reported WUE of 31 kg ha^{-1} mm^{-1} in 1995 and 4.5 kg ha^{-1} mm^{-1} 1996, most likely due to one month earlier planting date in 1995, allowing the crop to better utilize soil water prior to harvest. The authors of [13] found the WUE of pearl millet, in a study of various cultivars and plant populations in Niger, ranged from 7.6 to 8.7 kg ha^{-1} mm^{-1}. In Tunisia, [41] reported a pearl millet WUE range of 6.4 to 7.6 kg ha^{-1} mm^{-1} among four water stress treatments. The authors of [22] reported pearl millet DM values of 4000 to 8300 kg ha^{-1} and WUE from 13.8 to 17.9 kg ha^{-1} mm^{-1}, respectively, in India.

Whereas, in Akron, CO, among forage and grain, foxtail and proso millet cropping systems described by [33], DM production ranged from 562 to 4545 kg ha^{-1} with WUE ranging from 4.8 to 27.5 kg ha^{-1} mm^{-1} during the five-year study. Additionally, [20] also reported high WUE ranging between 35.7 and 41 kg ha^{-1} mm^{-1} under four levels of irrigation. Grain sorghum WUE has been reported between 11 to 49.2 kg ha^{-1} mm^{-1} among locations throughout the Central and Southern Great Plains U.S. [42,43].

In the studies of pearl millet and sorghum cited above, minimum or no-till was used by [27] and [43], however, tillage information was absent from the other studies. In this study, a WUE response to agronomic management was only observed from tillage practices (Figure 2), whereas WUE was not impacted in the same way in this study by the increasing DM and WU from irrigation treatments as previously observed in other studies [20,22,41]. In a review of soil management effects on WUE, [23] reported that in no-till, where plant residue is maintained on the soil surface, higher WUE is typically found due to reduced evaporative water loss (E) from soil, lower soil surface temperatures, and improved water infiltration. The authors of [23] analyzed nineteen studies of crop and soil management practices and synthesized their results into a graphical depiction of crop WUE responses to soil management and seasonal effects. They demonstrated that crop biomass and yield increased or decreased due to changes in soil management. In this study, pearl millet DM response to soil tillage reflects differences in biomass due to soil management changes as described by [23] (Figure 7). However, when, soil surface E is roughly approximated at the x-intercept of the DM/WU regression line (Figure 7) [38,44], the two-season average estimated E from tilled and no-till soil was approximately 114 and 131 mm, respectively. While these estimates are similar, the result is contrary to reports of less E from no-till soil [23]. Additionally, WU was not different between tillage or row spacing treatments in either year. Furthermore, the spread of DM responses observed, lack-of-fit to the DM/WU predicted line, and no differences in DM or WU detected between row space levels and WU between tillage levels indicates that WU was insensitive to changes in the management practices other than when water was added by irrigation (Figures 2, 7 and 8). The authors of [24] reported that pearl millet yield is largely independent of WU when LAI is low, as was observed in both seasons of this pear millet forage study, due to the majority of light being intercepted by the soil surface allowing water loss to E in the sparse canopied crop with LAI values < 2 [44]. In 2016, LAI was not different between the tillage levels, however, in 2017, LAI in the tilled treatment reached approximately 2 nearly 100 GDDs before no-till reached a LAI of approximately 2. As mentioned above, the N deficiency observed predominately in no-till explains slower growth, lower DM and WEU for both years even though till had lower early season plant count than in no-till during 2017. Thus, higher WUE in tilled soil was achieved as a result of earlier canopy development, more light interception and greater DM production.

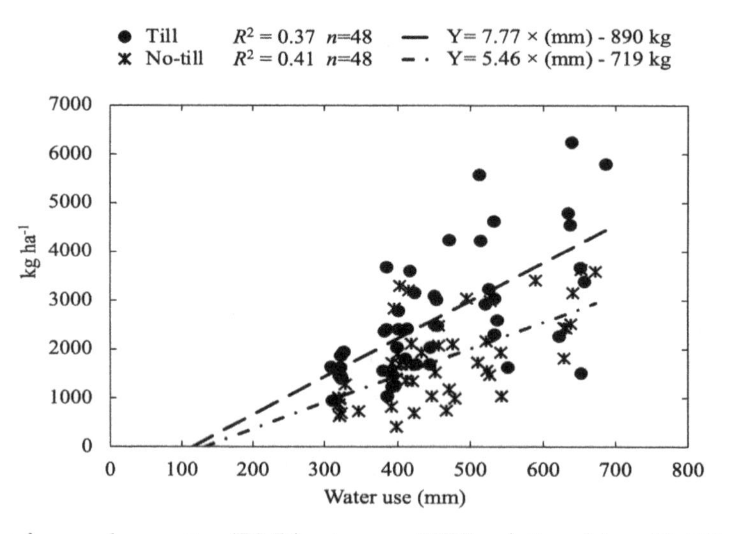

Figure 7. Pearl millet forage dry matter (DM)/water use (WU) relationship with tillage levels indicating management impact on DM production for 2016 and 2017 at Canyon, TX.

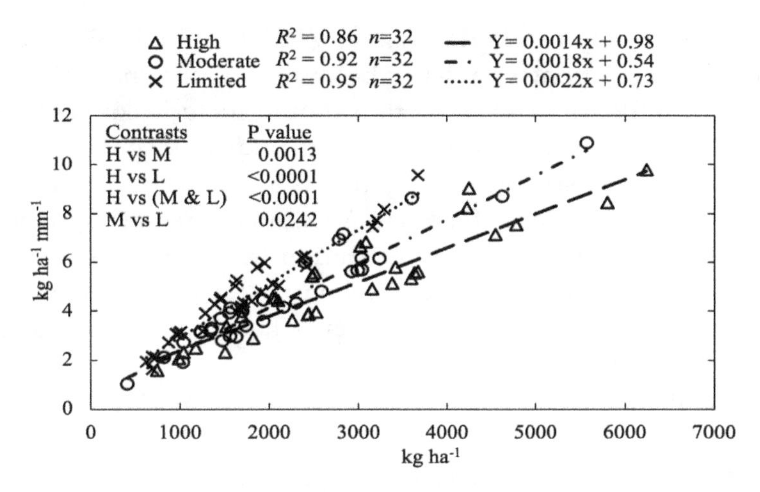

Figure 8. Direct relationship between water use efficiency (WUE) and pearl millet dry matter (DM). Slopes show trends of increasing WUE as DM increases for each irrigation level.

The proceeding discussion explaining WUE differences in the tilled treatment also helps explain the absence of main effect interactions and lack of WUE response to decreasing irrigation. Poor soil nutrient supply and high weed competition reduced pearl millet crop growth thereby obscuring any possible interactions among the main effects and responses typically reported from no-till or water stress studies. The authors of [20,22,41] found that pearl millet WUE increased as water stress also increased, which was not observed in this study. However, the relationship between WUE and DM [26,37,38] affords direct comparison of WUE at each irrigation level (Figure 8). Statistical contrast analysis of the slopes and intercepts for the respective irrigation levels [26] show that L had greater positive gain in WUE than was achieved in the H or M. This analysis suggests that WUE at the L level had a higher rate of return for accumulated DM than the other water levels. The authors of [22] demonstrated that in water stressed pearl millet WUE increased 23% from well-watered conditions. Analysis of their results demonstrates that pearl millet maintains physiological functions even when constrained by water stress. For example, the ratio of net photosynthesis (g CO_2 m^{-2} h^{-1}) to WU (mm) was greater in the limited water treatment than in the full water treatment (0.029, 0.002, respectively) [22]. This result can be explained by the biochemical pathway in warm season C_4 plants used to continue photosynthesis when vapor pressure deficit is high in low density canopies [45,46]. Furthermore, [44] explained that increased WUE this is the result of a greater proportional increase in WUE for low LAI values than when LAI is >2. Thus, the direct relationship between WUE and DM illustrates an ecophysiological response from pearl millet interacting with the crop environment.

Although irrigation applied lead to increased DM, WU did not well predict the behavior of pearl millet forage production across both years and all treatments. The production function $y = 6.68 \times x$ (mm) $- 837$ kg ha^{-1} ($R^2 = 0.31$, $n = 96$) cannot be readily utilized in other production situations that might have better weed and sol nutrient management. In contrast, for most DM/WU linear regressions, the coefficient of determination is much higher (>0.7) [33,42] and more confidence is given to estimating crop yield from WU given a set of WUE optimizing management strategies.

5. Conclusions

These results demonstrate the challenges of field study management to optimize WUE. Future research in the study region might explore agroecological strategies to mitigate poor soil nutrient supply and weed competition for pearl millet by practicing intercropping with legumes or agroforestry where possible [24]. Despite the factors that obscured the main effects and interactions, greatest average forage production was achieved with the highest irrigation level, however, highest WUE was attained in tilled soil due to greater LAI, light interception, and plant growth than in no-till. While the application of water increases forage production, low LAI values will increase estimated E and reduce WUE, especially without adequate nutrient application. Differences in DM were associated with changes in soil management, which ultimately resulted in higher WUE for tilled soil. The DM/WU pearl millet forage production function lacked strong correlation because of weeds and LAI < 2. An agronomic WUE optimization plan for pearl millet forage production should include conventional till, weed control, and proper nutrient management. Narrow is preferable to wide row spacing for greater LAI and light interception earlier in the growing season. If irrigation is available, water application should range from 400–600 mm of total water available for ET to maximize forage production, however, climatic demand will cause greater water loss from transpiration than if limited water is applied. Future research has an opportunity to conduct a more comprehensive review and meta-analysis of global pearl millet management practices used to improve water use efficiency and the effective use of water [47].

Author Contributions: Conceptualization, B.C., B.B., M.R.; Methodology, B.C., B.B., and M.R.; Formal analysis, B.C., M.D., and M.R.; Funding acquisition, B.B.; Investigation, B.C., and B.B.; Project administration, B.C., and B.B.; Resources, B.B.; Supervision, B.B., M.D., and M.R.; Validation, B.B., M.D., and M.R.; Visualization, B.C.; Writing—original draft, B.C.; Writing—review and editing, B.C., B.B., M.D., and M.R. All authors have read and agreed to the published version of the manuscript.

Acknowledgments: Research funding was provided by the Dryland Agriculture Institute, West Texas A&M University, Canyon, TX, USA. The authors also wish to express their gratitude for the time and efforts of two anonymous reviewers whose thoughts contributed to the improvement of the manuscript.

References

1. Farooq, M.; Hauuain, M.; Ul-Allah, S.; Siddique, K.H.M. Physiological and Agronomic Approaches for Improving Water-Use Efficiency in Crop Plants. *Agric. Water Manag.* **2019**, *219*, 95–108. [CrossRef]
2. Mbava, N.; Mutema, M.; Zengeni, R.; Shimelis, H.; Chaplot, V. Factors Affecting Crop Water Use Efficiency: A Worldwide Meta-Analysis. *Agric. Water Manag.* **2020**, *228*. [CrossRef]
3. Andrews, D.L.; Kumar, K.A. Pearl millet for food, feed, and forage. In *Advances in Agronomy*; Sparks, D.L., Ed.; Academic Press, Inc.: Cambridge, MA, USA, 1992; Volume 48, pp. 89–139.
4. Mason, S.C.; Maman, N.; Pale, S. Pearl Millet Production Practices in Semi-Arid West Africa: A Review. *Exp. Agric.* **2015**, *15*, 501–541. [CrossRef]
5. Samake, O.; Stomph, T.J.; Kropff, M.J. Integrated Pearl Millet Management in the Sahel: Effects of Legume Rotation and Follow Management on Productivity and Striga hermonthica infestation. *Plant Soil* **2006**, *286*, 245–257. [CrossRef]
6. Baumhardt, R.L.; Salinas-Garcia, J. Dryland agriculture in Mexico and the U.S. Southern Great Plains. In *Dryland Agriculture*, 2nd ed.; Agronomy Monograph No. 23; American Society of Agronomy, Crop Science Society of America, Soil Science Society of America: Madison, WI, USA, 2006.
7. Bhattarai, B.; Singh, S.; West, C.P.; Saini, R. Forage Potential of Pearl Millet and Forage Sorghum Alternatives to Corn under the Water-Limiting Conditions of the Texas High Plains: A Review. *Crop Forage Turfgrass Manag.* **2019**, *5*, 1–12. [CrossRef]
8. Machicek, J.A.; Blaser, B.C.; Darapuneni, M.; Rhoades, M.B. Harvesting regimes affect brown midrib sorghum-sudangrass and brown midrib pearl millet forage production and quality. *Agronomy* **2019**, *9*, 416. [CrossRef]
9. FAO. Production Quantities of Millet by Country. FAOSTAT, Crops. 2020. Available online: http://www.fao.org/faostat/en/#data/QC/visualize (accessed on 22 October 2020).
10. Ismail, S.M. Optimizing Productivity and Irrigation Water Use Efficiency of Pearl Millet as a Forage Crop in Arid Regions Under Different Irrigation Methods and Stress. *Afr. J. Agric. Res.* **2012**, *7*, 2509–2518. [CrossRef]
11. Sharma, B.; Kumari, R.; Kumari, P.; Meena, S.K.; Singh, R.M. Effect of Planting Pattern of Productivity and Water Use Efficiency of Pearl Millet in the Indian Semi-Arid Region. *J. Indian Soc. Soil Sci.* **2015**, *6*, 259–265. [CrossRef]
12. Silungwe, F.R.; Graef, F.; Bellingrath-Kimura, S.D.; Tumbo, S.D.; Kahimba, F.C.; Lana, M.A. The Management Strategies of Pearl Millet Farmers to Cope with Seasonal Rainfall Variability in a Semi-Arid Agroclimate. *Agronomy* **2019**, *9*, 400. [CrossRef]

13. Payne, W.A. Managing yield and water use of pearl millet in the Sahel. *Agron. J.* **1997**, *89*, 481–490. [CrossRef]
14. Maman, N.; Lyon, D.J.; Mason, S.C.; Galusha, T.D.; Higgins, R. Pearl millet and grain sorghum yield responses to water supply in Nebraska. *Agron. J.* **2003**, *95*, 1618–1624. [CrossRef]
15. Bhattarai, B.; Singh, S.; West, C.P.; Ritchie, G.L.; Trostle, C.L. Water Depletion Pattern and Water Use Efficiency of Forage Sorghum, Pearl Millet, and Corn Under Water Limiting Conditions. *Agric. Water Manag.* **2020**, *238*, 106206. [CrossRef]
16. Jahansouz, M.R.; Afshar, R.K.; Heidari, H.; Hashemi, M. Evaluation of Yield and Quality of Sorghum and Millet as Alternative Forage Crops to Corn under Normal and Deficit Irrigation Regimes. *Jordan J. Agric. Sci.* **2014**, *10*, 699–715. Available online: https://platform.almanhal.com/Files/2/94696 (accessed on 20 October 2020).
17. Ibrahim, A.; Abaidoo, R.C.; Fatandji, D.; Opoku, A. Hill Placement of Manure and Fertilizer Micro-Dosing Improves Yield and Water Use Efficiency in the Sahelian Low Input Millet-Based Cropping System. *Field Crops Res.* **2015**, *180*, 29–36. [CrossRef]
18. Manyame, C. On-Farm Yield and Water Use Response of Pearl Millet to Different Management Practices in Niger. Ph.D. Thesis, Texas A&M University, College Station, TX, USA, 2006. Available online: https://core.ac.uk/download/pdf/147132537.pdf (accessed on 20 October 2020).
19. Neal, J.S.; Fulkerson, W.J.; Hacker, R.B. Differences in Water Use Efficiency among Annual Forages Used by the Dairy Industry Under Optimum and Deficit Irrigation. *Agric. Water Manag.* **2011**, *98*, 759–774. [CrossRef]
20. Rostamza, M.; Chaichi, M.; Jahansouz, M.; Alimadadi, A. Forage Quality, Water Use and Nitrigen Utilization Efficiencies of Pearl Millet (*Pennisetum americanum* L.) Grown Under Different Soil Moisture and Nitrogen Levels. *Agric. Water Manag.* **2011**, *98*, 1607–1614. [CrossRef]
21. Geerts, S.; Raes, D. Deficit Irrigation as an on-farm Strategy to Maximize Crop Water Productivity in Dry Areas. *Agric. Water Manag.* **2009**, *96*, 1275–1284. [CrossRef]
22. Singh, B.R.; Singh, D.P. Agronomic and physiological responses of sorghum, maize, pearl millet, to irrigation. *Field Crops Res.* **1995**, *42*, 57–67. [CrossRef]
23. Hatfield, J.L.; Sauer, T.J.; Prueger, J.H. Managing Soils to Achieve Greater Water Use Efficiency: A review. *Agron. J.* **2001**, *93*, 271–280. [CrossRef]
24. Payne, W.A. Optimizing crop water use in sparse stands of pearl millet. *Agron. J.* **2000**, *92*, 808–814. [CrossRef]
25. Zhang, Z.; Whish, J.P.M.; Bell, L.W.; Nan, Z. Forage Production, Quality and Water-Use-Efficiency of Four Warm-Season Annual Crops at Three Sowing Times in the Loess Plateau Region of China. *Eur. J. Agron.* **2017**, *84*, 84–94. [CrossRef]
26. Nielsen, D.C.; Vigil, M.F.; Benjamin, J.G. Forage response to water use for dryland corn, millet, and triticale in the Central Great Plains. *Agron. J.* **2006**, *98*, 992–998. [CrossRef]
27. Unger, P.W. Alternative and opportunity dryland crops and related soil conditions in the Southern Great Plains. *Agron. J.* **2001**, *93*, 216–226. [CrossRef]
28. Web Soil Survey. 2019. Available online: https://websoilsurvey.sc.egov.usda.gov/App/HomePage.htm.
29. Allen, R.G. *REF-ET, Reference Evapotranspiration Calculator, Version 4.1*; Research and Extension Center, University of Idaho: Kimberly, ID, USA, 2016.
30. Ritchie, J.T.; Alagarswamy, G. Simulation of sorghum and pearl millet phenology. In *Modeling the Growth and Development of Sorghum and Pearl Millet*; Virmani, S.M., Tandon, H.L.S., Alagarswamy, G., Eds.; Research Bulletin No. 12; ICRISAT: Patancheru, Andhra Pradesh, India, 1989.
31. Olsen, J.E.; Hansen, P.K.; Berntsen, J.; Christensen, S. Simulation of above-ground suppression of competing species and competition tolerance in winter wheat varieties. *Field Crops Res.* **2004**, *89*, 263–280. [CrossRef]
32. Blaser, B.C.; Singer, J.W.; Gibson, L.R. Winter cereal canopy effect on cereal and interseeded legume productivity. *Agron. J.* **2011**, *103*, 1080–1185. [CrossRef]
33. Stephenson, R.J.; Posler, G.L. Forage yield and regrowth of pearl millet. *Trans. Kans. Acad. Sci.* **1984**, *87*, 91–97. [CrossRef]
34. Nielsen, D.C.; Vigil, M.F.; Benjamin, J.G. Evaluating decision rules for dryland rotation crop selection. *Field Crops Res.* **2011**, *120*, 254–261. [CrossRef]
35. Paris, P.; Di Matteo, G.; Tarchi, M.; Tosi, L.; Spaccino, L.; Lauteri, M. Precision Subsurface Drip Irrigation Increases Yield While Sustaining Water-Use Efficiency in Mediterranean Poplar Bioenergy Plantations. *For. Ecol. Manag.* **2018**, *409*, 749–756. [CrossRef]
36. Bouchard, A.; Vanasse, A.; Seguin, P.; Belanger, G. Yield and Composition of Sweet Pearl Millet as Affected by Row Spacing and Seeding Rate. *Agron. J.* **2011**, *103*, 995–1001. [CrossRef]

37. Musick, J.T.; Jones, O.R.; Stewart, B.A.; Dusek, D.A. Water-yield relationships for irrigated and dryland wheat in the U.S. Southern Plains. *Agron. J.* **1994**, *86*, 980–986. [CrossRef]
38. Ehlers, W.; Goss, M. *Water Dynamics and Plant Production*, 2nd ed.; CAB International: Boston, MA, USA, 2016.
39. Myers, R.J.K.; Foale, M.A. Row spacing and population density in grain sorghum—A simple analysis. *Field Crops Res.* **1981**, *4*, 147–154. [CrossRef]
40. Mohler, C.L. Enhancing the Competitive Ability of Crops. In *Ecological Management of Agricultural Weeds*; Liebman, M., Mohler, C.L., Staver, C.P., Eds.; Cambridge University Press: Cambridge, UK, 2004; pp. 269–305.
41. Nagaz, K.; Toumi, I.; Mahjoub, I.; Masmoudi, M.M.; Mechlia, N.B. Yield and Water-Use Efficiency of Pearl Millet (*Pennisetum glaucum* (L.) R. Br.) Under Deficit Irrigation with Saline Water in Arid Conditions of Southern Tunisia. *Res. J. Agron.* **2009**, *3*, 9–17.
42. Nielsen, D.C.; Vigil, M.F. Defining a Dryland Grain Sorghum Production Function for the Central Great Plains. *Agron. J.* **2017**, *109*, 1582–1590. [CrossRef]
43. Hao, B.; Xue, Q.; Bean, B.W.; Rooney, W.L.; Becker, J.D. Biomass Production, Water and Nitrogen Use Efficiency in Photoperiod-Sensitive Sorghum in the Texas High Plains. *Biomass Bioenergy* **2014**, *62*, 108–116. [CrossRef]
44. Ritchie, J.T. Efficient water use in crop production: Discussion on the generality of relationships between biomass production and evapotranspiration. In *Limitations to Efficient Water Use in Crop Production*; Tanner, H.M., Jordan, W.R., Sinclair, T.R., Eds.; ASA-CSSA-SSSA: Madison, WI, USA, 1983; pp. 29–44.
45. Sinclair, T.R.; Weiss, A. *Principles of Ecology in Plant Production*, 2nd ed.; CABI: Cambridge, MA, USA, 2010.
46. Hardwick, S.R.; Toumi, R.; Pfeifer, M.; Turner, E.C.; Nilus, R.; Ewers, R.M. The Relationship between Leaf Area Index and Microclimate in Tropical Forest and Oil Palm Plantation: Forest Disturbance Drives Changes in Microclimate. *Agric. For. Meteorol.* **2015**, *201*, 187–195. [CrossRef]
47. Blum, A. Effective Use of Water (EWU) and not Water-Use Efficiency (WUE) is the Target of Crop Yield Improvement under Drought Stress. *Field Crops Res.* **2009**, *112*, 119–123. [CrossRef]

12

The Efficiencies, Environmental Impacts and Economics of Energy Consumption for Groundwater-Based Irrigation in Oklahoma

Divya Handa [1], Robert S. Frazier [1], Saleh Taghvaeian [1,*] and Jason G. Warren [2]

1 Department of Biosystems and Agricultural Engineering, Oklahoma State University, Stillwater, OK 74078, USA; dhanda@ostatemail.okstate.edu (D.H.); robert.frazier@okstate.edu (R.S.F.)

2 Department of Plant and Soil Sciences, Oklahoma State University, Stillwater, OK 74078, USA; jason.warren@okstate.edu

Abstract: Irrigation pumping is a major expense of agricultural operations, especially in arid/semi-arid areas that extract large amounts of water from deep groundwater resources. Studying and improving pumping efficiencies can have direct impacts on farm net profits and on the amount of greenhouse gases (GHG) emitted from pumping plants. In this study, the overall pumping efficiency (OPE), the GHG emissions, and the costs of irrigation pumping were investigated for electric pumps extracting from the Rush Springs (RS) aquifer in central Oklahoma and the natural gas-powered pumps tapping the Ogallala (OG) aquifer in the Oklahoma Panhandle. The results showed that all electric plants and the majority of natural gas plants operated at OPE levels below achievable standard levels. The total emission from the plants in the OG region was 49% larger than that from plants in the RS region. However, the emission per unit irrigated area and unit total dynamic head of pumping was 4% smaller for the natural gas plants in the OG area. A long-term analysis conducted over the 2001–2017 period revealed that 34% and 19% reductions in energy requirements and 52% and 20% decreases in GHG emissions can be achieved if the OPE were improved to achievable standards for plants in the RS and OG regions, respectively.

Keywords: pumping plants; energy audit; life cycle assessment; greenhouse gas emission; center-pivot irrigation

---

1. Introduction

Irrigated agriculture around the world relies heavily on energy resources to extract freshwater and to convey it to application sites. This is especially the case in arid/semi-arid regions, where large amounts of irrigation water are required to sustain crop production. As a result, the availability and cost of energy are among major factors influencing the economic viability of irrigated agriculture in these regions. In addition, energy consumption for irrigation has major environmental consequences, mainly due to the emission of greenhouse gases [1,2]. Wang et al. [3] have reported that pumping groundwater for irrigation accounts for 3% of total emissions from agriculture in China. A similar study in Iran has found that groundwater pumping is responsible for 3.6% of total carbon emissions in the country [4]. In India, groundwater pumping is estimated to be the source of nearly 6% of India's total emissions [5]. In the U.S., carbon emissions due to pumping irrigation water have been reported to be about three million metric tons per year, with electric pumps responsible for 46% of the total emission, followed by diesel (32%) and natural gas (19%) [6].

Energy consumption and its associated costs and greenhouse gas emissions can be reduced by improving pumping efficiency [7]. In a study in central Tunisia, improving pumping efficiency was found to result in 33% cost reduction on average [8]. An average cost saving of 17% following efficiency improvement was also reported in [9] for an irrigated area in southeastern Spain. Pump efficiency is primarily dependent on operating conditions such as total dynamic head (TDH) and the condition of the pump. Any deviation from optimum conditions can lead to reduced efficiency and increased costs and emissions.

One deviation from optimum conditions is the change in TDH, caused by declines in groundwater levels. This is especially the case in irrigated areas that experience large declines due to increasing groundwater extraction. In the North China Plain, Qui et al. [10] estimated that groundwater declines from 1996 to 2013 have led to a 22% increase in energy consumption and a 42% increase in greenhouse gas emissions. Increases in groundwater depth will not only increase TDH and consequently energy use, but will also result in a gradual deviation from design parameters used in selecting the most efficient pump and hence a reduction in system efficiency.

Irrigated agriculture in Oklahoma has been facing similar energy-related challenges. In 2013, Oklahoma producers spent over USD 22 million to power more than 5,300 pumps [11]. Electricity was the main source of pumping energy, supplying water to 46% of all irrigated areas in the state. This was closely followed by natural gas, which powers pumps to irrigate 42% of all irrigated lands [11]. The remainder of irrigation pumps typically use diesel or propane units. In addition, Oklahoma producers who rely on groundwater resources have been experiencing a decline in water availability, reflected in a reduction in average well flow rates from 0.032 $m^3\ s^{-1}$ in 2008 to 0.026 $m^3\ s^{-1}$ in 2013 [11]. The groundwater decline has been significant in the Panhandle region due to increased usage, drought periods, and negligible recharge rates. Identifying energy consumption efficiencies and improved practices can have a considerable impact on the profitability of agricultural production in Oklahoma.

The goal of this study was to identify the overall efficiency and environmental impact of irrigation pumping in two agricultural regions of central and western Oklahoma that rely on two aquifers with significantly different depths to groundwater. The more specific objectives included: (i) to estimate the overall pumping efficiency for several pumping plants in each region; (ii) to conduct a life cycle assessment and calculate greenhouse gas emissions under existing and achievable efficiencies; and (iii) to investigate the impacts of changing groundwater depths and energy prices on the economics and emissions of irrigation pumping plants over a 17-year period. To the best of our knowledge, only a few previous studies have identified the overall efficiency of agricultural irrigation pumping. For example, Luc [8] has determined the efficiency of 18 electric pumps in central Tunisia. The lack of data in this field is most probably due to large human, technical, and financial resources required to carry out field evaluations of pumping plants. As a result, many previous life cycle assessment studies have assumed or approximated pumping efficiencies in their analysis as opposed to using measured values [2–4,7,10]. This study combines efficiencies measured through field audits with greenhouse gas analysis and uses the results to explore long-term effects of changing groundwater levels on energy costs and emissions. The results will give Oklahoma agricultural producers, water managers, and other decision makers an insight into current economics and the environmental footprint of irrigation pumping in the study areas, as well as potentials for improvement. Moreover, the results will be transferable to areas with similar agro-climatological and groundwater resources conditions.

2. Materials and Methods

2.1. Study Area

A total of 24 irrigation pumping plants in central and Panhandle regions of Oklahoma were tested between 2015 and 2018 with the aim of determining their energy consumption efficiencies, emissions, and expenses. Of the pumping plants evaluated, fourteen were located within the Ogallala (OG) aquifer and ten within the Rush Springs (RS) aquifer (Figure 1). The OG sites were all natural gas internal combustion powered and the RS sites were electricity-powered pumping plants. The Ogallala is one of the most important aquifers in Oklahoma, supplying more than 98% of total water demand in the Panhandle regions. It is classified as an unconfined bedrock aquifer composed of semi-consolidated clay, silt, sand, and gravel layers [12]. The maximum thickness of the aquifer is 213 m and the groundwater flow direction is toward the east/southeast, similar to the land surface elevation gradient. OG water quality is considered good, with an average pH of 7.3 and specific conductance of 0.64 decisiemens per metre (dS m^{-1}) [12]. RS is another important bedrock aquifer in the state and provides irrigation water to numerous fields in central Oklahoma. This aquifer is composed of Rush Springs sandstone on top of the Marlow formation. The maximum thickness of RS is 101 m, with groundwater moving in a south/southeast direction [12]. Similar to OG, the water quality of RS is good, with an average pH of 7.2 and specific conductance of 1.08 dS m^{-1} [12]. The depth to groundwater, however, is much larger in the OG (with a 2018 average of 57.1 m), and it has experienced a steady decline over the past several decades, while RS is shallower (with a 2018 average of 18.2 m) and is more sensitive to inter-annual variations in precipitation [11,12].

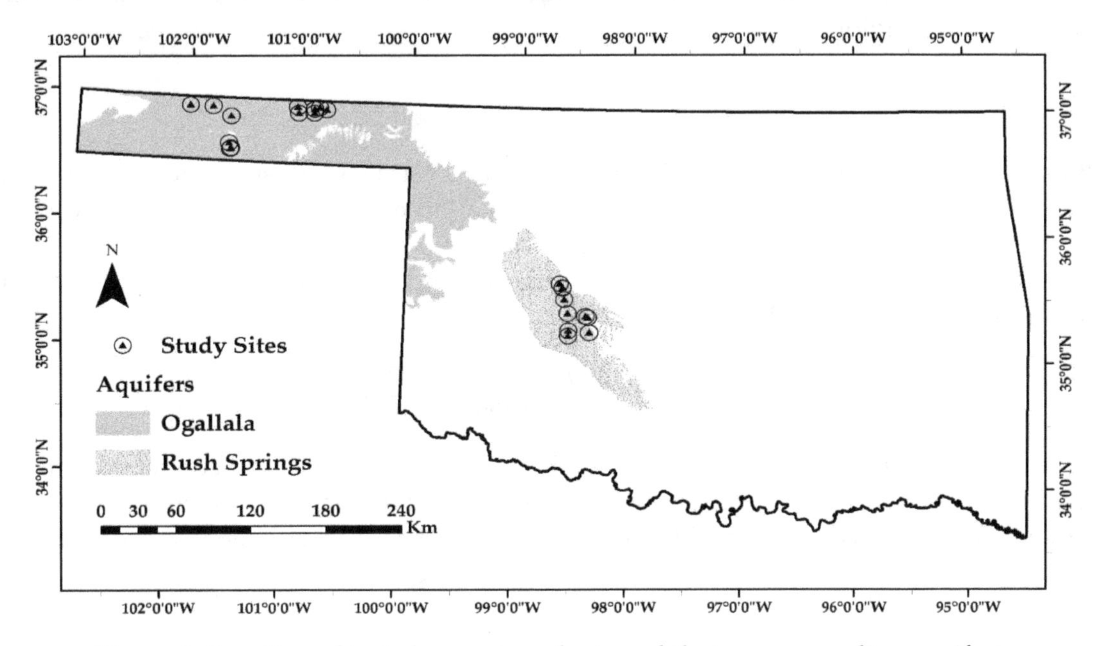

Figure 1. Location of tested pumping plants and their corresponding aquifers.

2.2. Energy Audits

The energy audits included determining several basic irrigation well and pump parameters such as groundwater depth (GWD), water pressure, and discharge rate. These parameters were then used to estimate the overall pumping efficiency (OPE), a widely used metric for assessing the efficiency of irrigation pumping plants. The OPE is the ratio of the output work the pump exerts on the water at the pump outlet, known as water horsepower (WHP), to the required energy input or energy horsepower (EHP) of the driving unit required to pump the measured water output [13], and is calculated as

$$OPE = \frac{Water\ output\ (WHP)}{Energy\ input\ (EHP)} \times 100 \quad (1)$$

The WHP (kW) can be determined as

$$WHP = \frac{Q \times TDH}{F} \quad (2)$$

where Q is the discharge rate ($m^3\ s^{-1}$), TDH is the total dynamic head (m) and F is a conversion factor equal to 0.102 ($m^4\ s^{-1}\ kW^{-1}$). In this study, Q was measured using an ultrasonic flow meter (Portaflow-C, Fuji Electric Co., Japan) on the discharge pipe from the pump. The accuracy of the ultrasonic flow meter was tested previously against a calibrated flow device and found to be acceptable [14].

The TDH is the total equivalent pressure that must be applied to the water column being pumped while also taking into account the losses due to friction [13]. It may be expressed as

$$\text{TDH} = \text{pumping lift} + \text{pressure head} \quad (3)$$

where pumping lift is the vertical distance between the pumping water level and center of the pump outlet (m) and pressure head is the pressure required at the pump outlet (m). The pumping lift was measured by lowering a water level meter (model 102, Solinst Canada Ltd., Canada) probe through an access hole in the pump base-plate while a pressure gauge close to the pump outlet was used to measure the pressure head [15].

The estimation procedure for energy horsepower (EHP) depends on the type of energy used and differs among electric motors in the Rush Springs aquifer region and natural gas engine driven pumps in the Ogallala aquifer region.

2.2.1. Electric Motors

In Oklahoma, electric motor driven irrigation pumps tend to be used where the ground water depth is less than 80 m and three-phase power is available. These pumps usually require less maintenance and operational activity than internal combustion engines. For electric motors, the energy input (kW) is the electrical power supplied to the motor and can be calculated using the following equation for a three-phase motor [16]:

$$EHP = \frac{V \times I \times PF \times 1.732}{1000} \quad (4)$$

where V is voltage (V), I is current (A), PF is the power factor, and 1.732 is a conversion factor. In this study V, I, and PF were measured using a three-phase electric meter. The current of each of the three legs was first measured individually and then averaged together. The voltage was measured across all three legs and also averaged.

2.2.2. Natural Gas Engines

The natural gas consumption of the internal combustion engines was measured by a rotary gas meter (Dresser Roots® Series B, General Electric, Boston, MA, USA). The gas meter was installed by first turning off the gas supply to the engine at the gas meter. Next, the main fuel line running to the intake manifold was disconnected and the rotary meter was installed in-line with this gas line, which was then reconnected to the engine. The engine was allowed to run until in steady state operating temperature. The irrigation water pump was also allowed to bring the entire irrigation system up to operating pressure (with water delivery from all nozzles). Then the engine and pump system were allowed to run for 30–45 min, at which time average fuel consumption readings and correction factors were recorded. Removing the rotary meter was the reverse of the installation procedure. The meter auto-corrects for gas pressure, density, and temperature. The display gives readings of cubic feet per minute, which were converted to mechanical power and used as the input fuel power.

The estimated OPE of the audited pumping plants was compared against the widely used Nebraska Pumping Plant Performance Criteria (NPPPC). According to the NPPPC, the OPE of accurately designed and appropriately maintained electricity- and natural gas-driven pumping plants should be 66% and 17%, respectively [16]. With the electric pumping plants, their OPE was also compared against thresholds proposed by Hanson [17] for different required corrective actions.

2.3. *Life Cycle Assessment*

The greenhouse gas (GHG) emissions for the pumping sites were calculated using the GREET® (GREET.NET version 2017, Argonne National Laboratory, Argonne, IL, USA) and the U.S. Environmental Protection Agency (EPA) Greenhouse Gas Inventory Guidance [18]. The GREET Life Cycle Assessment (LCA) tool is an online resource that is well established and supported in the LCA focus area. While the GREET model is basically a transportation LCA tool used to model vehicle emissions due to different energy sources (e.g., biofuels, natural gas, electricity, etc.), it can also provide a good approximation for stationary power systems. The GREET model examines fuels and energy production from raw materials (e.g., coal or crude oil), extraction and processing (the "well"), to storage (the "pump"). This technique in GREET is called "well to pump" (WTP).

In the case of electricity where the end-use is essentially emission-free for both stationary and vehicle uses, the WTP model can be used alone for stationary irrigation pumping plants without the need for any additional modification. The "U.S. Central and Southern Pains" Utility Mix category was chosen to represent the electrical grid generator composition for Oklahoma. In all cases, the emissions were calculated for 1000 h of irrigation system operation. In the case of natural gas, there are on-site emissions due to the engine combustion process. Therefore, the GREET WTP analysis (extraction, transmission, and distribution) was added to an established EPA stationary engine emissions calculation technique [18] to give an approximation of the total GHG emissions for stationary engines from raw material extraction to end-use combustion. In short, the electricity production generates its GHGs at different points from the natural gas engines. Both generate GHGs in preparing the fuel/energy, but the natural gas engines contribute large percentages of GHGs at the end-use (combustion) phase. Of course, current electrical production also generates combustion GHGs at the generation phase depending on the generator technology mix. GREET captures this in the electrical WTP technique.

The EPA end-use GHG estimation for the natural gas methodology employed was based on using the natural gas volume consumed (measured) during the audit field tests with a gas flow meter. The methodology is may be given as

$$\text{Em} = \text{Fuel} \times \text{HHV} \times \text{EF} \tag{5}$$

where Em is the mass of CO_2, CH_4, or N_2O emitted, Fuel is the mass or volume of fuel combusted, HHV is the fuel heat content (higher heating value) in units of energy per mass of fuel, and EF is the emission factor of CO_2, CH_4, or N_2O per energy unit. The HHV and EF values reported in [18] for natural gas combustion were used in this study. For the total GHG emissions from combustion, CO_2 equivalence factors of 25 for CH_4 and 298 for N_2O were applied. As with electric motor pumps, the emissions for natural gas pumping engines were reported for 1000 h of irrigation system operation. Readers should be aware that the LCA approach implemented in this study has uncertainties beyond those caused by errors in the input data, mainly due to assumptions and simplifications adopted in the procedure.

2.4. *Long-Term Trends*

Irrigated agriculture in the study area relies heavily on groundwater. Hence, it is of great importance to investigate the impacts of long-term fluctuations in groundwater levels on the efficiencies, emissions, and economics of irrigation pumping. The first step to conducting this analysis was to estimate variations in energy requirement in response to changes in groundwater depth for each of the studied aquifers (Ogallala and Rush Springs). Several previous studies have investigated energy required for pumping groundwater as a function of depth to groundwater [4,7,19,20]. These studies have used the following equation or a variation of it; it has been selected in the present study and applied to estimate annual energy requirements over the 17-year period from 2001 to 2017:

$$Energy = \frac{TDH \times M \times g}{3.62 \times 10^6 \times OPE} \tag{6}$$

where energy is in kWh, M is the total mass of groundwater pumped for irrigation (kg), g is the gravitational acceleration (9.8 m s^{-2}), and other parameters are as they were defined above.

Since actual long-term TDH data for audited systems were not available, this parameter was approximated through developing a linear regression model to predict TDH from groundwater depth based on the data collected during energy audits. The assumption was that GWD is by far the largest portion of TDH, especially since all tested center pivot systems were of mid-elevation spray application type and thus required significantly lower operating pressures compared to center pivots equipped with impact sprinklers. The close proximity of irrigation wells to irrigation systems meant that pressure losses during water conveyance were small too. Once this relationship was developed it was applied to the average annual GWD estimated during the 2001–2017 period based on readings reported by the Oklahoma Water Resources Board at 42 and 22 observation wells in the Ogallala and Rush Springs aquifers, respectively. The readings were made on an annual basis (in winter) using electric tapes. The observation wells were selected in a way as to to represent the entire aquifer. The number of wells with continuous GWD data dropped quickly for years before 2001, which is why the 17-year period from 2001 to 2017 was selected for analysis in this study.

The discharge rates obtained during the energy audits were averaged for each studied aquifer and used in obtaining M, assuming 1000 h of pump operation per year. The OPE was estimated in a similar fashion, assuming that the average OPE of audited systems in each region was a reasonable representative of the average OPE of all systems in that region. In addition, this average OPE was assumed to be an acceptable representative of average OPE over the long-term period considered (2001–2017). After obtaining annual energy requirements, the LCA analysis explained in the previous section was implemented to estimate variations in GHG emissions as impacted by fluctuations in groundwater depth.

2.5. Economic Analysis

The economic implications of improving OPE at studied pumping plants was also investigated. Irrigation energy costs can be one of the largest categories of costs a producer in Oklahoma will incur over a season, reaching about 22 million USD [11]. Improving OPE to the levels recommended by the NPPPC could decrease irrigation pumping costs. In this study, energy cost analysis was conducted in two parts. The first part focused on estimating the energy cost of pumping 1000 h per year under current efficiencies and potential savings if the OPE of each tested pumping plant was improved to NPPPC levels. Unit energy costs of 0.05 USD per kWh for electricity-powered and 3.30 USD per 1000 cubic feet (MCF) of fuel for natural gas-powered plants were used in the analysis based on costs reported in the U.S. Energy Administration Information web portal for Oklahoma for the year when the tests were conducted [21].

The second part examined changes in long-term (2001–2017) pumping costs for the average OPE in each aquifer region by taking into account annual variations in groundwater depth and energy costs. The cost of electricity in Oklahoma varied from 0.04 USD kWh^{-1} in 2002 to 0.06 USD kWh^{-1} in 2008 and 2014 [21]. It should be noted that the actual blended rates would be higher than the projected rates used here. Hence, the actual pumping costs for these plants will be larger than those estimated in the present study. Compared to the cost of electricity, greater inter-annual variations were observed in natural gas costs in Oklahoma, ranging from 2.94 USD MCF^{-1} in 2016 to 13.03 USD MCF^{-1} in 2008 [21]. Finally, the energy costs for all electric and natural gas pumping plants in Oklahoma were approximated by assuming the estimates for audited sites are statistically representative of all corresponding plants in the state. This assumption may not be accurate, but it provides an estimate of the magnitude of statewide energy expenses for irrigation pumping.

3. Results and Discussion

3.1. Energy Audits

The measured parameters showed significant differences among audited sites in the Rush Springs and Ogallala aquifer regions. The average static groundwater depths for instance, were 24.4 and 79.7 m for RS and OG, respectively. Barefoot [22] has tested 13 natural gas irrigation pumping plants in the Oklahoma Panhandle region (OG) and reported a similar average pumping lift of 80.4 m. The average dynamic GWD, measured 15 min after starting the pump, was 30.7 and 89.1 m for the same aquifers, respectively. The measured water pressure was larger for irrigation systems in RS, resulting in a smaller difference in TDH compared to GWD. The average TDH was 67.8 and 105.9 m for the RS and OG aquifers, respectively.

The difference in TDH was accompanied by a corresponding difference in input energy. With an average value of 270 kW (362 Hp), the input power requirement in the OG was nearly five times larger than that in the RS region, which had an average of 56 kW (75 Hp). This probably explains the preference of natural gas engines over electric motors as an energy source for powering OG pumping plants since large electric motors have specific wiring and utility constraints. The water discharge rates were similar in the two study regions, with average values of 36.2 and 36.0 $l\ s^{-1}$ for the RS and OG aquifers, respectively. The average discharge reported in [22] was 47.9 $l\ s^{-1}$ in the OG aquifer region, about 33% larger than the value found in the present study.

The overall pumping efficiency of the sites in the RS aquifer region (electricity-powered) varied from 24.9% to 62.6%. Of the ten pumping plants evaluated, seven had an OPE of less than 50%, which was proposed in [17] as the threshold below which repairing or replacing the pump should be considered. All of the systems had efficiencies smaller than the recommended OPE of 66% by the NPPPC standard. The average OPE for the RS region was 43.3%. The difference between the estimated OPE and the NPPPC standard implies that nearly 23% of electrical energy is wasted on average due to poor efficiency of the pumping plant in the RS region. The average OPE in this study compares well with the average OPE of 42.6% reported in [23] and 47.0% in [24] for pumping plants in the High Plains and Trans-Pecos areas of Texas. The range of efficiencies in [24] was also similar to that in this study, with values varying from 16.8% to 70.6%. However, DeBoer et al. [25] have reported a larger average OPE of 58% in for electricity-driven pumping plants in west central Minnesota, North Dakota, and South Dakota. The plants tested in DeBoer's study were fairly new, with 74% being less than six years old. The younger age of the pumps could be the cause of higher average efficiency.

The OPE of the natural gas-powered pumping plants in the OG aquifer region ranged from 5.7% to 21.4%. Out of 14 audited pumping plants, ten had an OPE less than the NPPPC recommended standard of 17% for natural gas internal combustion engines pumping plants. The average OPE for the OG region was 13.6%, close to average OPEs of 13.2%, 11.7%, and 13.1% as reported for natural gas-powered pumping plants in Oklahoma and Texas by Barefoot [22], Fipps et al. [23], and New and Schneider [24], respectively. The range of OPEs in [24] was 2.2–21.6%, similar to the range of OPEs estimated in the present study.

Linear regression analysis conducted on data collected at each site and the two sites combined revealed no significant relationship between OPE and the two parameters of TDH and discharge rate (p values larger than 0.37). This suggests that the performance of audited systems was impacted by other factors such as the type, age, and condition of the pumping plants. Small sample sizes of systems tested may have also contributed to the lack of correlation. Table 1 presents the average values of key parameters for tested plants in the two study areas.

Table 1. Average values of the main characteristics of studied pumping plants in the Rush Springs (RS) and Ogallala (OG) aquifer regions.

Parameter	RS	OG
Static groundwater depth (m)	24.4	79.7
Dynamic groundwater depth (m)	30.7	89.1
Total dynamic head (m)	67.8	105.9
Discharge ($l\ s^{-1}$)	36.2	36.0
Overall pumping efficiency (%)	43.3	13.6

3.2. Life Cycle Assessment

The LCA of electric motor pumps in the RS region examined the greenhouse gas (GHG) emissions at an electric generation station for 1000 h of pump operation. The total emissions from these pumping plants ranged from 28.4 to 52.9 and averaged 40.1 metric tons of equivalent CO_2 (t CO_2e) emissions. In order to facilitate comparison with other studies, it is useful to report emissions for the unit irrigated area and unit TDH. The pumping plants in the present study were all serving a standard-size center pivot system, with an irrigated area of about 50.8 ha per system. Hence, the average emission from electrical pumps per unit area of irrigated land and unit TDH would be 11.8 kg CO_2 e ha^{-1} m^{-1}. This is within the range of 4–93 kg CO_2 e ha^{-1} m^{-1} reported in [7] for a variety of crops irrigated by electrical pumps in the Haryana state of India.

LCA of natural gas driven pumps examined the emissions through a two-part analysis that used GREET WTP and EPA calculations for stationary engines. The first part of the analysis provided estimates for natural gas extraction, processing, storage, and transportation (off-site), while the second part resulted in emission estimates for end-use at the irrigation field (on-site). The total off-site GHG emissions estimated for natural gas pumping plants in the OG region averaged 11.0 and had a range of 6.0–17.6 t CO_2e. The on-site emissions varied from 26.7 to 78.1 and averaged 48.8 t CO_2e. The average total emission from off- and on-site analyses was 59.8 t CO_2e. This is equal to 11.4 kg CO_2e ha^{-1} m^{-1} when expressed in terms of emission per unit area of irrigated land and unit TDH. The maximum total emission was 95.7 t CO_2e and belonged to a pumping plant that had the fourth lowest OPE and the seventh largest groundwater depths, a combination resulting in the maximum amount of energy use.

On average, the total GHG emission from pumping plants in the OG region was 49% larger than that of the pumping plants in the RS region. However, the emission per unit irrigated land and unit TDH was 4% smaller. This indicates that TDH, which is greatly impacted by groundwater depth, plays a significant role in determining the GHG emissions from agricultural pumping plants. The emissions can also be reported per unit volume of water extracted from aquifers. This analysis resulted in emission ranges of 0.20–0.45 and 0.34–0.99 kg CO_2 e m^{-3} for pumping plants in the RS and OG regions, respectively. These estimates are similar to the range of 0.18–0.60 kg CO_2 e m^{-3} reported in [3] for all 31 provinces in China with variable proportions of electricity and diesel driven pumps.

3.3. Long-Term Trends

Examination of the observed groundwater depth data showed that the Rush Springs aquifer levels varied between 18.2 and 21.0 m below the ground surface over the 17-year period, with a net decline of 1.5 m (Figure 2a). On the other hand, the Ogallala aquifer GWD experienced a steady decline from 56.6 to 62.3 m below the ground surface (Figure 2b). The OG aquifer has significantly smaller recharge rates. As a result, no rise in water level was observed in the OG aquifer during wet periods in 2005, 2007–2009, and 2015–2017, while the RS aquifer experienced rises in groundwater level. The rate of decline in water levels was greater during the drought years of 2011–2014 compared to wet and normal years for both aquifers, an indication of increased pumping for irrigation during this dry period.

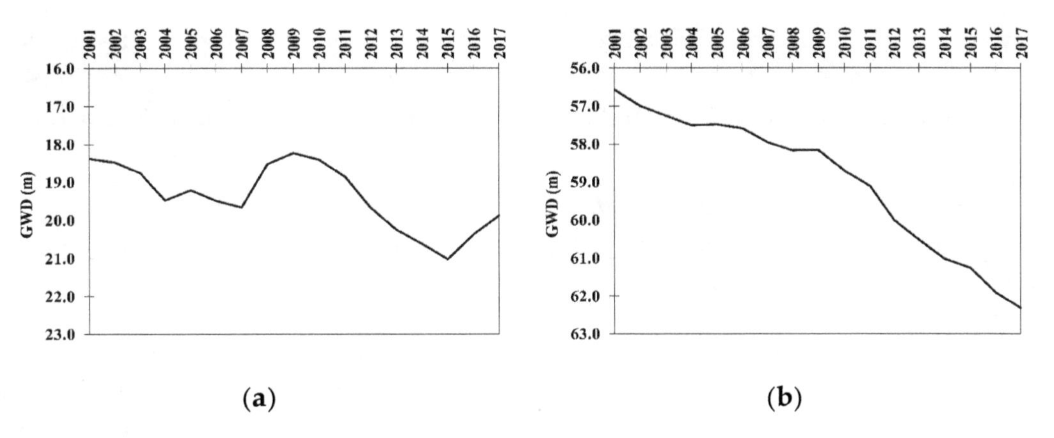

Figure 2. Annual variations in groundwater depth (GWD) for plants in the (**a**) Rush Springs, and (**b**) Ogallala aquifers.

The TDH and GWD measurements at audited sites were strongly correlated, with a Pearson coefficient of 0.88. The linear relationship developed based on TDH and GWD is presented below:

$$\text{TDH} = 0.67 \times \text{GWD} + 53.76, \quad (7)$$

This relationship was statistically significant ($p < 0.001$) and had a coefficient of determination (R^2) of 0.78, suggesting that over three-fourths of variability in TDH could be explained by changes in GWD. A similar approach was employed in [3], where the slope, intercept, and R^2 were 0.91, 21.75, and 0.62 for a linear relationship between pump lift and GWD.

As expected, the variations in energy requirement during the 2001–2017 period had a pattern similar to that of GWD in each aquifer region. In the case of RS, the energy requirement for 1000 h of pump operation per year varied from 53,721 to 55,247 kWh during the 17 years considered and had an average of 54,344 kWh. The energy requirement was much larger at OG and increased over time, with a range of 233,175–242,980 kWh and an average of 237,277 kWh (Figure 3). This was more than four times larger than the average in RS. When considering energy requirements per unit volume of pumped water, the RS and OG regions had average rates of 0.42 and 1.84 kWh m^{-3}, respectively. These values are similar to energy use rates of 0.21 to 0.66 kWh m^{-3} reported in [3] across all provinces in China.

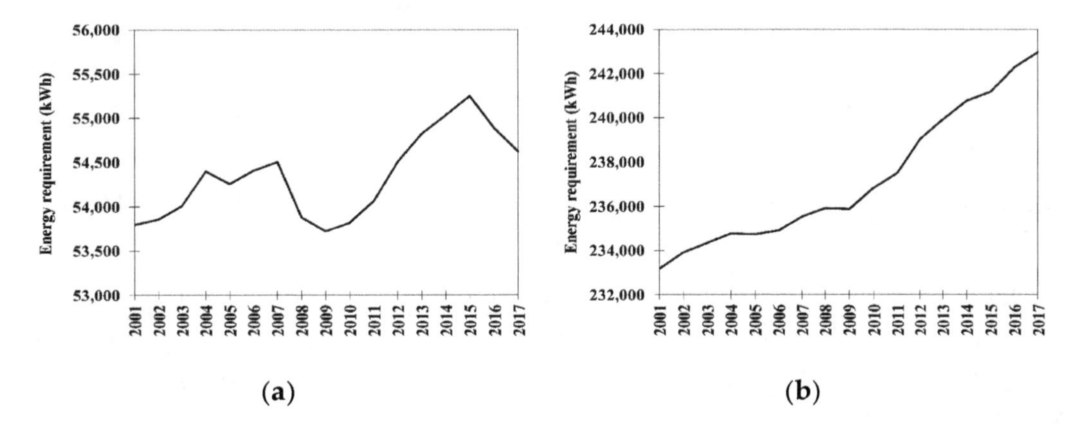

Figure 3. Annual variations in energy requirements for 1000 h of operation per year for plants in the (**a**) Rush Springs, and (**b**) Ogallala aquifers.

In OG, the increase in energy requirements due to increases in GWD over the 17-year period was 4% of the initial (2001) amount. Qiu et al. [10] have reported a significantly larger increase of

22% in energy use in China between 1996 and 2013. However, the rate of groundwater level decline in their study was 0.6 m $year^{-1}$, two times larger than the drop rate of 0.3 m $year^{-1}$ observed in the present study. The results also revealed that improving the OPE in each region to achievable levels recommended by the NPPPC would result in 34% and 19% reductions in average energy requirements in the RS and OG regions, respectively.

The increase in energy requirements in the OG aquifer region over the long-term period resulted in a continuous increase in total GHG emissions from 51.7 t CO_2e in 2001 to 53.9 t CO_2e in 2017. In the RS aquifer region, the total GHG emissions varied between 38.3 t CO_2e in 2009 and 39.4 t CO_2e in 2015, the year that marked the end of a severe drought that caused significant declines in groundwater levels. Apart from the groundwater level, the OPE had a large influence on the energy use rate and emissions. Improving the OPE of electricity-powered pumping plants in the RS aquifer region to the NPPPC recommended standard of 66% could on an average reduce total GHG emissions by nearly 52%. Similarly, improving the OPE of natural gas-powered pumping sites from an average of 13.75% to the NPPPC recommended 17% level in the OG aquifer region could potentially reduce emissions by 20%. In India, it has been reported that improving electric pumping system efficiency (OPE) to 51% from an existing 34.7% level could lead to a decline of 32% in CO_2 emissions [7].

3.4. *Economic Analysis*

Based on the audit results, the current seasonal pumping cost for 1000 h was highly variable among pumping plants, with an average of 2827 USD for electricity-powered pumping sites in the RS aquifer region. The pumping expense was also variable among natural gas-powered pumping sites, with an average of 3042 USD, which was only 8% larger than that of electric plants. Such a small difference despite significant differences in energy use is due to low natural gas prices during the study period. It should be also noted that producers may run their pumps longer than 1000 h per year depending on crop type, weather conditions, and well yield. A longer operating hour will result in a linear increase in energy consumption costs. A significant potential for reduction in pumping costs was estimated if OPE was improved to meet NPPPC standards. The average saving for electric pumps was 35% of the current pumping costs. With the prices at the time of study, this amount of saving was equal to 1000 USD or one USD per every hour of pumping. For natural gas-powered pumping plants, the average saving was 23% of existing pumping costs, equal to 711 USD (71 cents per every hour of pumping). As mentioned before, this study was conducted when natural gas prices were among the smallest in the past several years in Oklahoma. Higher costs would have resulted in larger dollar values of saving. The results obtained here are consistent with the findings of Hardin and Lacewell [26] who reported significant decreases in fuel costs and increases in farm profits if the OPE of irrigation pumping plants in the Texas High Plains was improved to achievable levels.

Our long-term analysis showed that pumping cost for electric plants ranged from 2052 USD in 2002 to 3219 USD in 2014 and averaged 2788 USD for 1000 h of pumping per year (Figure 4). In the case of natural gas, inter-annual fluctuations were significantly larger. The pumping costs for these plants varied between 2429 USD in 2016 and 10,482 USD in 2008 and averaged 6490 USD for 1000 h of pumping per year (Figure 4). According to the most recent Farm and Ranch Irrigation Survey (FRIS) conducted by the U.S. Department of Agriculture, 3456 electricity-powered and 1354 natural gas-powered pumping plants are used for irrigation in Oklahoma [27]. Assuming that these plants have pumping costs similar to those estimated in this study, the total annual cost of irrigation pumping in Oklahoma can be approximated at 9.6 and 8.8 million USD for electric and natural gas plants, respectively. These estimates are similar to the total energy expenditure of 9.5 and 10.5 million USD reported for the same two sources of energy in [27], respectively.

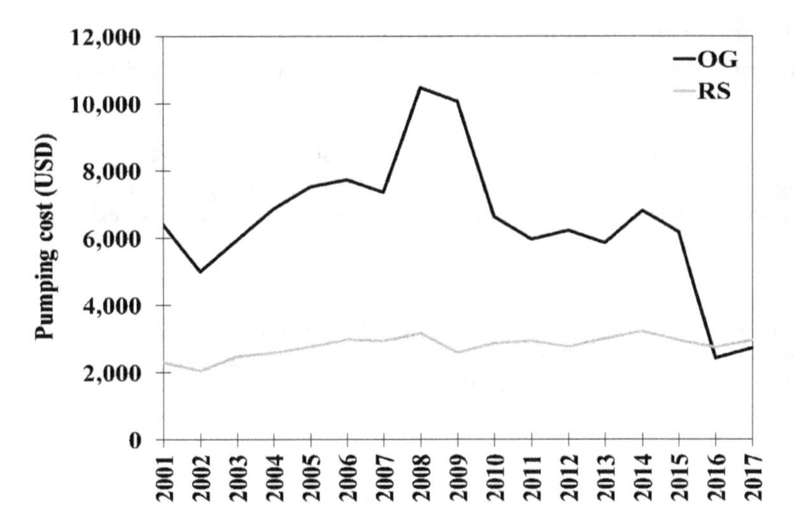

Figure 4. Annual variations in pumping cost for 1000 h of operation per year for the Ogallala and Rush Springs aquifers.

4. Conclusions

The future of irrigated agriculture is largely dependent on its financial and environmental sustainability. In arid/semi-arid regions where large amounts of irrigation water is required to meet crop demand and where available water resources are usually limited or difficult to extract, energy consumption for irrigation pumping can play a significant role in both environmental and financial sustainability of agricultural production. In this study, the efficiencies, environmental footprint, and economics of irrigation pumping were investigated in two Oklahoma areas that rely heavily on groundwater resources. The first area was in central Oklahoma where electric pumps are used to extract water from the Rush Springs aquifer and the second area was in the Oklahoma Panhandle where natural gas-powered pumps remove water from the declining Ogallala aquifer. Field visits were conducted in the period from 2015 to 2018 to collect required data for analysis. The results confirmed significant differences among the regions in terms of depth to groundwater. It was also revealed that all electric pumps and the majority of natural gas plants had an overall pumping efficiency below the standard rates achievable under field conditions. The range and average OPE obtained in this study were similar to those reported in previous studies, including those that were conducted a few decades ago.

The life cycle assessment results showed a significant difference in total emissions, with natural gas-powered plants emitting 49% more greenhouse gases when considering the entire process from the extraction and transportation of natural gas to its combustion at the pump site. However, the average emission expressed per unit irrigated area and unit total dynamic head of pumping was slightly smaller for the natural gas plants in the OG region. In the same region, the energy required for pumping increased by 4% between 2001 and 2017 due to continuous declines in OG water levels. The results showed that improving the OPE to achievable standards could have resulted in 34% and 19% reductions in average energy consumption during the 2001–2017 period in the RS and OG regions, respectively. The reductions in GHG emissions would have been 52% and 20% for the same study areas, respectively.

The cost of irrigation pumping was also estimated for the pumping plants tested and for the two regions over the 2001–2017 period. Compared to RS, large inter-annual variations in the cost of pumping were found for agricultural producers in the OG, mainly due to the large fluctuations in the natural gas price. Assuming that the pumping plants audited in the present study represent all electric and natural gas plants in Oklahoma, the statewide energy expenses were estimated and found to be in good agreement with those reported in surveys conducted by the U.S. Department of Agriculture. The results of this study highlight the need for regular evaluation of pumping plant efficiencies to identify systems with poor performance and taking corrective actions to improve farm profitability and reduce environmental consequences of energy consumption for irrigation extraction.

Author Contributions: Conceptualization, R.S.F. and S.T.; Methodology, R.S.F. and S.T.; Formal analysis, D.H., R.S.F., and S.T.; Investigation, D.H., R.S.F., S.T., and J.G.W.; Writing—original draft preparation, D.H., R.S.F., S.T., and J.G.W.; Writing—review and editing, D.H., R.S.F., S.T., and J.G.W.; Project administration, R.S.F.; Funding acquisition, R.S.F.

Acknowledgments: The authors are grateful to Mr. Chris Stoner and Mr. Donald Sternitzke from the Oklahoma Natural Resources Conservation Service, U.S. Department of Agriculture, for their support. We are also thankful to all agricultural producers who collaborated with us on this project.

References

1. Khan, M.A.; Khan, M.Z.; Zaman, K.; Naz, L. Global estimates of energy consumption and greenhouse gas emissions. *Renew. Sustain. Energy Rev.* **2014**, *29*, 336–344. [CrossRef]
2. Pradeleix, L.; Roux, P.; Bouarfa, S.; Jaouani, B.; Lili-Chabaane, Z.; Bellon-Maurel, V. Environmental impacts of contrasted groundwater pumping systems assessed by life cycle assessment methodology: Contribution to the water-energy nexus study. *Irrig. Drain.* **2015**, *64*, 124–138. [CrossRef]
3. Wang, J.; Rothausen, S.G.; Conway, D.; Zhang, L.; Xiong, W.; Holman, I.P.; Li, Y. China's water–energy nexus: Greenhouse-gas emissions from groundwater use for agriculture. *Environ. Res. Lett.* **2012**, *7*, 014035. [CrossRef]
4. Karimi, P.; Qureshi, A.S.; Bahramloo, R.; Molden, D. Reducing carbon emissions through improved irrigation and groundwater management: A case study from Iran. *Agric. Water Manag.* **2012**, *108*, 52–60. [CrossRef]
5. Shah, T. Climate change and groundwater: India's opportunities for mitigation and adaptation. *Environ. Res. Lett.* **2009**, *4*, 035005. [CrossRef]
6. Follett, R.F. Soil management concepts and carbon sequestration in cropland soils. *Soil Tillage Res.* **2001**, *61*, 77–92. [CrossRef]
7. Patle, G.T.; Singh, D.K.; Sarangi, A.; Khanna, M. Managing CO_2 emission from groundwater pumping for irrigating major crops in trans indo-gangetic plains of India. *Clim. Chang.* **2016**, *136*, 265–279. [CrossRef]
8. Luc, J.P.; Tarhouni, J.; Calvez, R.; Messaoud, L.; Sablayrolles, C. Performance indicators of irrigation pumping stations: Application to drill holes of minor irrigated areas in the Kairouan plains (Tunisia) and impact of malfunction on the price of water. *Irrig. Drain.* **2006**, *55*, 85–98. [CrossRef]
9. Mora, M.; Vera, J.; Rocamora, C.; Abadia, R. Energy efficiency and maintenance costs of pumping systems for groundwater extraction. *Water Resour. Manag.* **2013**, *27*, 4395–4408. [CrossRef]
10. Qiu, G.Y.; Zhang, X.; Yu, X.; Zou, Z. The increasing effects in energy and GHG emission caused by groundwater level declines in North China's main food production plain. *Agric. Water Manag.* **2018**, *203*, 138–150. [CrossRef]
11. Taghvaeian, S. Irrigated Agriculture in Oklahoma. Oklahoma Cooperative Extension, Publication BAE-1530. 2014. Available online: http://pods.dasnr.okstate.edu/docushare/dsweb/Get/Document-9561/BAE-1530web.pdf (accessed on 10 October 2018).
12. Oklahoma Water Resources Board. Oklahoma Groundwater Report. Beneficial Use Monitoring Program. 2017. Available online: https://www.owrb.ok.gov/quality/monitoring/bump/pdf_bump/Reports/GMAPReport.pdf (accessed on 10 October 2018).
13. Brar, D.; Kranz, W.L.; Lo, T.H.; Irmak, S.; Martin, D.L. Energy conservation using variable-frequency drives for centerpivot irrigation: Standard systems. *Trans. ASABE* **2017**, *60*, 95–106.
14. Masasi, B.; Frazier, R.S.; Taghvaeian, S. Review and Operational Guidelines for Portable Ultrasonic Flowmeters. Oklahoma Cooperative Extension, Publication BAE-1535. 2017. Available online: http://pods.dasnr.okstate.edu/docushare/dsweb/Get/Document-10723/BAE-1535web.pdf (accessed on 10 October 2018).
15. Frazier, R.S.; Taghvaeian, S.; Handa, D. Measuring Depth to Groundwater in Irrigation Wells. Oklahoma Cooperative Extension, Publication BAE-1538. 2017. Available online: http://pods.dasnr.okstate.edu/docushare/dsweb/Get/Document-10865/BAE-1538web.pdf (accessed on 10 October 2018).

16. Ross, E.A.; Hardy, L.A. *National Engineering Handbook; Irrigation Guide*; USDA: Beltsville, MD, USA, 1997.
17. Hanson, B. Improving pumping plant efficiency does not always save energy. *Calif. Agric.* **2002**, *56*, 123–127. [CrossRef]
18. U.S. Environmental Protection Agency. *Greenhouse Gas Inventory Guidance*; United States Environmental Protection Agency: Washington, DC, USA, 2016.
19. Rothausen, S.G.; Conway, D. Greenhouse-gas emissions from energy use in the water sector. *Nat. Clim. Chang.* **2011**, *1*, 210–219. [CrossRef]
20. Shahdany, S.M.H.; Firoozfar, A.; Maestre, J.M.; Mallakpour, I.; Taghvaeian, S.; Karimi, P. Operational performance improvements in irrigation canals to overcome groundwater overexploitation. *Agric. Water Manag.* **2018**, *204*, 234–246. [CrossRef]
21. U.S. Energy Information Administration. Available online: https://www.eia.gov/ (accessed on 10 October 2018).
22. Barefoot, A.D. *Investigation of Factors Affecting Energy for Irrigation Pumping*; Oklahoma Water Resources Research Institute: Stillwater, OK, USA, 1980.
23. Fipps, G.; Neal, B. *Texas Irrigation Pumping Plant Efficiency Testing Program*; Texas A & M University System: College Station, TX, USA, 1995.
24. New, L.; Schneider, A.D. *Irrigation Pumping Plant Efficiencies, High Plains and Trans-Pecos Areas of Texas*; Texas Agricultural Experiment Station, Texas A & M University System: College Station, TX, USA, 1988.
25. DeBoer, D.W.; Lundstrom, D.R.; Wright, J.A. Efficiency analysis of electric irrigation pumping plants in the upper Midwest, USA. *Energy Agric.* **1983**, *2*, 51–59. [CrossRef]
26. Hardin, D.C.; Lacewell, R.D. Implication of improved irrigation pumping efficiency for farmer profit and energy use. *J. Agric. Appl. Econ.* **1979**, *11*, 89–94. [CrossRef]
27. U.S. Department of Agriculture 2013 Farm and Ranch Irrigation Survey. Available online: https://www.nass.usda.gov/Publications/AgCensus/2012/Online_Resources/Farm_and_Ranch_Irrigation_Survey/ (accessed on 10 October 2018).

13

Assessment of Landsat-Based Evapotranspiration using Weighing Lysimeters in the Texas High Plains

Ahmed A. Hashem [1,2,3], Bernard A. Engel [3,*], Vincent F. Bralts [3], Gary W. Marek [4], Jerry E. Moorhead [5], Sherif A. Radwan [2] and Prasanna H. Gowda [6]

1 College of Agriculture, Arkansas State University, 422 University Loop W, Jonesboro, AR 72401, USA; ahashem@astate.edu
2 Agricultural Engineering Department, Suez Canal University, Kilo 4.5 Ring Road, Ismailia 41522, Egypt; Sherifabdelhak@hotmail.com
3 Agricultural & Biological Engineering Department, Purdue University, 225 South University Street, West Lafayette, IN 47907, USA; bralts@purdue.edu
4 USDA-ARS Conservation and Production Research Laboratory, 300 Simmons Road, Unit 10, Bushland, TX 79012, USA; Gary.Marek@usda.gov
5 Lindsay Corporation, 8948 Centerport Blvd, Amarillo, TX 79108, USA; Jed.moorhead@lindsay.com
6 USDA, ARS Southeast Area, 141 Experimental Station Road, Stoneville, MS 38776, USA; prasanna.gowda@usda.gov
* Correspondence: engelb@purdue.edu.

Abstract: Evapotranspiration (ET) is one of the largest data gaps in water management due to the limited availability of measured evapotranspiration data, and because ET spatial variability is difficult to characterize at various scales. Satellite-based ET estimation has been shown to have great potential for water resource planning and for estimating agricultural water use at field, watershed, and regional scales. Satellites with low spatial resolution, such as NASA's MODIS (Moderate Resolution Imaging Spectroradiometer), and those with higher spatial resolution, such as Landsat (Land Satellite), can potentially be used for irrigation water management purposes and other agricultural applications. The objective of this study is to assess satellite based-ET estimation accuracy using measured ET from large weighing lysimeters. Daily, seven-day running average, monthly, and seasonal satellite-based ET data were compared with corresponding lysimeter ET data. This study was performed at the USDA-ARS Conservation and Production Research Laboratory (CPRL) in Bushland, Texas, USA. The daily time series Landsat ET estimates were characterized as poor for irrigated fields, with a Nash Sutcliff efficiency (NSE) of 0.37, and good for monthly ET, with an NSE of 0.57. For the dryland managed fields, the daily and monthly ET estimates were unacceptable with an NSE of −1.38 and −0.19, respectively. There are various reasons for these results, including uncertainties with remotely sensed data due to errors in aerodynamic resistance surface roughness length estimation, surface temperature deviations between irrigated and dryland conditions, poor leaf area estimation in the METRIC model under dryland conditions, extended gap periods between satellite data, and using the linear interpolation method to extrapolate daily ET values between two consecutive scenes (images).

Keywords: BEARS; bushland; climate; evapotranspiration; groundwater management; irrigation water management; Ogallala aquifer region; remote sensing; lysimeter ET assessment

---

1. Introduction

Evapotranspiration (ET) is an important component of the water cycle and the surface energy balance, where water changes phase from liquid to vapor, and a change in energy takes place [1]. The development of reliable long-term estimates of ET is needed to improve agricultural water use efficiency (WUE). WUE is defined as the assimilated carbon amount as biomass/grain produced per consumed unit of water by a crop. Paulson [2] reported that climate elements and wind conditions affect regional and seasonal ET estimates. For the irrigation scheduling decision-making process, knowledge of ET variability is important for proper water resources management. This is especially true in arid regions, where crop water requirements exceed rainfall, and irrigation is essential for crop production.

To sustain agricultural production in regions that are dependent on irrigation, blue water (surface water) is extracted from streams and aquifers (groundwater) to help sustain the crop and to meet water demand. Over time, farmers are required to dig deeper wells and extract water from deeper aquifers to ensure adequate water supply for their irrigation needs [3]. Unfortunately, the recharge rate of some aquifers is not sufficient to meet this demand, and historical declines in underground water table have occurred due to extensive water withdrawal [4]. In some cases, this has resulted in land subsidence and localized infrastructure damage [3]. ET is the main driver in irrigation water management and planning. The more accurate the ET estimates, the more likely associated management strategies are to achieving potential crop yields. ET is also an important component in estimating soil water to facilitate improved water use efficiency.

Remote sensing-based ET models are effective for crop water requirement estimations at the field and regional scales [5]. Several ET algorithms have been developed to utilize airborne and satellite data for irrigation scheduling and management purposes. ET can be measured over a surface using the Bowen ratio (BR), the eddy covariance (EC), and lysimeter systems at the field-scale. In all of these cases, spatial variability does not apply, since each method represents a very small scale measurement (typically less than 150 m in an agricultural setting). In addition, these methods only provide a single, averaged value that may not adequately capture the variability across a region. Satellite-based ET models produce regional-scale crop water use [6]. Many remote sensing-based ET algorithms have been developed, assessed, and widely used for estimating regional ET [6–9].

Park et al. [10], Jackson [11], and Choudhury et al. [12] reported that regional and watershed-scale spatially distributed ET can be better represented using remote sensing compared to traditional ET estimation methods. There are typically two approaches that are used to provide remote sensing ET estimates: (1) the land surface energy balance (EB) approach, and (2) the reflectance-based crop coefficient (Kc) and reference ET approach. The first approach is based on ET being a change of the state of water, using available energy in the environment for vaporization [13].

Multiple satellite platforms are available to use with energy balance ET models, such as Land Satellite (Landsat), The Advanced Spaceborne Thermal Emission and Reflection Radiometer (ASTER), Geostationary Operational Environmental Satellite (GEOS), moderate resolution imaging spectroradiometer (MODIS), and others, based on surface energy balance [14]. However, the spatiotemporal resolution is complex and cannot be combined with only one satellite. For daily estimates of ET with a high spatial resolution (field-scale), results depend on data assimilation, model applicability, and accuracy [15]. ET calibration based on remote sensing ET estimates has been conducted for a single source method such as the Surface Energy Balance Algorithm for Land (SEBAL) and the Mapping Evapotranspiration with Internalized Calibration model (METRIC) [16,17], and two source methods such as the Two-Source Energy Balance model (TSEB) [18,19]. Energy balance models based on thermal remote sensing include TSEB [18,20], SEBAL [8], METRIC [6,17], and several others [7,21].

A detailed assessment of existing EB models [7,22] stated that daily ET estimates varied 3 to 35% in comparison to Bowen ratio and EC ET measurements. Error sources included (a) modeling uncertainties, and (b) measurement errors and discrepancies in model-measurement scales. Other studies that assessed multiple airborne high-resolution sensor platforms indicated good agreement with measured ET [23].

The BEAREX08 experiment [15] was a robust remote sensing experiment that involved mass balance measurements of ET using four weighing lysimeters [24,25]. A wide array of instrumentation was installed for supportive data, including neutron probe (NP) access tubes for soil water measurements, multi-level canopy temperature measurements, and above and below canopy irradiance measurements. The purpose of these measurements was to provide as many measured data as possible to reduce estimations and provide more points of comparison. These data were used to address the lack of studies where EB model estimates of ET are compared with mass balance measurements, reducing uncertainty sources in EB models. Gowda et al. [7] reported that few remotely sensed ET estimates are used in irrigation scheduling and in-field management due to the absence of daily data with field-scale resolution. Many methods have been proposed to overcome this issue, such as use of infrequent, high spatial resolution data, including Landsat with 60 m spatial resolution, 16-day temporal resolution, in combination with lower spatial resolution and more frequent data, such as the Moderate Resolution Imaging Spectroradiometer (MODIS) with 1000 m spatial resolution and 1-day temporal resolution. The Disaggregated Atmosphere-Land Exchange Inverse (DisALEXI) model demonstrated this concept [14,26,27], but no evaluation of this approach for management has been successful thus far.

Spatial and temporal resolution problems could be resolved using aircraft. However, cost, data processing, and lack of experienced users have prevented their widespread use in providing such platforms as potential imagery sources for water management. The EB method typically provides an instantaneous value of ET, and interpolated to daily ET using the evaporative fraction method, [28], or reference ET approach [6,29]. Each ET estimation approach has its uncertainties; however, measurement data quality assessment and quality control reduce such modeling uncertainties [30,31]. Multi-time scale remotely-sensed ET estimation accuracy is essential for water management and irrigation planning. Extensive studies have been conducted to assess the METRIC model performance on irrigated fields [32–35]; however, none of these studies evaluated the model performance under dryland conditions for extended periods with various crops and temporal resolutions [36]. This study assessed daily interpolated ET estimation accuracy using the linear interpolation method on dryland and irrigated lysimeters for a ten year period with various crops at multiple time scales. Such assessment is crucial for water management policies, researchers, hydrologists, and irrigators when using such technologies for real time irrigation decisions.

The objective of this research was to assess the daily, seven-day running average, monthly, and seasonal satellite-based ET estimation accuracy versus large weighing lysimeter ET measurements in the Texas High Plains under irrigated and dryland conditions during the growing and non-growing seasons.

2. Materials and Methods

2.1. Study Site

Data used in this study from 2001–2010 were obtained at the USDA-ARS Conservation and Production Research Laboratory (CPRL) in Bushland, Texas, USA (35.19° N, 102.10° W). Four square fields were selected for this study; each field was ~4.7 ha. Four large precise weighing lysimeters were installed towards the center of each field [24,25]. Two lysimeters were managed as irrigated (NE and SE), and the other two lysimeters were managed as dryland (NW and SW). The irrigated lysimeter field was equipped with a linear-move irrigation system with Nelson sprinklers (Nelson Irrigation Corporation, Walla Walla, WA, USA) [37]. Irrigation scheduling was performed using neutron probe data, and in 2001 50% of crop water requirements (CWR) was replenished for cotton (deficit irrigation), and 100% of CWR was added for the remaining study period. Crop management data were collected and summarized by the CPRL [38,39].

Leaf area index (LAI) data were collected from the study site between 2001 and 2010. Plant samples were collected from locations within fields near the lysimeters biweekly during the growing season through destructive plant sampling. Leaves were separated from stems, and the average LAI values were obtained from at least three samples. Plant samples were not collected from the lysimeters, as the sampling was destructive and would impact lysimeter ET measurements. A digital scanning bed leaf area meter (model LI-3100) was used to measure the leaf area index of the leaf samples. The LAI values were calculated as the ratio of the upper side leaf area (m^2) to the ground area (m^2) [22,40].

For analysis simplicity, one lysimeter was selected for each management condition (irrigated and dryland). Statistical assessment was achieved for the estimated ET values for the dryland lysimeter (NW) and the irrigated lysimeter (NE). The soil characteristics for the study field are deep, well-drained Pullman silty clay loam (fine, mixed, superactive, thermic torrertic paleustoll) [40]. The local climate is classified as semi-arid, with large daily air temperature variations. Cotton, soybean, grain, and silage sorghum, sunflower, and cotton were the predominant crops for the research fields during the ten year study period [38,41].

2.2. *Bushland Evapotranspiration and Agricultural Remote Sensing (BEARS)*

Bushland Evapotranspiration and Agricultural Remote Sensing (BEARS) is an image processing and geographic information system (GIS) software developed by researchers at the USDA ARS CPRL in Bushland, TX, used for deriving hourly, daily, and seasonal ET maps, and other energy exchanges between land and atmosphere using Landsat 5,7, and 8 [42]. It is an open-source Java software, Version 1.0.1 available for download at: https://data.nal.usda.gov/dataset/bushland-evapotranspiration-and-agricultural-remote-sensing-system-bears-software. The software allows for custom models and equations but provides the option to select one of five default energy balance-based ET methods: Mapping Evapotranspiration at High Resolution with Internalized Calibration (METRIC), Surface Energy Balance Algorithm for Land (SEBAL), Surface Energy Balance System (SEBS), Two-Source Model (TSM), and Simplified Surface Energy Balance (SSEB) [42].

2.3. *Image Analysis*

In this study, the METRIC model was used to analyze Landsat satellite imagery, producing ET time-series datasets for the study period. The METRIC model description and required inputs based on Landsat satellite datasets are summarized in the literature [6,41]. Several hourly and daily outputs were obtained at the completion of each image analysis. Four daily outputs were obtained including daily ET (mm day^{-1}), evaporation fraction (EF) (unitless), leaf area index (LAI) (m^2 m^{-2}), and normalized difference vegetation index (NDVI) (unitless). Hashem et al. [41] assessed hourly estimation accuracy for ET (mm hr^{-1}), net radiation (Rn) (W m^{-2}), soil heat flux (Go) (W m^{-2}), and surface air temperature (Ts) (°C). A detailed description of the hourly assessment procedure was summarized [38,41].

2.4. *Landsat Satellite Dataset and Processing*

Landsat 5 TM cloud-free images were selected for analysis, which were obtained through Earth Explorer (https://earthexplorer.usgs.gov/), with Paths 30 and 31 and Row 36 from 2001 to 2010. A total of 129 images were analyzed with a spatial resolution of 30 m and a temporal resolution of 16 days. The number of clear images vary from year to year based upon geographic location. For this study site, 2001 had the lowest number of annual clear images, with six images, and 2008 had the highest number of clear images with 16 images, as summarized in Table 1. Processed image date and day of the year (DOY) are summarized in Table 1.

Table 1. Landsat image dates and corresponding day of year (DOY).

	2001 (Cotton-Limited Irrigation)	2002 (Cotton-Limited Irrigation)	2003 (Soybean)	2004 (Soybean)	2005 (Sorghum)	2006 (Forage Corn)	2007 (Forage Sorghum)	2008 (Cotton)	2009 (Sunflower)	2010 (Cotton)
#	Date (DOY)	Date (DOY)	Date (DOY)	Date (DOY)	Date (DOY)	Date (DOY)	Date (DOY)	Date (DOY)	Date (DOY)	Date (DOY)
1	March 12 (71)	January 26 (26)	January 13 (13)	February 17 (48)	January 25 (25)	January 18 (28)	January 8 (8)	January 18 (18)	January 13 (13)	April 29 (119)
2	May,22 (142)	February 11 (42)	April 10 (100)	March 20 (80)	February 3 (34)	February 13 (44)	February 25 (56)	February 19 (50)	January 20 (20)	June 25 (176)
3	June 16 (167)	March 3 (90)	May 5 (125)	March 27 (87)	March 7 (66)	April 18 (108)	March 4 (63)	March 22 (82)	January 29 (29)	July 18 (199)
4	June 23 (174)	May 9 (129)	May 28 (148)	April 21 (112)	June 18 (169)	May 20 (140)	March 29 (88)	April 7 (95)	February 5 (36)	August 3 (215)
5	July 9 (190)	May 18 (138)	June 24 (205)	May 14 (135)	June 27 (178)	June 5 (156)	June 8 (159)	May 2 (123)	February 21 (52)	August 12 (224)
6	July 25 (206)	June 10 (161)	July 17 (198)	May 30 (151)	July 20 (201)	July 23 (204)	July 26 (207)	May 18 (139)	March 18 (77)	August 19 (231)
7	August 19 (231)	June 19 (170)	September 17 (260)	December 1 (336)	August 30 (242)	August 8 (220)	August 11 (223)	June 3 (155)	April 3 (93)	September 4 (247)
8	September 27 (270)	July 21 (202)	September 26 (269)	December 12 (352)	September 22 (265)	August 24 (236)		June 10 (162)	June 22 (173)	September 29 (272)
9	October 13 (286)	September 23 (266)	October 19 (292)		October 1 (274)	September 18 (261)		July 21 (203)	July 8 (189)	October 15 (288)
10	November 7 (311)		November 29 (333)		October 24 (297)	September 25 (268)		August 6 (219)	August 16 (228)	November 16 (320)
11	December 9 (343)				November 18 (315)	October 11 (284)		August 22 (235)	November 4 (308)	November 23 (327)
12	December 12 (359)					October 27 (300)		September 30 (274)	November 20 (324)	December 12 (355)
13						November 28 (332)		October 25 (299)		December 25 (359)
14								November 1 (306)		
15								November 17 (322)		
16								December 28 (363)		

2.5. Landsat ET Gap Filling

Linear interpolation methods were used to estimate the satellite-based ET during the gap period using the satellite-based daily evaporation fraction between consecutive images. The gap between consecutive images was split equally into two periods of eight days; the first eight days used the first image evaporation fraction value, and the second eight-day period used the next image evaporation fraction value.

To estimate the actual ET time series, the evaporation fraction (k_c) for each day was multiplied by the reference ET [43]. The daily reference ET was estimated using the ASCE–2005 [44] standardized reference evapotranspiration equation. Hourly ET was calculated for the study period, and the daily ET was obtained by accumulating the hourly ET in a given day. The average frequency and the maximum gap periods (days) were calculated using Equations (1) and (2), respectively. The year of 2006 had the longest average frequency with 52 days, and 2008 had the shortest average frequency with 23 days. The longest and shortest maximum gap period was 184 days and 40 days for 2004 and 2008, respectively, as shown in Table 2. Areas of interest (AOIs) of 3 by 3 grids (30 m spatial resolution), with a total of 9 pixels and a surface area of 8100 m^2 each, were created, and each lysimeter was located at the center of each AOI. ArcGIS 10.2.2 model builder tool software was utilized to extract the average ET for each AOI.

$$Average\ frequency = \frac{Number\ of\ days\ in\ the\ growing\ season\ (April\ to\ October)}{Number\ of\ clear\ images\ used\ in\ the\ analysis} \quad (1)$$

$$Average\ frequency = \frac{Number\ of\ days\ in\ the\ entire\ year\ (January\ to\ December)}{Number\ of\ clear\ images\ used\ in\ the\ analysis} \quad (2)$$

Table 2. Average frequency and maximum gap each year for the study site.

Year	Average Frequency (Days)	Maximum Gap (Days)	Dates of Maximum Gap	DOY of Maximum Gap
2001	30	70	March 12–May 22	DOY 71 To DOY 142
2002	41	99	September 23–December 31	DOY 266 To DOY 365
2003	37	86	January 13–April 10	DOY 13 To DOY 100
2004	46	184	May 30–December 1	DOY 151 To DOY 336
2005	33	102	March 7–June 18	DOY 66 To DOY 169
2006	28	63	February 13–April 18	DOY 44 To DOY 108
2007	52	141	August 19–December 31	DOY 231 To DOY 365
2008	23	40	June 10–July 21	DOY 162 To DOY 203
2009	30	56	August 16–November 11	DOY 228 To DOY 315
2010	28	118	January 1–April 29	DOY 1 To DOY 119

The years 2004 and 2007 were omitted from the analysis due to limited clear image availability. Day of the year (DOY).

2.6. Dry and Wet Pixel Determination

Obtaining the most accurate wet (cold) and dry (hot) agricultural pixels is a significant step that is essential to produce accurate ET maps [34,45]. Wet and dry pixels represent the extreme conditions per scene. The manual selection method was used, along with the surface temperature and NDVI histogram distribution, to obtain these pixels for each image. NDVI and surface temperature threshold values are summarized in Table 3 to obtain the hot and dry pixel characteristics. The hot pixel is defined as bare agricultural soil with high temperature and low ET value, and the cold pixel is defined as cultivated agricultural soil with low temperature and high ET value [6,46]. The histogram distribution for surface temperature and NDVI help to determine the range of temperature and NDVI values per image.

Table 3. Hot and cold pixel conditions.

Pixel	Constraint		Condition
	T_s	NDVI	
Hot (dry)	High	≤0.2	Bare agricultural soil
Cold (Wet)	Low	≥0.7	Cultivated agricultural soil

2.7. Statistical Analysis

Statistical comparisons were performed including multiple statistical coefficients, including the coefficient of determination (R^2), regression line slope and intercept, root mean square error (*RMSE*), percent of root mean square error (% *RMSE* error), mean bias error (*MBE*), and Nash Sutcliff efficiency (*NSE*), between the measured and the calculated satellite-based ET. A detailed description of each coefficient and its range and interpretation of values were provided [38,41,47]. The mean bias error (*MBE*) and % *MBE* provided a means to determine the deviation between the measured and satellite-based estimates, with *MBE* = 0 indicating no bias in the estimation within the satellite data. The *MBE* and % *MBE* were calculated using Equations (3) and (4), respectively.

$$MBE = \frac{\sum_{i=1}^{n} (y_t - \hat{y}_t)}{n} \quad (3)$$

where y_t = the *i*th measured value, $\hat{y}_t$ = the *i*th simulated value, and n = the total number of observations.

$$\%MBE = \frac{MBE}{\overline{x}_{obs}} \times 100 \quad (4)$$

where $\overline{x}_{obs}$ is mean of measured values.

3. Results

3.1. Dryland Lysimeter ET Estimation

The northwest dryland lysimeter daily measured ET, Landsat ET, and seven-day running average values were plotted from 2001 to 2005, as shown in Figure 1a, and from 2006 to 2010 as shown in Figure 1b. The seven-day running average was used for visual comparison. The 1:1 graph between the daily measured and calculated Landsat ET can be seen in Figure 2, and the daily and monthly summary statistics are summarized in Table 4. The monthly average ET values were plotted, as shown in Figure 3. Detailed growing and non-growing season summary statistics are reported in Table 5.

Table 4. Daily and monthly summary statistics for the NW dryland lysimeter with Landsat ET.

		Daily	Monthly
RMSE (mm)		1.8	1.2
% *RMSE* error		144.3	105.7
NSE		−1.38	−0.19
Measured average ET (mm d^{-1})		1.3	1.1
Landsat average ET (mm d^{-1})		1.7	1.4
Regression	R^2	0.01	-
	Slope	0.09	-

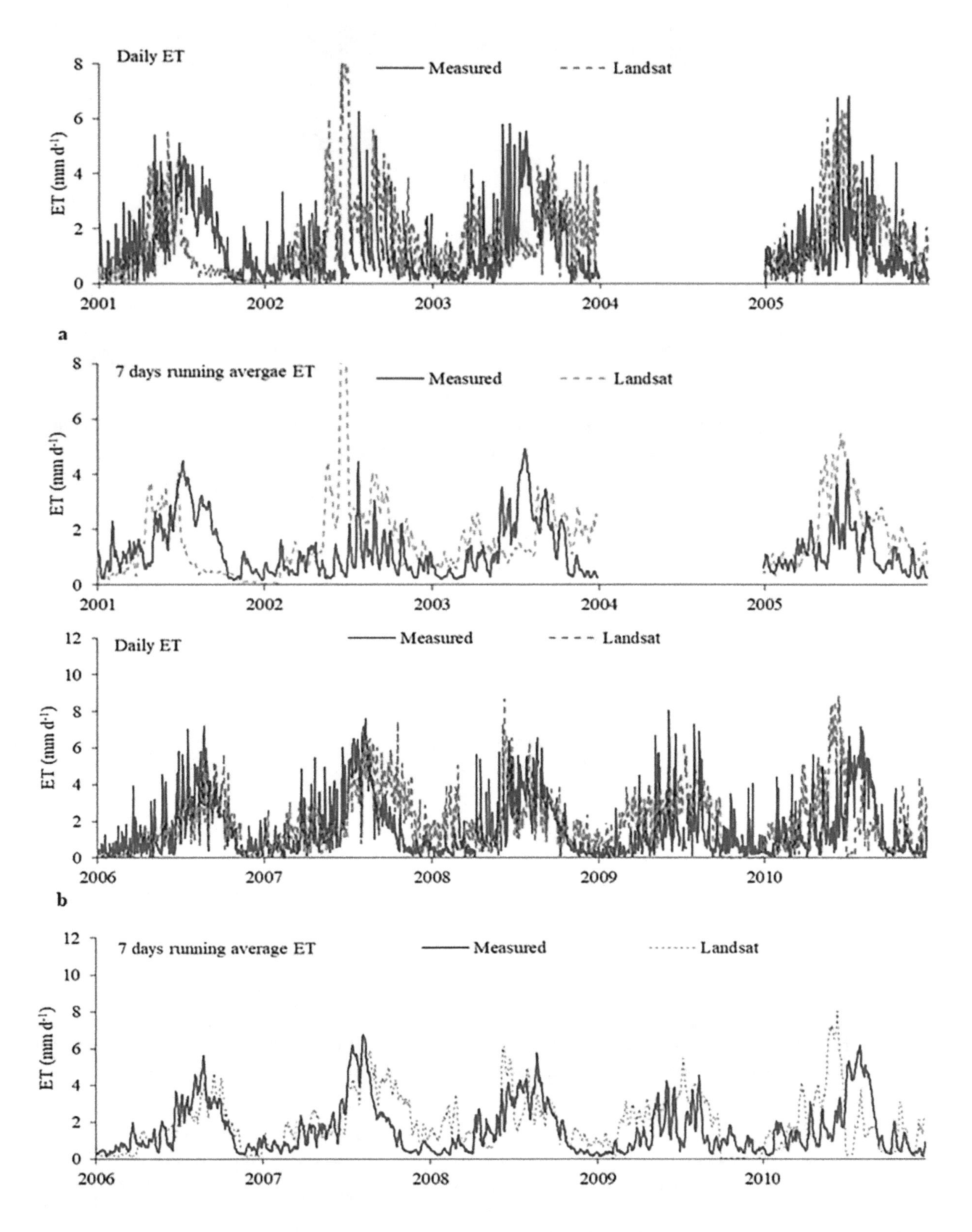

Figure 1. Daily and seven-day running average measured and Landsat evapotranspiration (ET) from (**a**) 2001–2005, and (**b**) 2006–2010 for the northwest (NW) dryland lysimeter. The years 2004 and 2007 had limited clear remote sensing observations during the growing season.

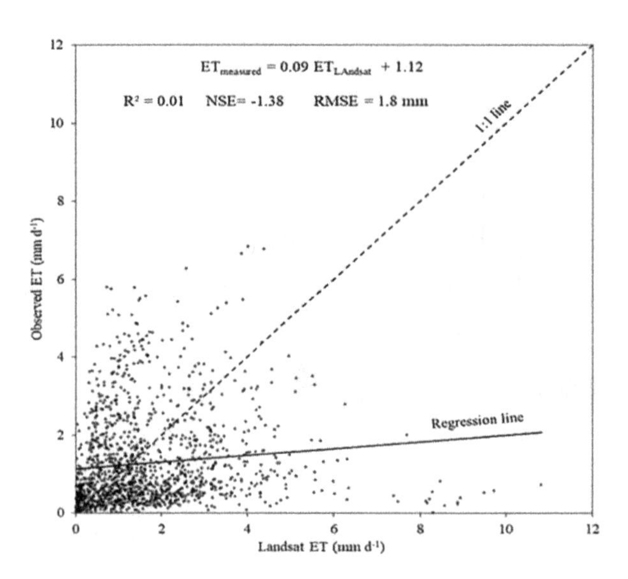

Figure 2. Daily 1:1 graph of measured and Landsat ET for NW dryland lysimeter. The years 2004 and 2007 were omitted from the analysis due to limited clear image availability.

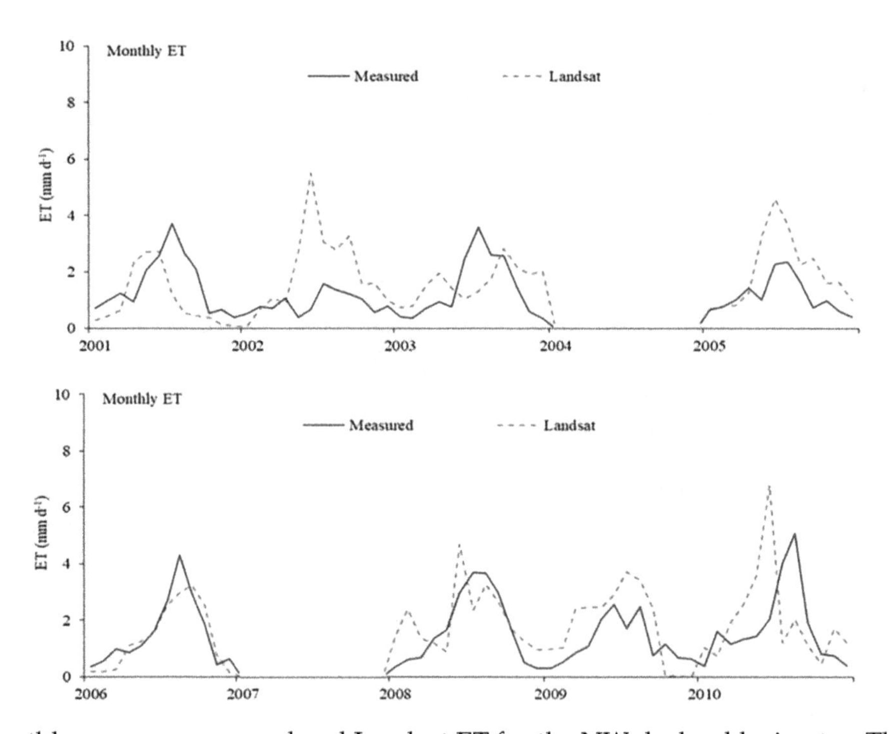

Figure 3. Monthly average measured and Landsat ET for the NW dryland lysimeter. The years 2004 and 2007 had limited clear remote sensing observations during the growing season.

Table 5. Seasonal summary statistics for the NW dryland lysimeter with Landsat ET.

Crop		RMSE (mm)	% RMSE Error	Measured Average ET (mm d^{-1})	Landsat Average ET (mm d^{-1})
Cotton	2001 GS	1.7	83.4	2.0	1.1
	2001 NG	1.2	123.1	0.9	0.9
Fallow	2002 GS	-	-	-	-
	2002 NG	2.5	288.4	0.9	2.0

Table 5. *Cont.*

Crop		RMSE (mm)	% RMSE Error	Measured Average ET (mm d^{-1})	Landsat Average ET (mm d^{-1})
Sorghum	2003 GS	1.9	71.4	2.7	1.7
	2003 NG	1.4	171.0	0.8	1.5
Fallow	2005 GS	-	-	-	-
	2005 NG	1.7	143.6	1.2	2.0
Sorghum	2006 GS	1.3	48.9	2.7	2.6
	2006 NG	0.8	104.5	0.8	0.7
Cotton	2008 GS	1.8	70.5	2.5	2.6
	2008 NG	1.5	177.6	0.8	1.4
Fallow	2009 GS	-	-	-	-
	2009 NG	1.8	144.3	1.2	1.
Soybean	2010 GS	3.3	99.1	3.3	2.0
	2010 NG	2.2	215.9	1.0	2.0

GS: growing season, NG: non-growing season. The years 2004 and 2007 were omitted from the analysis due to limited clear image availability.

The measured mean ET was 1.3 mm d^{-1}, and the mean Landsat ET was 1.7 mm d^{-1}. Due to the daily ET value deviations, the summary statistics provided a poor match during the comparison period with an R^2 value of 0.01, NSE value of −1.38, RMSE of 1.8, and an RMSE error of ~144.3%, which is considered a high error value. These statistical parameters indicated that there was no correlation between measured and calculated values [48].

The growing season dryland measured LAI was plotted versus calculated Landsat estimates for the days where Landsat images were available, as shown in Figure 4. It can be clearly seen that the Landsat LAI underestimated the LAI during the study period for all cultivated crops. Dryland conditions in Bushland, TX typically have much less vegetation than irrigated areas. The lower amounts of plant biomass allow the soil background to be more prominent at the Landsat spatial resolution. The increased soil background can skew the reflectance and LAI, causing lower values. Another potential source of errors is due to uncertainties with the remotely sensed datasets under dryland conditions, due to surface roughness length and ET extrapolation methods that have been incorporated in the METRIC model [32,38,49]. Chavez et al. [7] reported that the METRIC model estimation error was (0.7 ± 0.9 mm d^{-1}), explaining that the variations were due to errors associated with the surface roughness length and aerodynamic resistance. Another potential source of uncertainties is that the METRIC model uses a SURFACE NDVI vs. LAI relationship, where it is generated by fitting a generalized equation to six LAIs compared to NDVI functions [35], defined in the MODIS LAI backup [45], indicating a higher source of LAI estimation under dryland conditions.

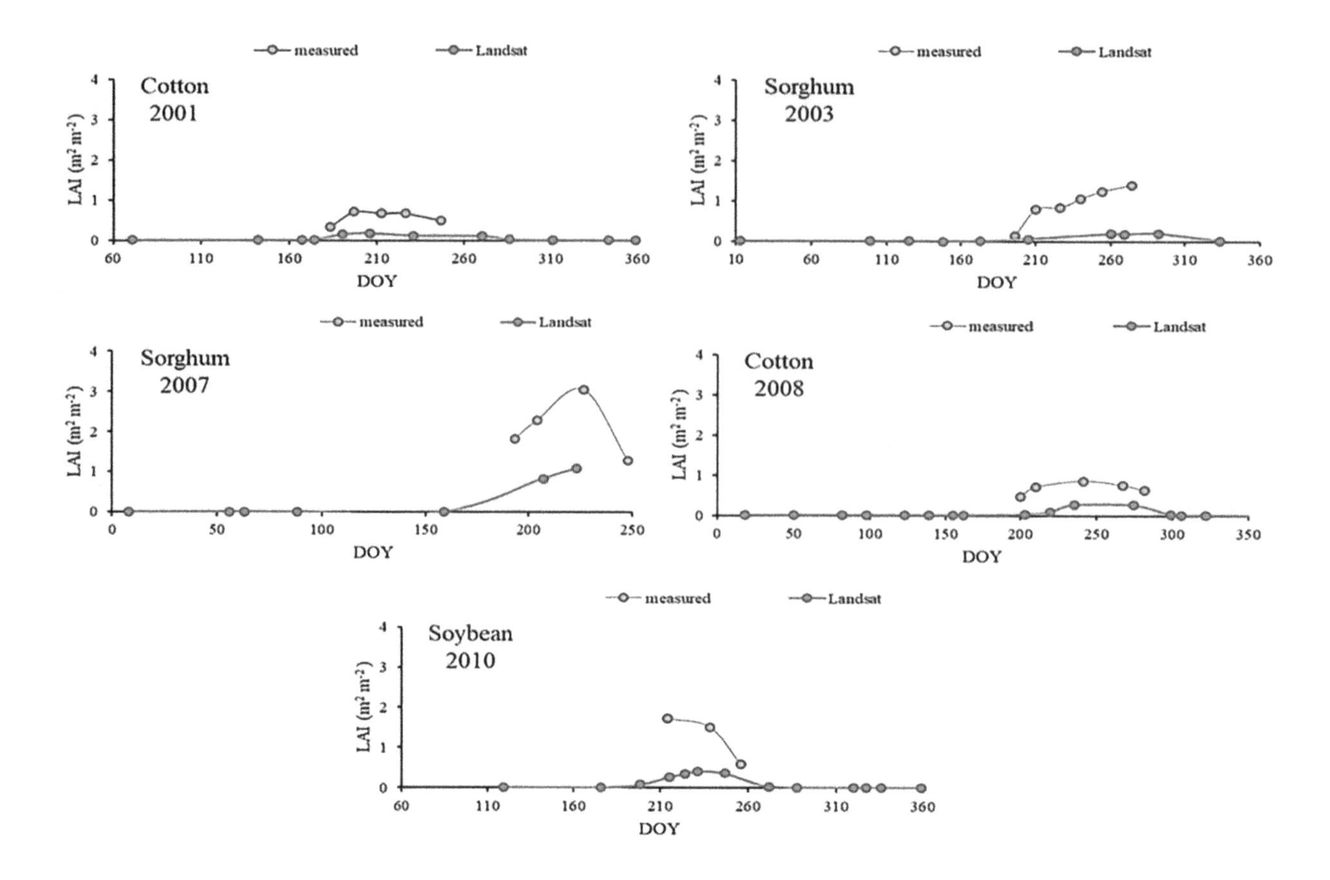

Figure 4. Measured and calculated Landsat leaf area index (LAI) for the NW dryland lysimeter. The year 2004 graph omitted because Landsat LAI values were not available.

3.2. Irrigated Lysimeter ET Estimation

Figure 5a,b presents the northeast irrigated lysimeter daily measured ET, Landsat ET, and seven-day running average from 2001 to 2010. The 1:1 graph between the daily measured and calculated Landsat ET is shown in Figure 6, and the daily and monthly summary statistics are summarized in Table 6. The growing and non-growing season summary statistics are reported in Table 7. The daily 1:1 graph is shown in Figure 6, and the monthly ET is shown in Figure 7.

Table 6. Daily and monthly summary statistics for the NE irrigated lysimeter with Landsat ET.

		Daily	Monthly
RMSE (mm)		2.1	1.5
% RMSE error		86.4	56.7
NSE		0.37	0.57
Measured average ET (mm d^{-1})		2.4	1.9
Landsat average ET (mm d^{-1})		2.4	1.9
Regression	R^2	0.38	-
	Slope	0.86	-

The years 2004 and 2007 were omitted from the statistical analysis due to limited clear image availability.

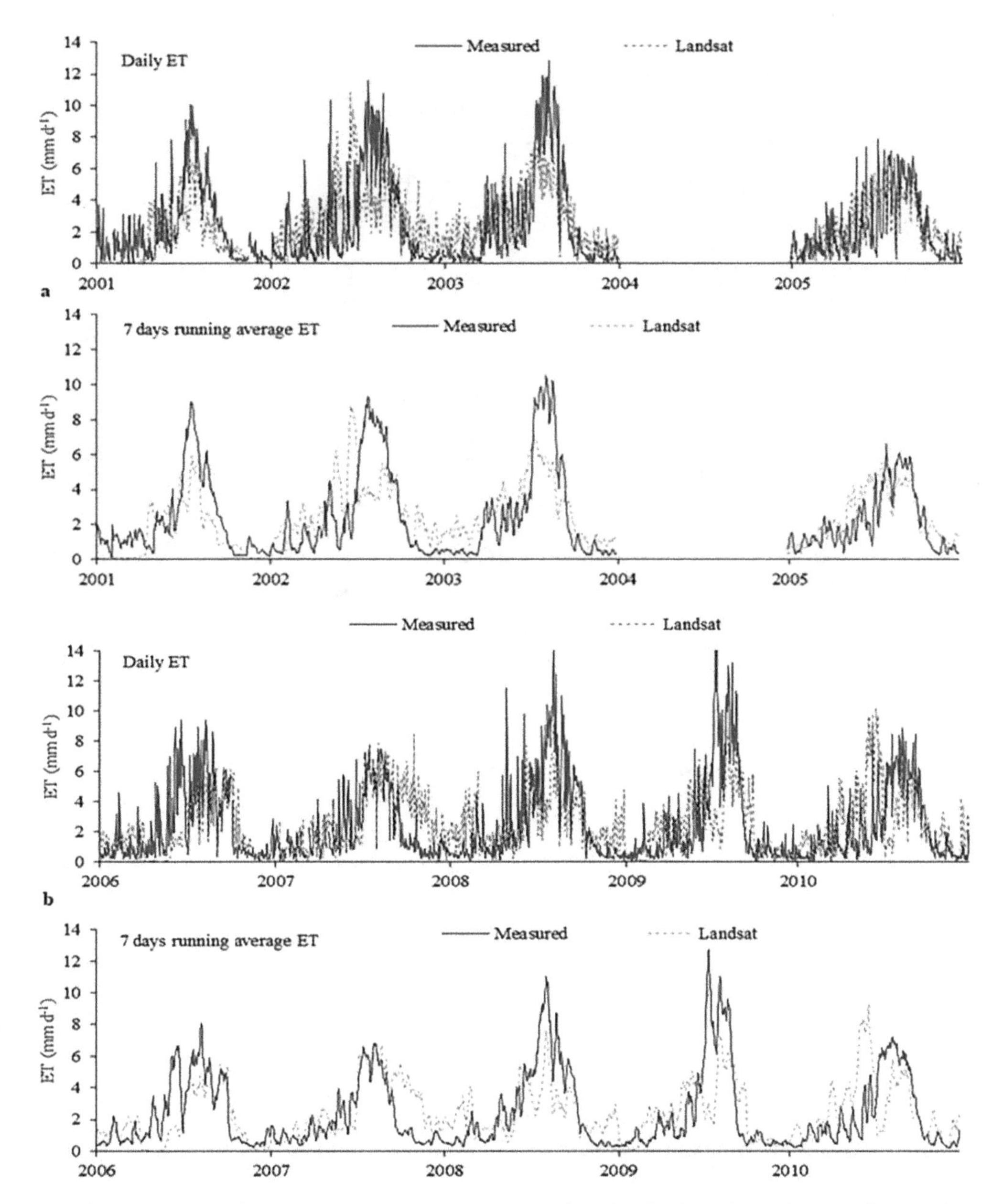

Figure 5. Daily and seven-day running average measured and calculated Landsat ET from (**a**) 2001–2005, and (**b**) 2006–2010 for the northeast (NE) irrigated lysimeter. The years 2004 and 2007 had limited clear remote sensing observations during the growing season.

Table 7. Seasonal summary statistics for the NE irrigated lysimeter with Landsat ET.

Crop		RMSE (mm)	%RMSE Error	Measured Average ET (mm d^{-1})	Landsat Average ET (mm d^{-1})
Cotton	2001 GS	2.0	66.2	3.1	2.1
	2001 NG	1.4	132.1	1.1	0.5
Cotton	2002 GS	3.5	81.9	4.2	4.0
	2002 NG	1.8	156.3	1.1	1.9
Soybean	2003 GS	2.7	55.2	4.9	4.0
	2003 NG	1.4	156.0	0.9	1.8

Table 7. *Cont.*

Crop		RMSE (mm)	%RMSE Error	Measured Average ET (mm d^{-1})	Landsat Average ET (mm d^{-1})
Sorghum	2005 GS	1.8	53.4	3.4	3.6
	2005 NG	1.1	114.2	1.0	1.3
Forage corn	2006 GS	2.7	62.3	4.3	3.1
	2006 NG	1.3	138.4	0.9	0.9
Cotton	2008 GS	2.7	56.1	4.9	3.3
	2008 NG	1.8	186.9	0.9	1.8
Sunflower	2009 GS	3.9	75.7	5.1	3.8
	2009 NG	1.4	139.9	0.9	1.4
Cotton	2010 GS	3.5	87.8	3.9	3.9
	2010 NG	1.8	185.5	0.9	1.8

GS: growing season, NG: non-growing season. The years 2004 and 2007 were omitted from the analysis due to limited clear image availability.

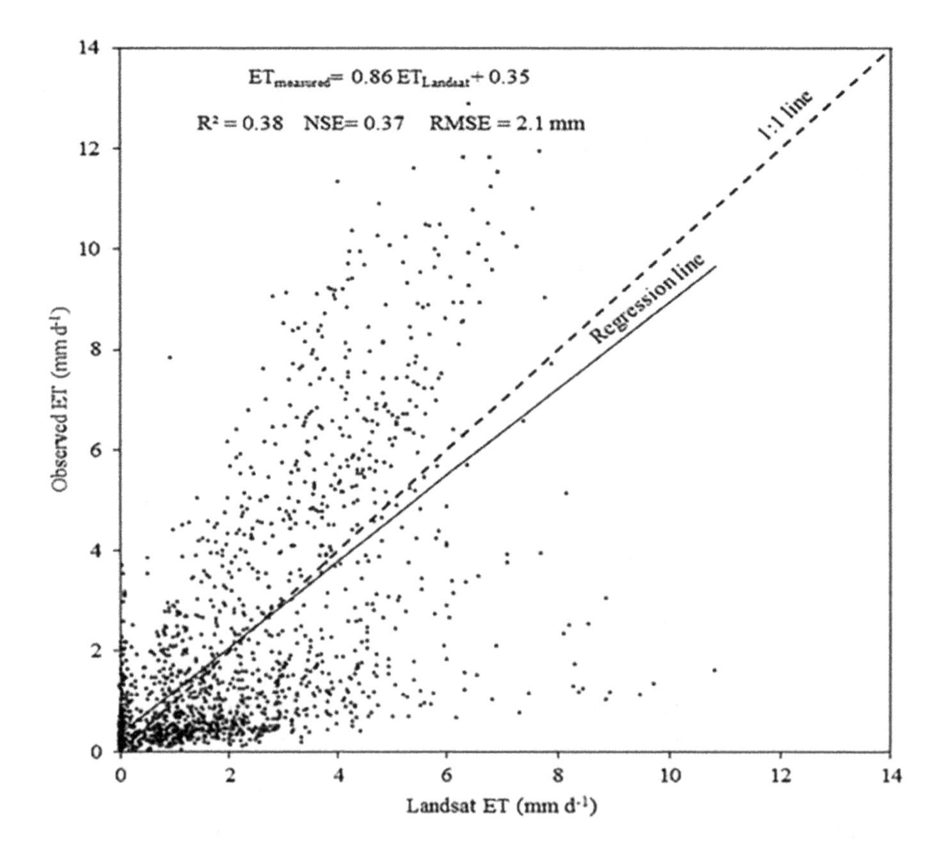

Figure 6. Daily 1:1 graph of measured and Landsat ET for the NE irrigated lysimeter. The years 2004 and 2007 were omitted from the analysis due to limited clear image availability.

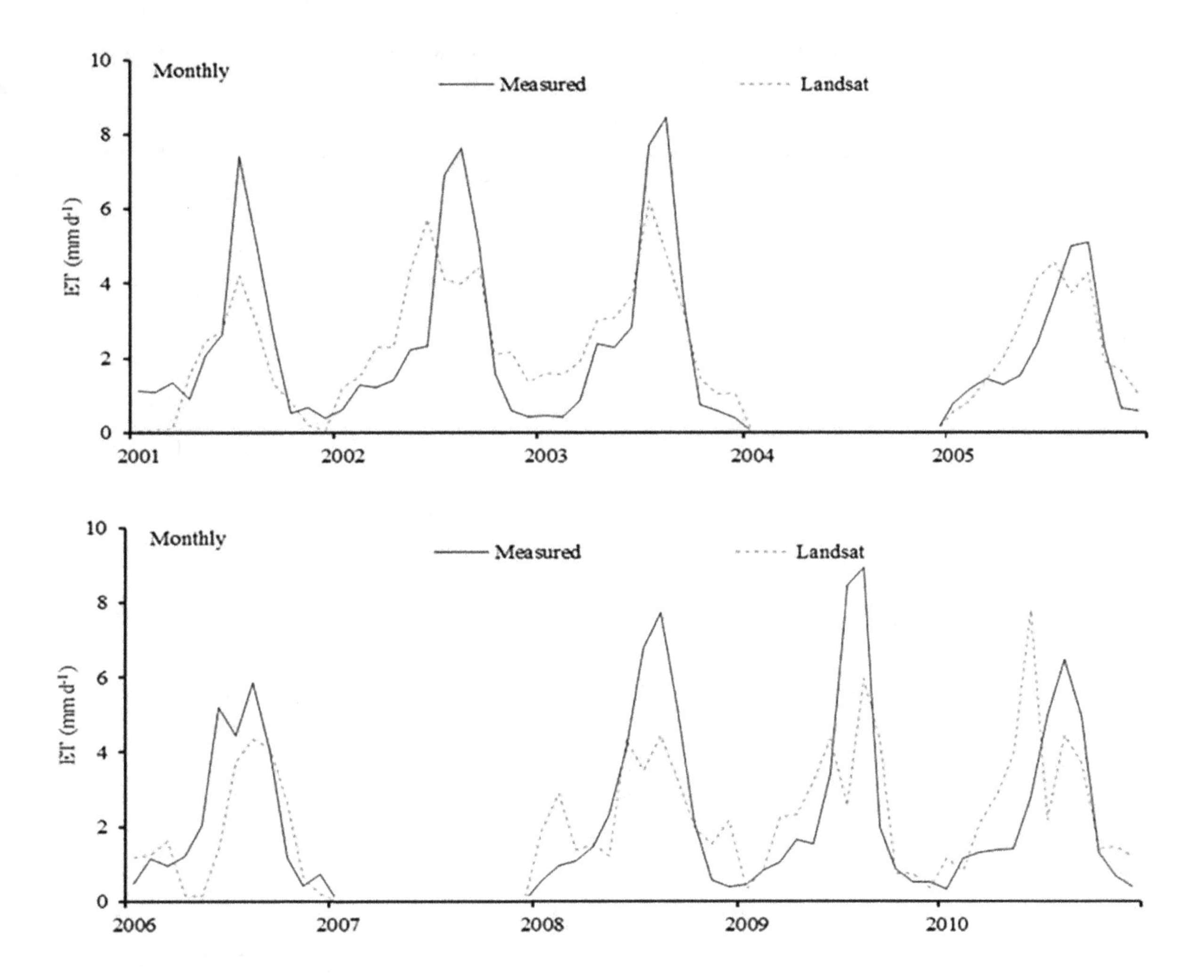

Figure 7. Monthly average measured and Landsat estimated ET for the NE irrigated lysimeter. The years 2004 and 2007 had limited clear remote sensing observations during the growing season.

The daily mean ET was 2.4 mm d^{-1} and 2.4 mm d^{-1} for the Landsat and measured, respectively. The summary statistics improved with the irrigated field and provided a weak correlation with an R^2 value of 0.38, NSE of 0.37, RMSE of 2.1, and RMSE ~86.4%.

The growing season irrigated measured LAI was plotted versus Landsat estimates for the days where Landsat images were available during the year, as shown in Figure 8. Landsat better estimated LAI under irrigated conditions compared to the dryland conditions. Consequently, higher NDVI values were obtained [38], producing higher LAI values for irrigated fields, and resulted in better ET estimates.

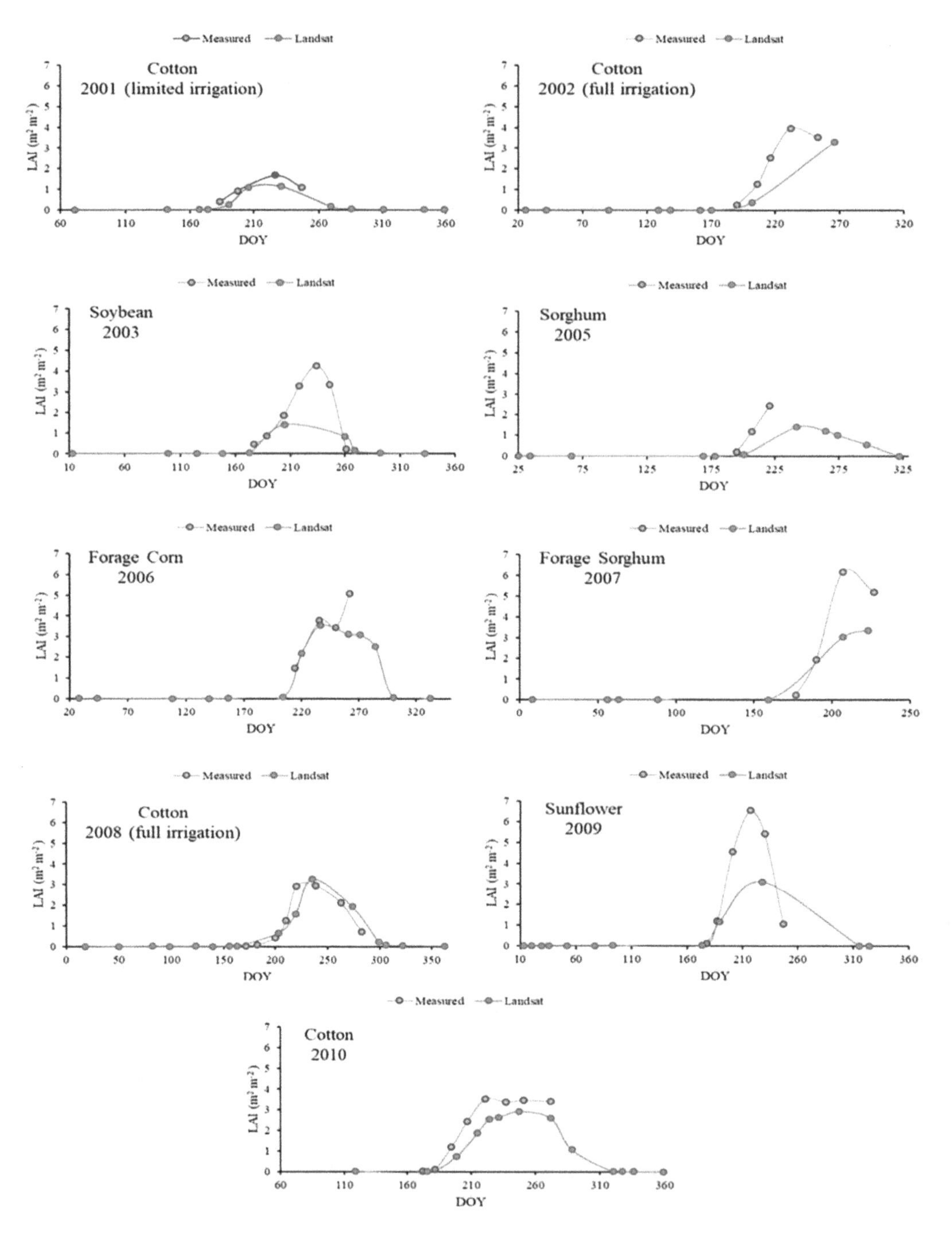

Figure 8. Measured and Landsat leaf area index (LAI) for the NE irrigated lysimeter. The year 2004 graph omitted because Landsat LAI values were not available.

4. Discussion

4.1. Dryland Daily ET Comparison

The relationship between measured and Landsat ET for the dryland lysimeter showed significant deviation with periods of both over and underestimation of ET throughout the year for the entire study period. The satellite-based LAI was assessed versus the measured LAI (Figure 4), and the LAI assessment summary is summarized in Hashem [38]. The daily time series ET deviations were related to errors in LAI estimation [38,41,47], where Landsat LAI estimates were significantly lower than measured LAI during the growing season for the dryland lysimeter. The higher the NDVI values, the more the LAI values increase, resulting in greater ET values.

In 2002, 2005, and 2009, the lysimeter field was fallow, and Landsat overestimated ET in each of the three years (Figure 1), and these results agree with Allen et al. [46]. Cotton was cultivated in 2001 and 2008, and Landsat estimates of ET closely matched the measured ET at the beginning of each year. However, towards the end of 2001, Landsat significantly underestimated ET due to low NDVI values and, consequently, underpredicting LAI [38]. ET data in 2004 and 2007 were omitted from the analysis, as the Landsat data overestimated the ET compared to the measured ET due to the large gap period in Landsat data and linear interpolation method used to fill the gap. In 2003, when sorghum was cultivated, the satellite-based-ET overestimated measure ET in both the beginning and towards the end of the year, and underestimated towards the middle of the growing season.

A detailed statistical analysis was performed for the growing and non-growing seasons (Table 5), where the growing season was defined as the days between planting and harvest. Monthly statistics showed better statistical performance, with monthly RMSE and NSE of −0.19 and 1.2 compared to values of −1.38 and 1.8 mm for the daily assessment [38,46,50]. The RMSE during the growing season was greater compared to the non-growing season (Table 5), with values almost double for the growing season compared to the non-growing season due to low measured ET values during the non-growing season. Hence, there was less variation between the measured and satellite-based ET values. However, the %RMSE error was higher during the non-growing season than the growing season, and these results agree with Allen et al. [46].

The satellite was able to distinguish between bare soil and vegetation in the field, providing useful information on when the field was fallow versus when a crop was growing. However, the overall LAI estimation from Landsat was lower than the measured LAI for all cultivated crops during this study under dryland conditions. Potential reasons for the LAI undercalculations are the water stress during the growing season producing low NDVI values under dryland conditions, uncertainties with aerodynamic resistance surface roughness length [36], long gap periods, and using the linear interpolation method to generate daily ET time series [38].

4.2. *Irrigated Daily ET Comparison*

The relationship between measured and Landsat ET for the irrigated lysimeter provided overall better agreement compared to the dryland field [38,41,51,52]. The Landsat ET estimates were closely matched most of the year, except the middle of the growing seasons, during the peak crop water requirements. The satellite-based approach underestimated ET toward the middle of the growing season for cotton and soybeans, and overestimated the ET early and late during the growing season.

A detailed statistical analysis was performed for the daily and monthly ET (Table 6). The irrigated daily ET estimates were considered poor with an NSE of 0.37, RMSE of 2.1 mm d^{-1}, and %RMSE of 86.4%. However, there was a statistical improvement with the monthly ET values with an NSE of 0.57, RMSE of 1.5 mm d^{-1}, and % RMSE of 56.7%. Similar to the dryland lysimeter, the RMSE during the growing season was greater compared to the non-growing season (Table 7), with values almost double for the growing season compared to the non-growing season due to low ET measured values during the non-growing season. Hence, there was less variation between the measured and satellite-based ET values. However, the %RMSE error was higher during the non-growing season than the growing season, and these results agree with Allen et al. [46]. Allen et al. [6] illustrated that the use of reference ET considers the advective effects on a METRIC model performance, which can make the METRIC ET overestimate the ET from irrigated fields, exceeding daily net radiation in arid and semi-arid conditions. Allen et al. [46] reported that daily ET had the largest differences due to ET fluctuating the most during the growing season, and the monthly and season ET lumped most of the daily variations [38,46], and this is in agreement with the current study results.

Similarly, for the dryland lysimeter, the deviation between Landsat and measured ET was related to higher LAI estimation [32,36,38,47], advective condition effects under irrigated conditions [6, 36], and extended gap periods [38]. In addition, most of the studies conducted evaluated the ET on the current scene (image) days with minimal EB closure errors [36,49,53], and no studies evaluated the extrapolated daily ET assessment for dryland conditions with clumped crops [36]. However, the irrigated field difference magnitude was far less than for the dryland field.

Landsat LAI estimates were better for the irrigated lysimeter, as the METRIC model performance was affected with the wet and cold pixel determination, and the METRIC model performed better with full canopy (full irrigated), compared to dryland (partial canopy) [32,36,38,49]. As irrigated fields produced more vegetation vigor, higher NDVI values were obtained [36,38], and consequently higher estimates of LAI were obtained and resulted in better estimates of Landsat ET for areas managed under irrigated conditions (Figure 8) [32,36,38,41].

The overall Landsat LAI estimation somewhat matched the measured LAI for most cultivated crops from 2001 to 2010 (2004 and 2007 omitted due to large gaps in clear Landsat data during the growing season). Three indicators that the satellite imagery was able to differentiate between irrigated (full canopy) and dryland fields (partial canopy) as well as identify the growing season were as follows:

1. The estimated LAI values for the irrigated field were much larger than that of the dryland field.
2. All LAI values were zero in the beginning and end of each year, and this reflects that the field was bare soil. However, there were LAI values recorded during the growing season for the same field, providing useful information on when the field was fallow versus when a crop was growing.
3. The magnitudes of NDVI values for irrigated fields were higher than those for the dryland fields [36,38].

The reason behind this is likely that LAI is better estimated for the irrigated field than the dryland field [36,38] due to more vegetation coverage, resulting in higher NDVI values and consequently ET values. In 2007 and 2009, when forage sorghum and sunflowers were cultivated, respectively, the LAI estimated using Landsat was slightly lower than the measured LAI for the irrigated lysimeter.

5. Conclusions

Remote sensing-based ET estimation is considered a promising tool for irrigation water management. However, uncertainties associated with satellite-based ET estimation still exist, especially with various remotely sensed platforms due to variations in spatial and temporal resolution. In this study, satellite-based ET was evaluated using Landsat under semi-arid conditions in Texas under irrigated and dryland conditions.

Ten years of lysimeter measured ET data were used in this study. The Landsat-based ET overestimated the measured ET early and late in the growing season and underestimated ET during the peak of the growing season. The daily and monthly ET for the dryland lysimeter was unacceptable with negative NSE (−1.38 and −0.19), indicating there was no correlation between the estimates and measured ET; however, the daily and monthly ET for the irrigated lysimeter values showed better statistics with an NSE of 0.37 and 0.57, respectively. Seasonal ET showed more variations during the growing season compared to the non-growing season, because higher ET values were estimated during the growing season.

Under dryland conditions, there was significant LAI underestimation compared to the measured LAI values due to water stress during the growing season. LAI plays a significant role in evapotranspiration; where greater values of NDVI were obtained, consequently greater LAI was obtained under irrigated conditions, resulting in more ET for irrigated conditions. There are several reasons behind uncertainties of LAI and ET estimation, including the following: (1) METRIC model uncertainties with partial canopy estimates, (2) dryland plants' rapid modification of LAI based on available soil water (partial canopy), and (3) uncertainties with aerodynamic resistance surface roughness length as well as surface temperature deviations between irrigated and dryland conditions.

Extended gap periods are another significant challenge, and the selection of the filling method can account for ET estimation errors. In this study, gap periods reached up to 184 days in 2004, and the minimum was in 2008 with 40 days. The linear interpolation method was utilized to extrapolate the daily ET estimates between every two consecutive images in this study.

More satellite-based ET assessment under arid and semi-arid conditions is required, where the magnitude and frequency of precipitation are erratic, and irrigation is the only source under arid conditions to replenish crop water needs. With advances in remote sensing, more frequent satellite imagery will be available, with high spatial resolutions. Other extrapolation methods should be considered to generate daily time-series ET datasets. This would likely improve overall ET estimation accuracy by improving the overall spatial and temporal resolution.

Future research opportunities that include the assessment of ET relationship with crop physiology, yield, and yield components (number of flowers, grain quality, etc.) would provide potential information on crop response under dryland and irrigated conditions. Economic analysis of commodity market prices would be another research project due to groundwater decline in the Ogallala aquifer.

Author Contributions: Writing—original draft, A.A.H.; methodology, A.A.H. and P.H.G.; software, J.E.M., G.W.M., and P.H.G.; resources, B.A.E. and P.H.G.; review and editing, B.A.E., G.W.M., V.F.B., S.A.R., and J.E.M.; supervision, B.A.E. All authors have read and agreed to the published version of the manuscript.

Acknowledgments: The authors express their sincere thanks to (1) the Egyptian government general mission scholarship administrated by the Egyptian Cultural and Education Bureau, Washington, DC, for partially supporting this research; (2) the Purdue Research Foundation and the Agricultural and Biological Engineering Department for funding support during this research; and (3) the USDA-ARS at Bushland, Texas, USA for sharing the lysimeter data and data analysis.

References

1. Sellers, P.J.; Randall, D.A.; Collatz, G.J.; Berry, J.A.; Field, C.B.; Dazlich, D.A.; Zhang, C.; Collelo, G.D.; Bounoua, L. A revised land surface parameterization (SiB2) for atmospheric GCMs. Part I: Model formulation. *J. Clim.* **1996**, *9*, 676–705. [CrossRef]
2. Paulson, R.W. *Evapotranspiration and Droughts. National Water Summary 1988–1989: Hydrologic Events and Floods and Droughts*; US Government Printing Office: Washington, DC, USA, 1991; Volume 2375, pp. 1–147.
3. Famiglietti, J.S. The global groundwater crisis. *Nat. Clim. Chang.* **2014**, *4*, 945–948. [CrossRef]
4. Yaeger, M.A.; Massey, J.H.; Reba, M.L.; Adviento-Borbe, M.A.A. Trends in the construction of on-farm irrigation reservoirs in response to aquifer decline in eastern Arkansas: Implications for conjunctive water resource management. *Agric. Water Manag.* **2018**, *208*, 373–383. [CrossRef]
5. Gowda, P.H.; Chávez, J.L.; Colaizzi, P.D.; Evett, S.R.; Howell, T.A.; Tolk, J.A. Remote Sensing Based Energy Balance Algorithms for Mapping ET: Current Status and Future Challenges. *Trans. ASABE* **2007**, *50*, 1639–1644. [CrossRef]
6. Allen, R.G.; Tasumi, M.; Morse, A.; Trezza, R.; Wright, J.L.; Bastiaanssen, W.; Kramber, W.; Lorite, I.; Robison, C.W. Satellite-Based Energy Balance for Mapping Evapotranspiration with Internalized Calibration (METRIC)—Applications. *J. Irrig. Drain. Eng.* **2007**, *133*, 395–406. [CrossRef]
7. Gowda, P.H.; Chavez, J.L.; Colaizzi, P.D.; Evett, S.R.; Howell, T.A.; Tolk, J.A. ET mapping for agricultural water management: Present status and challenges. *Irrig. Sci.* **2007**, *26*, 223–237. [CrossRef]
8. Bastiaanssen, W.; Pelgrum, H.; Wang, J.; Ma, Y.; Moreno, J.; Roerink, G.; van Der Wal, T. A remote sensing surface energy balance algorithm for land (SEBAL). *J. Hydrol.* **1998**, 213–229. [CrossRef]
9. French, A.N.; Hunsaker, D.J.; Thorp, K.R. Remote sensing of evapotranspiration over cotton using the TSEB and METRIC energy balance models. *Remote. Sens. Environ.* **2015**, *158*, 281–294. [CrossRef]
10. Park, A.B.; Colwell, R.N.A.; Meyers, V.F. Resource Survey by Satellite; sci-ence fiction coming true. In *Yearbook of Agriculture*; US Government Printing Office: Washington, DC, USA, 1968; pp. 13–19.
11. Jackson, R.D. Remote Sensing of Vegetation Characteristics for Farm Management. *1984 Tech. Symp. East* **1984**, *475*, 81–97. [CrossRef]
12. Choudhury, B.; Idso, S.; Reginato, R. Analysis of an empirical model for soil heat flux under a growing wheat crop for estimating evaporation by an infrared-temperature based energy balance equation. *Agric. For. Meteorol.* **1987**, *39*, 283–297. [CrossRef]

13. Su, H.; McCabe, M.; Wood, E.F.; Su, Z.; Prueger, J.H. Modeling Evapotranspiration during SMACEX: Comparing Two Approaches for Local- and Regional-Scale Prediction. *J. Hydrometeorol.* **2005**, *6*, 910–922. [CrossRef]
14. Anderson, M.C.; Kustas, W.P.; Norman, J.M.; Hain, C.R.; Mecikalski, J.R.; Schultz, L.; González-Dugo, M.P.; Cammalleri, C.; D'Urso, G.; Pimstein, A.; et al. Mapping daily evapotranspiration at field to continental scales using geostationary and polar orbiting satellite imagery. *Hydrol. Earth Syst. Sci.* **2011**, *15*, 223–239. [CrossRef]
15. Evett, S.R.; Kustas, W.P.; Gowda, P.H.; Anderson, M.C.; Prueger, J.H.; Howell, T.A. Overview of the Bushland Evapotranspiration and Agricultural Remote sensing EXperiment 2008 (BEAREX08): A field experiment evaluating methods for quantifying ET at multiple scales. *Adv. Water Resour.* **2012**, *50*, 4–19. [CrossRef]
16. Bastiaanssen, W.G.M.; Noordman, E.J.M.; Pelgrum, H.; Davids, G.; Thoreson, B.P.; Allen, R. SEBAL Model with Remotely Sensed Data to Improve Water-Resources Management under Actual Field Conditions. *J. Irrig. Drain. Eng.* **2005**, *131*, 85–93. [CrossRef]
17. Tasumi, M.; Allen, R.G. Satellite-based ET mapping to assess variation in ET with timing of crop development. *Agric. Water Manag.* **2007**, *88*, 54–62. [CrossRef]
18. Norman, J.; Kustas, W.; Humes, K. Source approach for estimating soil and vegetation energy fluxes in observations of directional radiometric surface temperature. *Agric. For. Meteorol.* **1995**, *77*, 263–293. [CrossRef]
19. Colaizzi, P.D.; Evett, S.R.; Howell, T.A.; Gowda, P.H.; Shaughnessy, S.A.; Tolk, J.A.; Kustas, W.P.; Anderson, M.C. Two-Source Energy Balance Model: Refinements and Lysimeter Tests in the Southern High Plains. *Trans. ASABE* **2012**, *55*, 551–562. [CrossRef]
20. Kustas, W.P.; Norman, J.M. Evaluation of soil and vegetation heat flux predictions using a simple two-source model with radiometric temperatures for partial canopy cover. *Agric. For. Meteorol.* **1999**, *94*, 13–29. [CrossRef]
21. Kalma, J.D.; McVicar, T.R.; Matthew, F. UNSW Faculty of Engineering Matthew Francis McCabe Estimating Land Surface Evaporation: A Review of Methods Using Remotely Sensed Surface Temperature Data. *Surv. Geophys.* **2008**, *29*, 421–469. [CrossRef]
22. Moorhead, J.E.; Marek, G.W.; Gowda, P.H.; Lin, X.; Colaizzi, P.D.; Evett, S.R.; Kutikoff, S. Evaluation of Evapotranspiration from Eddy Covariance Using Large Weighing Lysimeters. *Agronomy* **2019**, *9*, 99. [CrossRef]
23. Anderson, M.; Norman, J.; Kustas, W.; Houborg, R.; Starks, P.; Agam, N. A thermal-based remote sensing technique for routine mapping of land-surface carbon, water and energy fluxes from field to regional scales. *Remote Sens. Environ.* **2008**, *112*, 4227–4241. [CrossRef]
24. Marek, T.H.; Schneider, A.D.; Howell, T.A.; Ebeling, L.L. Design and Construction of Large Weighing Monolithic Lysimeters. *Trans. ASAE* **1988**, *31*, 477–484. [CrossRef]
25. Howell, T.A.; Schneider, A.D.; Dusek, D.A.; Marek, T.H.; Steiner, J.L. Calibration and Scale Performance of Bushland Weighing Lysimeters. *Trans. ASAE* **1995**, *38*, 1019–1024. [CrossRef]
26. Norman, J.M.; Anderson, M.C.; Kustas, W.P.; French, A.N.; Mecikalski, J.; Torn, R.; Diak, G.R.; Schmugge, T.J.; Tanner, B.C.W. Remote sensing of surface energy fluxes at 101-m pixel resolutions. *Water Resour. Res.* **2003**, *39*, 39. [CrossRef]
27. Anderson, M.C.; Norman, J.M.; Mecikalski, J.R.; Otkin, J.A.; Kustas, W.P. A climatological study of evapotranspiration and moisture stress across the continental United States based on thermal remote sensing: 2. Surface moisture climatology. *J. Geophys. Res. Space Phys.* **2007**, *112*, 112. [CrossRef]
28. Anderson, M.; Kustas, W.P.; Norman, J.M. Upscaling Flux Observations from Local to Continental Scales Using Thermal Remote Sensing. *Agron. J.* **2007**, *99*, 240–254. [CrossRef]
29. Colaizzi, P.D.; Evett, S.R.; Howell, T.A.; Tolk, J.A. Comparison of Five Models to Scale Daily Evapotranspiration from One-Time-of-Day Measurements. *Trans. ASABE* **2006**, *49*, 1409–1417. [CrossRef]
30. Allen, R.G.; Pereira, L.S.; Howell, T.A.; Jensen, M.E. Evapotranspiration information reporting: I. Factors governing measurement accuracy. *Agric. Water Manag.* **2011**, *98*, 899–920. [CrossRef]
31. Allen, R.G.; Pereira, L.S.; Howell, T.A.; Jensen, M.E. Evapotranspiration information reporting: II. Recommended documentation. *Agric. Water Manag.* **2011**, *98*, 921–929. [CrossRef]

32. Chávez, J.L.; Gowda, P.H.; Howell, T.A.; Garcia, L.A.; Copeland, K.S.; Neale, C.M.U. ET Mapping with High-Resolution Airborne Remote Sensing Data in an Advective Semiarid Environment. *J. Irrig. Drain. Eng.* **2012**, *138*, 416–423. [CrossRef]
33. Mkhwanazi, M.; Chávez, J.L.; Rambikur, E.H. Comparison of Large Aperture Scintillometer and Satellite-based Energy Balance Models in Sensible Heat Flux and Crop Evapotranspiration Determination. *Int. J. Remote Sens. Appl.* **2012**, *2*, 24.
34. Morton, C.G.; Huntington, J.L.; Pohll, G.M.; Allen, R.G.; McGwire, K.C.; Bassett, S.D. Assessing Calibration Uncertainty and Automation for Estimating Evapotranspiration from Agricultural Areas Using METRIC. *JAWRA J. Am. Water Resour. Assoc.* **2013**, *49*, 549–562. [CrossRef]
35. Trezza, R.; Allen, R.; Tasumi, M. Estimation of Actual Evapotranspiration along the Middle Rio Grande of New Mexico Using MODIS and Landsat Imagery with the METRIC Model. *Remote Sens.* **2013**, *5*, 5397–5423. [CrossRef]
36. Chávez, J.L.; Gowda, P.H.; Howell, T.A.; Copeland, K.S. Evaluating Three Evapotranspiration Mapping Algorithms with Lysimetric Data in the Semi-arid Texas High Plains. In Proceedings of the 28th annual international irrigation show, San Diego, CA, USA, 9–11 December 2007.
37. Dusek, D.A.; Howell, T.A.; Schneider, A.D.; Copeland, K.S. Bushland weighing lysimeter data acquisition systems for evapotranspiration research. *Am. Soc. Agric. Eng.* **1987**, *1061*, 81–2506.
38. Hashem, A.A. Irrigation Water Management Using Remote Sensing and Hydrologic Modeling. Ph.D. Thesis, Purdue University, West Lafayette, IN, USA, 2018.
39. Marek, G.W.; Gowda, P.; Evett, S.R.; Baumhardt, R.L.; Brauer, D.; Howell, T.A.; Marek, T.H.; Srinivasan, R. Calibration and Validation of the SWAT Model for Predicting Daily ET over Irrigated Crops in the Texas High Plains Using Lysimetric Data. *Trans. ASABE* **2016**, *59*, 611–622. [CrossRef]
40. Marek, G.W.; Gowda, P.; Evett, S.R.; Baumhardt, R.L.; Brauer, D.K.; Howell, T.A.; Marek, T.H.; Srinivasan, R. Estimating Evapotranspiration for Dryland Cropping Systems in the Semiarid Texas High Plains Using SWAT. *JAWRA J. Am. Water Resour. Assoc.* **2016**, *52*, 298–314. [CrossRef]
41. Hashem, A.A.; Engel, B.; Bralts, V.F.; Marek, G.W.; Moorhead, J.E.; Rashad, M.; Radwan, S.; Gowda, P.H. Landsat Hourly Evapotranspiration Flux Assessment using Lysimeters for the Texas High Plains. *Water* **2020**, *12*, 1192. [CrossRef]
42. Moorhead, J.E.; Gowda, P.H.; Ponder, B.A.; Brauer, D.K. Bushland Evapotranspiration and Agricultural Remote Sensing System (BEARS) software. In Proceedings of the Managing Global Resources for a Secure Future, Tampa, FL, USA, 22–25 October 2017.
43. Allen, R.G.; Pereira, L.S.; Raes, D.; Smith, M. *Crop Evapotranspiration—Guidelines for Computing Crop Water Requirements*; FAO Irrigation and Drainage Paper 56: Rome, Italy, 1998; ISBN 92-5-104219-5.
44. Blonquist, J.; Allen, R.; Bugbee, B. An evaluation of the net radiation sub-model in the ASCE standardized reference evapotranspiration equation: Implications for evapotranspiration prediction. *Agric. Water Manag.* **2010**, *97*, 1026–1038. [CrossRef]
45. Allen, R.G.; Burnett, B.; Kramber, W.; Huntington, J.L.; Kjaersgaard, J.; Kilic, A.; Kelly, C.; Trezza, R. Automated Calibration of the METRIC-Landsat Evapotranspiration Process. *JAWRA J. Am. Water Resour. Assoc.* **2013**, *49*, 563–576. [CrossRef]
46. Allen, R.G.; Irmak, A.; Trezza, R.; Hendrickx, J.M.H.; Bastiaanssen, W.G.M.; Kjaersgaard, J. Satellite-based ET estimation in agriculture using SEBAL and METRIC. *Hydrol. Process.* **2011**, *25*, 4011–4027. [CrossRef]
47. Hashem, A.A.; Engel, B.A.; Marek, G.W.; Moorhead, J.E.; Rashad, M.; Flanagan, D.C.; Radwan, S.; Bralts, V.F.; Gowda, P.H. Evaluation of SWAT Soil Water Estimation Accuracy Using Data from Indiana, Colorado, and Texas. *Trans. ASABE* **2020**, *5*, 1539–1559.
48. Moriasi, D.N.; Arnold, J.G.; van Liew, M.W.; Bingner, R.L.; Harmel, R.D.; Veith, T.L. Model Evaluation Guidelines for Systematic Quantification of Accuracy in Watershed Simulations. *Trans. ASABE* **2007**, *50*, 885–900. [CrossRef]
49. Madugundu, R.; Al-Gaadi, K.A.; Tola, E.; Hassaballa, A.A.; Patil, V.C. Performance of the METRIC model in estimating evapotranspiration fluxes over an irrigated field in Saudi Arabia using Landsat-8 images. *Hydrol. Earth Syst. Sci.* **2017**, *21*, 6135–6151. [CrossRef]
50. Senay, G.B.; Friedrichs, M.; Singh, R.K.; Velpuri, N.M. Evaluating Landsat 8 evapotranspiration for water use mapping in the Colorado River Basin. *Remote Sens. Environ.* **2016**, *185*, 171–185. [CrossRef]

51. Moorhead, J.E.; Marek, G.W.; Colaizzi, P.D.; Gowda, P.; Evett, S.R.; Brauer, D.; Marek, T.H.; Porter, D.O. Evaluation of Sensible Heat Flux and Evapotranspiration Estimates Using a Surface Layer Scintillometer and a Large Weighing Lysimeter. *Sensors* **2017**, *17*, 2350. [CrossRef] [PubMed]
52. Gowda, P.H.; Howell, T.A.; Paul, G.; Colaizzi, P.D.; Marek, T.H.; Su, Z.; Copeland, K.S. Deriving Hourly Evapotranspiration Rates with SEBS: A Lysimetric Evaluation. *Vadose Zone J.* **2013**, *12*, 1–11. [CrossRef]
53. Numata, I.; Khand, K.; Kjaersgaard, J.; Cochrane, M.A.; Silva, S.S. Evaluation of Landsat-Based METRIC Modeling to Provide High-Spatial Resolution Evapotranspiration Estimates for Amazonian Forests. *Remote Sens.* **2017**, *9*, 46. [CrossRef]

14

Calibration and Global Sensitivity Analysis for a Salinity Model used in Evaluating Fields Irrigated with Treated Wastewater in the Salinas Valley

Prudentia Zikalala [1,*], Isaya Kisekka [2] and Mark Grismer [2]

[1] Department of Land, Air, and Water Resources, University of California, One Shield Avenue, Davis, CA 95616, USA

[2] Department of Land, Air, and Water Resources & Biological and Agricultural Engineering, University of California, One Shield Avenue, Davis, CA 95616, USA; ikisekka@ucdavis.edu (I.K.); megrismer@ucdavis.edu (M.G.)

* Correspondence: pgzikalala@ucdavis.edu.

Abstract: Treated wastewater irrigation began two decades ago in the Salinas Valley of California and provides a unique opportunity to evaluate the long-term effects of this strategy on soil salinization. We used data from a long-term field experiment that included application of a range of blended water salinity on vegetables, strawberries and artichoke crops using surface and pressurized irrigation systems to calibrate and validate a root zone salinity model. We first applied the method of Morris to screen model parameters that have negligible influence on the output (soil-water electrical conductivity (EC_{sw})), and then the variance-based method of Sobol to select parameter values and complete model calibration and validation. While model simulations successfully captured long-term trends in soil salinity, model predictions underestimated EC_{sw} for high EC_{sw} samples. The model prediction error for the validation case ranged from 2.6% to 39%. The degree of soil salinization due to continuous application of water with electrical conductivity (EC_w) of 0.57 dS/m to 1.76 dS/m depends on multiple factors; EC_w and actual crop evapotranspiration had a positive effect on EC_{sw}, while rainfall amounts and fallow had a negative effect. A 50-year simulation indicated that soil water equilibrium ($EC_{sw} \leq$ 2dS/m, the initial EC_{sw}) was reached after 8 to 14 years for vegetable crops irrigated with EC_w of 0.95 to 1.76. Annual salt output loads for the 50-year simulation with runoff was a magnitude greater (from 305 to 1028 kg/ha/year) than that in deep percolation (up to 64 kg/ha/year). However, for all sites throughout the 50-year simulation, seasonal root zone salinity (saturated paste extract) did not exceed thresholds for salt tolerance for the selected crop rotations for the range of blended applied water salinities.

Keywords: treated wastewater irrigation; salinization; model simulation; global sensitivity analysis

---

1. Introduction

Salinization of soils, groundwater, and surface waters from irrigation is a well-known problem often associated with the decline of ancient civilizations dependent on irrigated agriculture around the world, such as Mesopotamia [1]. Today, the salinity problem associated with irrigation in low rainfall regions continues to have numerous grave economic, social, and political consequences. For example, there is a high economic cost associated with salinity; the US Bureau of Reclamation spends $32 million annually to limit salt additions to the Colorado River and the Natural Resource Natural Resource Conservation Service-US Department of Agriculture (USDA-NRCS) spends some $13 million annually to control salinity in irrigation programs across the upper Colorado River Basin [2]. Simultaneously, as competition for available freshwater resources intensifies and use of treated wastewater (recycled water) having greater salinity grows to meet agricultural water demands, a key question inevitably remains, is long-term use of recycled water sustainable? In the 1980s, discussions about "sustainable agriculture" raised questions about changes in soil quality. Soil-water salinity is a transient condition whereby salts concentrate in the soil following root water uptake by plants as well as water loss by evaporation at the soil surface. Subsequent irrigation or rainfall can dilute the soil-water salinity or the solutes can be removed from the system by leaching to subsurface drains, or through deep percolation below the root zone. In areas having fine-textured soils overlying a shallow water table, additional root zone salinization can occur through capillary rise from the saline water table [3–5]. Salinity risks also increase when saline water is used for irrigation and when poor fertilizer and poor irrigation management are combined.

Salinity hazards caused by irrigation depend on the type of salts, soil, and climatic conditions, crop species, and the amount, quality, and frequency of water applications [6]. Increased irrigation efficiency through adoption of advanced irrigation technologies such as micro-irrigation and sprinkler systems may result in less water used in fields but may also decrease the leaching required to maintain satisfactory root zone salinity during the growing season. While advanced irrigation technologies are beneficial for increasing water productivity and protecting groundwater resources from pollutant leaching, the low leaching fractions may lead to soil salinization. In addition, surface runoff pickup of salts and leaching enable accumulated field salts to degrade river and groundwater resources. These trade-offs also suggest that refined guidelines for use of treated wastewater for irrigation are needed and could be aided through root zone salinity modeling.

Groundwater salinization is occurring in aquifers along the California coast [7] and is especially critical in the Salinas Valley of the Central Coast as seawater intrusion threatens groundwater supplies critical for irrigation of high-value fruit and vegetable crops. As a means to limit seawater intrusion, tertiary treated wastewater was made available for agricultural use in the Salinas Valley since 1998 as an alternative or supplement to groundwater and concerns are growing about possible root zone salinization in fields receiving recycled water. Accumulation of salinity in the crop root zone progressively decreases yields. For example, during a 13-year field experiment in the Castroville area, Platts and Grismer [8] observed an upward trend in soil electrical conductivity (EC) and chloride (Cl) indicating a soil salinization threat and a possible growing Cl toxicity threat to crop production in the Salinas Valley. The range of increase in EC in the root zone for sites irrigated with blended well and treated wastewater was 18 to 63% and an increase in Cl ranged from 48 to 510%. Moreover, agricultural return flows account for an estimated 33% of annual recharge to groundwater in the Valley [9]. In a geochemical analysis, Vengosh and others [10] suggested that 3–10 mm/year of vertical seepage associated with agriculture adversely affects the Valley's groundwater quality. On the other hand, Platts and Grismer [11] found that annual winter rainfall of roughly 250 mm was required to adequately leach accumulated salts associated with recycled water use for irrigation in the Valley. Moreover, from the lower Salinas River at Gonzales to the estuary, salinity is listed under EPA 303d indicating that salinity in the Salinas Valley threatens sensitive surface water supply and ecosystems.

Root zone soil-water models have been developed in an effort to gain both an understanding of the complex processes associated with soil–water–chemistry dynamics in the root zone and to provide guidance for water managers and growers. Dynamic soil-water models quantify many physical-chemical-biological interactions in irrigated agricultural systems and enable predictions to assess spatio-temporal changes in soil salinity during and between growing seasons [12]. Soil EC is one of nineteen measures advocated by the Soil Health Institute [13] as a measure of agricultural sustainability and is a critical output parameter from these models. Further, Maas and Hoffman [14] and Rhoades and others [15] developed crop-threshold EC values to assure successful use of saline water for irrigation. Important factors in such models include daily rain depths and evapotranspiration (ET) demands, soil properties, crops and irrigation method, water application depths and quality, and chemical factors such as salt precipitation and dissolution rates within the root zone. Understanding and predicting how root zone salinity changes in time under different irrigation methods and cropping systems provides insight into possible groundwater and surface water salinization. Several models have been developed to estimate the soil water balance of the soil–plant-atmosphere system, Decision Support System for Agrotechnology Transfer (DSSAT) for example is best suited to simulate process-based crop growth and development; the model does not include a salinity module and this model uses a "tipping bucket" water balance approach for soil hydrologic and water redistribution processes [16]. HYDRUS-1D simulates water flow in soils using the numerical solutions of the Richards equation, however, its simulation of crop-related processes is limited. Moreover, for long-term, multi-cropping simulations, HYDRUS-1D requires loose coupling with an external crop model for estimations of evapotranspiration. As such, we elected to use a simple daily time-step soil salinity model based on soil-water storage in four rootzone quarters and applicable to long records of meteorological data. This enables us to take into account a number of site-specific factors including soil properties, rainfall patterns, crop type, and irrigation methods to establish the effect of these factors on long-term soil salinity.

While dynamic soil-water processes can be simulated at multiple time scales, a daily time step is typically deployed because it represents the time scale at which rainfall, water application, and ET information is more readily available and because many of the actual root zone processes occur within hours to a day timeframe. While [11] employed a daily soil-water balance model to examine the effects of various hydrologic factors on soil salinity over the 13-year study, they did not include root zone soil-water chemistry, upward flow from shallow water tables and fertilizer management processes. Isidoro and Grattan [17] developed a root zone soil salinity and water balance model with root-water-extraction assumptions similar to foundational models of the past [18]. We extend this model to further account for drainage under shallow groundwater conditions to enable inclusion of saline water table effects on root zone soil water and salinity as similarly described in Reference [19].

Due in part to the greater capillary rise in finer-textured soils, greater upward flow rates from shallow water tables has been found in loamy soils than in sandy soils having small capillary rise, or in clayey soils with very slow permeability [20]. Crop water use from shallow water tables is controlled by its depth and water quality, crop type, growth stage and salt tolerance, and water application frequency and depth as affected by the vadose zone hydraulic conductivity [4,5,19,21–25]. In the Salinas Valley study area, groundwater depths range from 7.5 to 11.6 m (24.4 to 38 feet) below ground surface with maximum groundwater depths occurring in the fall following the summer irrigation season. In the 13-year study by Reference [8], they noted that observed Cl accumulation in the soil profile may have resulted from upward flow of saline groundwater.

Any modeling effort is a representation that necessarily simplifies reality; however, simulations enable investigation of "what if?" questions. Previous analyses by References [8,11] of overall soil salinity changes and leaching during the 13-year field study suggested that root zone salinity levels could be managed by winter rains when irrigating with blended recycled water. However, they underscored that more detailed process analyses were required to better elucidate what applied water salinity levels were tolerable. Here, we seek to quantify (model) long-term (50 years) trends in soil salinity within the Castroville area of the Salinas Valley when shallow groundwater is present as affected by irrigation with recycled water of varying salinity. Further, model applicability to a region, requires model calibration and validation for the study-site conditions so we explore use of a new two-step process that first identifies the critical model parameters and then focuses on those to calibrate the model. Use of models enables a comparatively inexpensive and environmentally-safe technique to evaluate the long-term effects of various agricultural management scenarios on soil salinity while also providing an aid for water managers considering these complex processes.

We use observations from the 13-year field experiment evaluating the long-term effects from use of varying fractions of recycled water (i.e., salinity) for vegetable and strawberry production on soil salinity in the Castroville area building on the previous efforts by Platts and Grismer [8,11]. Study objectives were to:

- Perform global sensitivity analysis of the modified Isidoro and Grattan [17] root zone salinity model to find the parameters most sensitive to model outputs:

- Complete a model calibration and validation using parameters to which model outputs are most sensitive as a guide: and
- Predict long-term (five decades) root zone salinity, salt output load with deep percolation and salt output load with surface runoff from fields using treated wastewater for irrigation.

2. Materials and Methods

2.1. Study Area

The study area in Castroville, overlies two main aquifers referred to as the "180-foot" and "400-foot" aquifers, respectively. These were formed from fluvial sands and gravels associated with the old Salinas River channel and possible delta conditions. Above and between the two aquifers are deposits of blue clay overlying the "180-foot" aquifer range from 8-m thick at Salinas to more than 30-m thick at Castroville [26–28]. The typical overlying soil profile in the study area is comprised of Pacheco sandy to clay loams as summarized in Table 1.

Table 1. Average soil profile texture variations in the study region [29].

Texture	Depth (m)	Textural Fractions (%)				Conductivity K_s (mm/d)	Bulk Density (kg/m^3)
		Sand	Clay	Silt	OM		
Clay loam	0–0.6	20–45	27–35	20–53	2	121–363	1660
Sandy loam	0.6–0.9	35–70	15–27	3–44	0.5	363–1218	1640
Loam	0.9–1.2	30–50	20–27	28–50	0.5	363–1218	1640
Silty clay loam	1.2–3	15–20	27–35	45–58	0.5	36–121	1700

We assembled the base data for the model using the estimates for saturated hydraulic conductivity, organic matter content, soil texture, and bulk density from SoilWeb [29] an interactive webtool to access detailed soil survey data (SSURGO). We then determined saturation water content (Ts), wilting point (Twp), and field capacity (Tfc) according to soil texture using artificial neural networks techniques implemented in Hydrus-1D [30]. Meteorological records for 1983 to 2014 were taken from the local (California Irrigation Management Information System (CIMIS) station number 19 [31] and the average monthly rainfall and grass-reference ET_o are shown in Figure 1. Average monthly reference ET_o exceeded average monthly rainfall during April through November, annual ET_o ranged from 862.3 mm to 1072.6 mm ($\pm$5.3%). Rainfall was concentrated from November to March (87% of annual rainfall) with annual rainfall depths ranging from 134.5 mm to 1026 mm ($\pm$47.1%) as shown in Figure 2.

The main crops grown in the project area are cool-season vegetables (lettuce, broccoli, cauliflower, artichoke, cabbage, spinach, celery) and strawberries. In 1995, the Monterey County Water Resources Agency (MCWRA) passed an ordinance prohibiting extraction of groundwater between sea level and −76.2 m in Salinas and Castroville. In 1998, Monterey County Water Recycling Projects (MCWRP) began delivering recycled water to 486 hectares (12,000 acres) in the northern Salinas Valley. Crop rotations and management practices at the eight sites of the study area are listed in Table A1; the crops include lettuce, broccoli, cauliflower, cabbage, celery, spinach, artichokes, and strawberries. Drip irrigation was used at the control site and vegetable crops were established with sprinklers for 20 to 30 days. At site 2, sprinkler irrigation switched to drip after plant establishment in 2002 while site 3 used sprinklers for vegetables and drip for strawberry. Sites 4 and 5 used sprinklers and drip and sites 6 and 7 used sprinklers for plant establishment and followed by furrow irrigation.

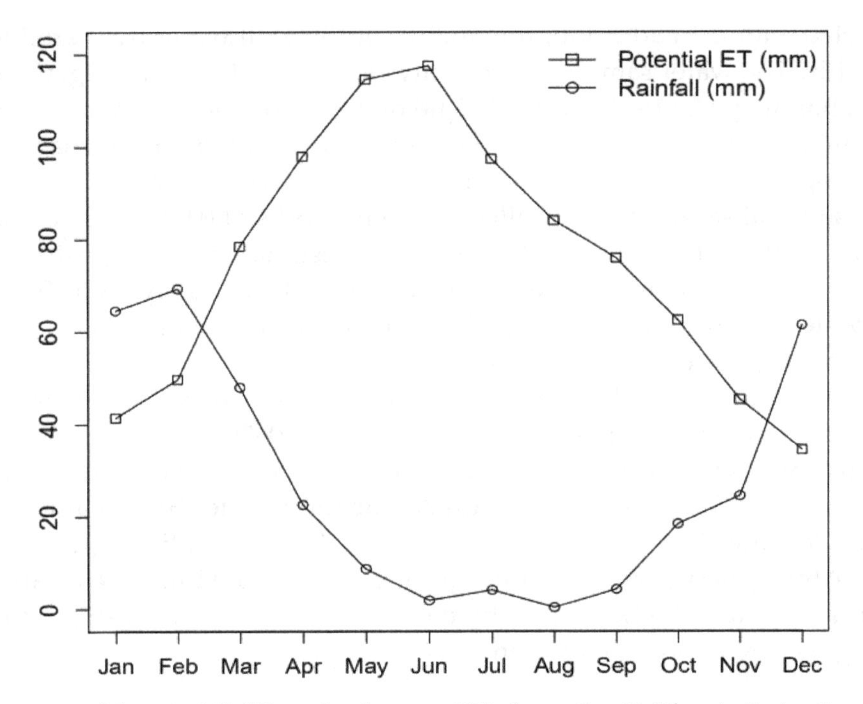

Figure 1. Mean monthly rainfall (P) and reference ET_o from the California Irrigation Management Information System (CIMIS) station 19 in Castroville, CA.

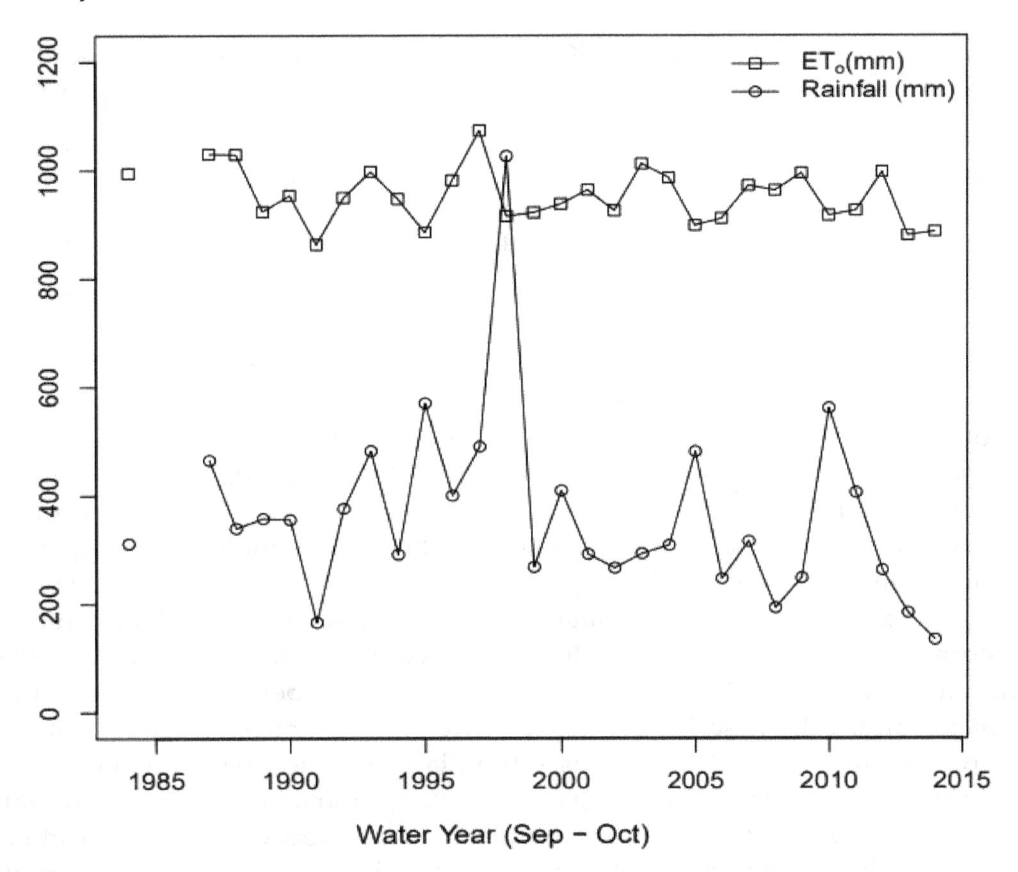

Figure 2. Annual rainfall and reference ET_o from CIMIS station 19 in Castroville, CA.

Tertiary treated wastewater effluent from Monterey Regional Water Pollution Control Agency, (MRWPCA) was sampled on a weekly basis to determine the levels of salt present in it before blending with the supplemental well water used to meet peak irrigation demand. Monthly delivery system sampling confirmed the quality of the water received by growers after supplemental well water was added to the recycled water. In addition, the quality of the well water delivered to the control site was sampled monthly. The water samples were analyzed for pH, EC_W, Na, Mg, Cl and K (potassium) by an accredited laboratory run by MRWPCA. The one control and seven test sites were randomly distributed throughout Castroville, USA area and were chosen based on soil characteristics, drainage systems, types of crops grown (lettuce, cole crops and strawberries), irrigation method and farming practices. At each site, soil samples were collected from depths of 0.03 to 0.30 m, 0.30 to 0.61 m and 0.61 to 0.91 m at four different locations within 1 m of a designated global positioning system (GPS) point. Sample analysis was done by an independent accredited lab (Valley Tech, Tulare, CA, USA) and included pH, soil water electrical conductivity (EC_{sw}), extractable cations B (boron), Ca, Mg, Na and K, and extractable anions Cl, NO_3 (nitrate) and SO_4 (sulphate).

Irrigation water salinity varied between sites and years as recycled water was blended with groundwater (2000–2009) or diverted Salinas River water (2010–2012) and the average applied water EC (EC_W) at the different sites for these time periods are summarized in Table 2. Tertiary treated wastewater in the study had on average Sodium Adsorption Ratio (SAR) value of 5.58, containing 192.1 mg/L of Na, 246.1 mg/L of Cl, and EC of 1.4 dS/m. The rain salinity (EC_P) values were taken from the National Atmospheric Deposition Program station in the Pinnacles National Park located ~322 km east of the study site. The EC_P varies by month ranging from 0.001 dS/m to 0.004 dS/m with May having the highest ion deposits with rain.

Table 2. Average electrical conductivity (EC) of applied water (EC_W) at different sites from 2000–2012.

Site #	2000–2009		2010–2012	
	% Recycled Water	EC_W (dS/m)	% Recycled Water	EC_W (dS/m)
Control	0	0.63	0	0.78
2	46	0.75	92	1.12
3	94	1.52	98	1.19
4	58	0.94	96	1.17
5	93	1.51	100	1.21
6	70	1.14	90	1.09
7	96	1.56	90	1.17

2.2. Root Zone Salinity Model

We coupled crop growth and soil water models applied across the root- and vadose-zones to simulate both upward flow from shallow water tables as well as downward percolation to the groundwater (see Appendix B for a detailed description of the model). These were combined with a root zone salinity model and used to predict root zone soil salinity (see model configuration in Figure 3). The two driving criteria for model selection included the simplification required to describe the processes mathematically without losing the detail needed to develop realistic results, and the model reliance on readily available input data. The Isidoro and Grattan [17] daily time-step model uses a closed-form solution of Reference [32] to describe vertical unsaturated water movement in the root zone and unsaturated zone (Equations (A1) to (A4)). A number of closed-form formulas have been proposed to empirically describe the dependence of unsaturated hydraulic conductivity and water content on pressure head [33–37]. We used the Clapp and Hornberger equation [32] to extend vertical flow through a continuous soil profile to compute the movement of water and salt across the entire vadose zone to the account for shallow groundwater flow processes. Thus, two additional layers from the root zone to the groundwater table were added for both unsaturated and saturated zones.

The crop component of the model includes crop development stages (Table A2), root growth (Equation (A8)), root water uptake and water stress response functions (Equations (A5)–(A7)). Evapotranspiration (ET) includes a combination of two separate processes whereby water is lost from the soil surface by evaporation and from the root zone by crop transpiration. The crop ET is calculated as the product of the reference ET_o and the estimated crop coefficient (K_c) that depends on crop characteristics, vegetative growth state, canopy cover, and height as well as surface-soil properties (Equations (A5)–(A7). In each layer (k), the actual crop ET can be lower than potential ETc(k) due to water stress, which depends on the soil water content and the sensitivity of the crop to low water contents, accounted for through the crop-specific parameter *p*: the ratio of readily available soil water (RAW) to total available water (TAW) (p = RAW/TAW) [38].

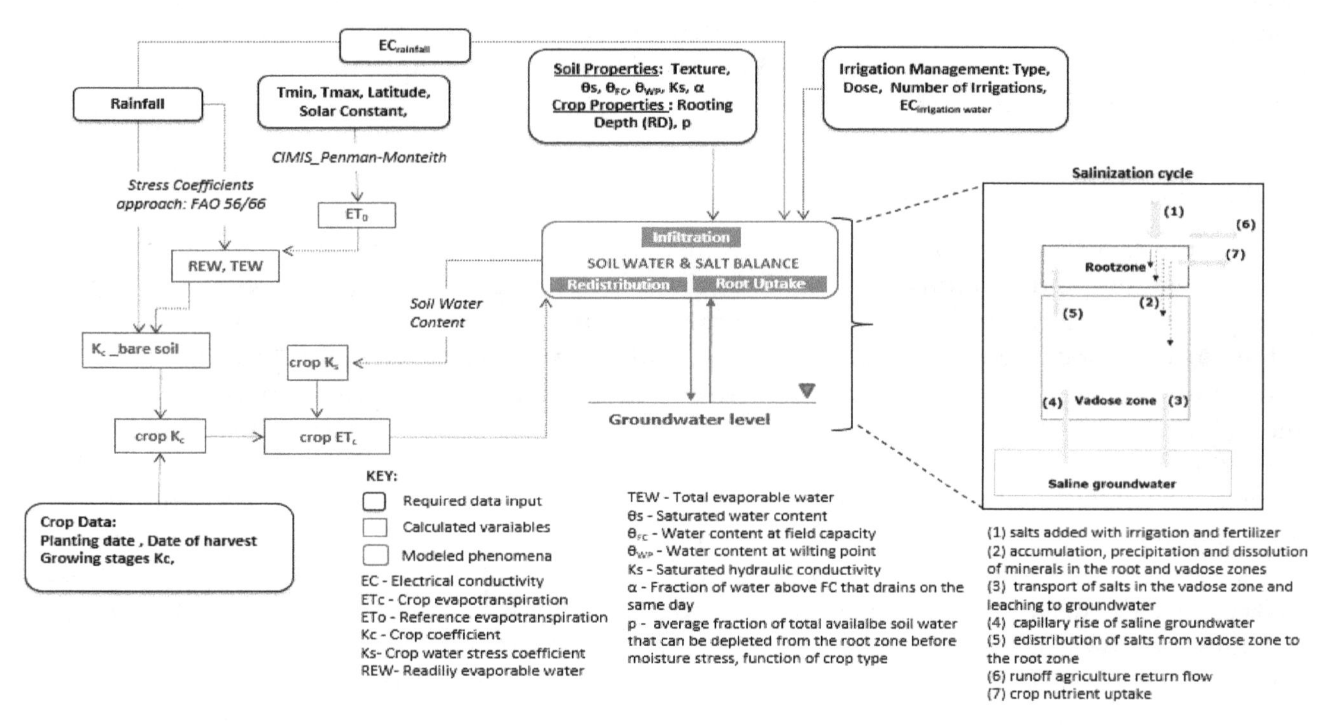

Figure 3. Main components of the soil–water and salinity balance model across the soil–water–plant–atmosphere–aquifer continuum.

The model domain consists of a one-dimensional vertical 7.7-m soil profile, representing the crop root zone and the unsaturated zone that overly a fixed saturated (water table) zone. The domain is discretized so that the clay loam root zone is divided into four quarters of equal depth to enable determination of plant water uptake fractions. The unsaturated zone below the root zone is divided into two layers: a sandy-loam layer immediately below the root zone and a silty clay-loam between the sandy-loam and the capillary fringe. Both upward flow from and downward flow to shallow groundwater is possible in the model. Surface runoff depths were calculated using the Soil Conservation Service (SCS) runoff method [39,40] see Equations (A9) and (A10). Equations (A11) and (A12) detail the soil–water balance simulations.

Salt balance calculations were performed in conjunction with the soil–water balance assuming complete mixing of water entering each layer with that already stored in that layer (Equations (A13)–(A19)). The soil–water EC is used as a salinity indicator, implicitly assuming there is a unique relationship between EC and total dissolved solids (TDS), and that EC behaves as a non-reactive (conservative) solute. Salinity of irrigation water (EC_W) and precipitation (EC_P) are input values. The mass of salts in layer k (Z(k)) is estimated from the product $EC_{SW}(k)$ W(k), where EC_{SW} is the electrical conductivity of the soil water in that layer.

Plant water uptake was assumed to be a descending extraction pattern that depends on irrigation frequency such that greater uptake is at the top quarter of the root zone [18,41–43]. Plant growth and root development parameters are summarized in Table A2 and Equation (A8) [38,44–47] and we assumed that strawberries were planted on 1.3 m wide raised beds as is common in the region. The model was calibrated for soil salinity generation due to dissolution and a dissolution rate is used to account for these processes. Irrigation and rainwater salinity are specified by the user and the model neglects plant root uptake of salts. While preferential flow and irrigation non-uniformity may also be important features, they were beyond the scope of this model.

Each simulation extended for a period of 13 and 50 water years (1 October to 30 September) and more importantly the model simulates carry over effect from one year to the next. The surface boundary conditions of rainfall, irrigation and ET_0 were specified daily together with irrigation and rain water EC. The lower boundary conditions were specified as fixed water table depth and groundwater salinity ECgw; though fixed water table depths are unlikely in the field, water table fluctuations are assumed to be dampened by capillary rise and evaporation from the water table. The model is written in R to make it more widely accessible to water managers and possibly growers.

2.3. Calibration and Sensitivity Analyses

Parameter sensitivity analyses provide insight into those model parameters that are most critical towards approximating measured results and are often used to help focus field sampling or measurement efforts and/or refinement of modeled processes. Here, we take a different approach and first used a global sensitivity analysis that considers variations within the entire variability space of the input factors. We used the elementary effect (EE) method for screening important input factors among the 33 factors initially considered important. Finally, the variance-based "Sobol" method was used with those factors determined to be significant from the Morris screening method for factor fixing and to identify those factors which, left free to vary over their range of uncertainty, make no significant contribution to the variance of the output results of interest. We applied the modified Isidoro and Grattan model to the 13 years of soil salinity observations up to 91.4 cm below the ground surface. Measured data from the control site, irrigated only with available groundwater, was used for calibration and one of the other eight test sites using predominantly recycled water (94–98% recycled/freshwater blend) was used for model validation so as to bracket possible model predictions. The calibrated model was then used to assess the long-term (50 years) salinity outcomes of variable management strategies including cropping patterns, irrigation technology, and irrigation water quality.

We use the Morris and Sobol's methods to support model calibration as shown in Figure 4. The sensitivity analysis was used to address the following questions:

- What input parameters or factors cause the largest variation in the output?
- Are there any factors whose variability has a negligible effect on the output?
- Are there interactions that amplify or dampen the variability induced by individual factors?

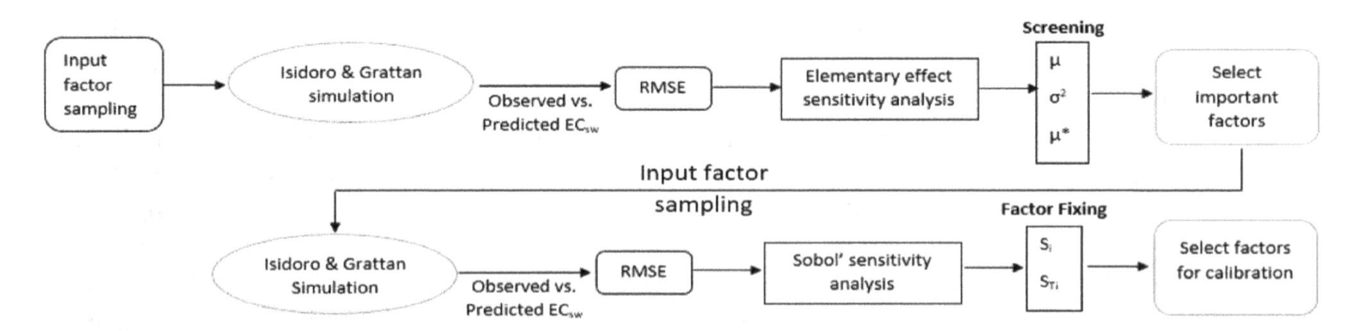

Figure 4. Workflow for calibration of the soil-water water and salt balance model.

Sensitivity Analysis Library (SALib version 1.1.3) an open source library written in Python, was used for performing the sensitivity analyses [48]. The model input variables considered for the sensitivity analysis are listed in Table 3 and parameters showing the greatest sensitivity were selected for calibration.

Table 3. Model parameters used in the sensitivity analysis. (Note: rz—root zone layers 1–4; 5—unsaturated zone layer 5; 6—unsaturated zone layer 6).

Property	Model Code	Model Units	Range	
			Min.	Max.
Soil Hydraulic Parameters				
Saturated water content	Tsrz	cm^3/cm^3	0.439	0.486
	Ts5	cm^3/cm^3	0.357	0.37
	Ts6	cm^3/cm^3	0.36	0.38
Water content at field capacity	Tfcrz	cm^3/cm^3	0.324	0.367
	Tfc5	cm^3/cm^4	0.240	0.320
	Tfc6	cm^3/cm^5	0.250	0.270
Water content at wilting point	Twprz	cm^3/cm^3	0.154	0.177
Residual water content	Tr5	cm^3/cm^5	0.11	0.14
	Tr6	cm^3/cm^5	0.066	0.08
Saturated hydraulic conductivity	Ksrz	cm/day	12.1	36.3
	Ks5	cm/day	36.3	121.8
	Ks6	cm/day	3.6	12.1
Fraction of excess water drained the first day	Arz	%	0.81	0.83
	A5	%	0.85	0.87
	A6	%	0.8	0.82
Runoff Curve Number for fallow periods	CNf	-	91	94
Growing season Curve Number	CNc	-	88	91
Capillary fridge height above the water table	H_d	cm	83	183
Depth to groundwater table	Hwt	M	7.45	11.57
Depth of surface soil layer subjected to drying by evaporation	Ze	mm	100	150
Plant Parameters				
Root water uptake for layers 1–3	RWU_1	%	60	71
	RWU_2	%	20	30
	RWU_3	%	6	7
Rooting depth of lettuce, broccoli, cabbage and cauliflower	ZrL	cm	30	50
	ZrBrc	cm	40	60
	ZrCabb	cm	50	80
	ZrCau	cm	40	70
Fraction of total available water that can be depleted from the root zone before moisture stress for lettuce	pL	-	0.3	0.7
Fraction of total available water that can be depleted from the root zone before moisture stress for broccoli, cabbage, and cauliflower	pBCC	-	0.45	0.7
Soil Chemical Parameter				
Rate of dissolution at 2.5–30.5 cm depth	k1	dSm^{-1}/day	0	0.014
Rate of dissolution at 30.5–61 cm depth	k2	dSm^{-1}/day	0	0.022
Salinity at shallow water table	ECgw	dSm^{-1}	0.35	1.58
Initial ECsw	ECsw	dSm^{-1}	0.29	4.1

Different crops have different water uptake patterns, but all take water from wherever it is most readily available within the rooting depth. The root zone water-uptake pattern depends on irrigation frequency. With infrequent irrigations, the typical extraction fractions by root zone layer is 40–30–20–10%. For frequent drip or sprinkler irrigation, the water uptake fractions are skewed towards greater uptake from the upper root zone, or a 60–30–7–3% uptake pattern [18]; this pattern is assumed in many classical analysis of saline soils [42]. Some have suggested use of an exponential model that specifies a greater proportion of uptake near the soil surface, that is, uptake fractions of 71–20–6–3% [41,43]. Ranges for the crop related data, including crop coefficients (Kc), rooting depths (Zr), and average fractions of total available water that can be depleted from the root zone before moisture stress (p), were taken from the Food and Agriculture Organization of the United Nations (FAO)-56 [38]. Estimates of strawberry crop coefficients were found in Reference [47].

All field sites considered were situated on Pacheco clay, clay-loam, and sandy loam soils with ranges of soil texture, available water content, bulk density, organic matter content, and saturated hydraulic conductivity (Ks) taken from SSURGO soil surveys as summarized previously in Table 1. Soil hydraulic properties required for model application were inferred from the soil survey information. We fitted that information to the van Genuchten model using a Multiobjective Retention Curve Estimator (MORE) based on the Multiobjective Shuffle Complex Evolution Metropolis (MOSCEM-AU) algorithm implemented in HYDRUS-1D [30]. The fraction "α" of the excess water that drained the first day ($0 < \alpha < 1$) was calculated from the soil texture in the layer through an empirical relation obtained to match the results presented in [49]. Grismer [20] provides a relationship between capillary fringe heights (Hd) and saturated intrinsic permeability for different soil textures. Groundwater levels were taken from regular measurements by the MCWRA at well 13S/02E-32E05 about 5 km west of study area. Estimated range of groundwater salinity was taken from Reference [50].

The primary output variable of concern was the soil–water EC_{sw} as determined for root zone layers up to 0.91 m deep. As the Isidoro and Grattan model is a dynamic model, the "output" term in the sensitivity analysis does not refer to the range of spatial and temporal distribution of EC_{sw} but to a summary variable. In this case, the root mean square error (RMSE) that is obtained as a scalar function of the simulated time series output EC_{sw} values. As such, for calibration, the objective function minimized the RMSE associated with model prediction.

2.3.1. Model Parameter Screening-Elementary Effects Method

The elementary effects (EE) method is an effective way of screening for important input factors contained in a model [51]. The fundamental concept of this method involves deriving measures of global sensitivity from a set of local derivatives, or elementary effects, sampled on a grid throughout the parameter space [52]. It is based on one-at-a-time (OAT) analysis, in which each parameter Xi is perturbed along a grid of size Δi to create a trajectory through the parameter space. For a given value of X, the elementary effect of the ith input factor is defined as:

$$\mathrm{EEi} = \frac{[Y(X_1, X_2, \ldots, X_{i-1} + X_i + \Delta, \ldots, X_k) - Y(X_1, X_2, \ldots, X_k)]}{\Delta} \tag{1}$$

where $Y(X_1, X_2, \ldots, X_k)$ is a prior point in the trajectory and $X = X_1, X_2, \ldots, X_k$ is any selected value in the parameter space such that the transformed point is still in the parameter range for each index $i = 1, \ldots, k$. The sensitivity measures μ and σ are the mean and the standard deviation of the distribution of EEi proposed by Morris. Mean parameter (μ) assesses the overall influence of the factor on the output parameter of interest; σ assesses the extent to which parameters interact. Thus, a small σ value implies that the effect of Xi is almost independent of the values taken by other factors; on the other hand, a large σ indicates that a factor is interacting with others because its sensitivity changes across the variability space. Campolongo and others [53] suggest that μ^* is a good proxy of the total sensitivity index, a measure of the overall effect of a factor on the output parameter inclusive of interactions. We analyzed μ^* for all input factors to screen out non-influential factors, and then performed a variance-based analysis with the remaining important factors. Once trajectories are sampled, the resulting r elementary effects per input are available, the statistics μ, $\sigma 2$ and μ^* for each factor are computed as:

$$\mu_i = \frac{1}{r}\sum_{j=1}^{r} EE_i^j \tag{2}$$

$$\mu_i^* = \frac{1}{r}\sum_{j=1}^{r} |EE_i^j| \tag{3}$$

$$\sigma_i^* = \frac{1}{r-1}\sum_{j=1}^{r} (EE_i^j - \mu)^2 \tag{4}$$

2.3.2. Factor Fixing-Sobol's Variance Method

Sobol's sensitivity analysis is a global-variance based method. Sensitivity measures are based on the decomposition of the model output variance to individual parameters and the interaction between parameters [54,55]. Variance-based sensitivity analysis relies on three principles:

- input factors are assumed to be stochastic variables of the model that induce a distribution in the output space;
- the variance of the output distribution is a good proxy of its uncertainty; and
- the contribution to the output variance from a given input factor is a measure of sensitivity.

Contribution to total output variance by individual input factors and their interaction can be written using an ANOVA high-dimensional model representation (HDMR) decomposition [51]:

$$V(Y) = \sum_{i}^{k} V_i + \sum_{1 \le i < j \le k} V_{ij} + \ldots + V_{12\ldots k} \tag{5}$$

where $V(Y)$ is the total or unconditional variance of the output, the conditional variance; V_i is the conditional variance or first-order effect of X_i on Y; V_{ij} is the joint effect of X_i and X_j minus the first order-effects for the same factors. Several variance-based indices can be defined; the first order index represents the main contribution effect of each input factor to the output variance can be determined from:

$$S_i = \frac{V_i}{V(Y)} \tag{6}$$

The total order index, S_T, a measure of the overall contribution to output variance from an input factor considering its direct effect and its interactions with all other factors and is determined from:

$$S_{Ti} = 1 - \frac{V_{\sim i}}{V(Y)} \tag{7}$$

where $V_{\sim i}$ is the conditional variance with respect to all the factors but one, i.e., $X_{\sim i}$. The condition $S_{Ti} = 0$ is necessary and sufficient for X_i to be a noninfluential factor on the output. That is, if $S_{Ti} \cong 0$, then X_i can be fixed at any value within it range of uncertainty without appreciably affecting the value of the output variance $V(Y)$. Here, we calculated all of the indices to determine the factors that can be fixed in the calibration process. A recommended sampling technique uses sequences of quasi-random numbers generating n, 2k matrix of random numbers where n is called a base sample and k is the number of input factors. This scheme allows for n (k + 2) model evaluations. We evaluated up to n = 12 and found no changes occurred after n = 10 and concluded that it was sufficient.

2.4. Long-Term Salinity Indicators

Salt output loads with deep percolation were used to assess the potential for groundwater resources deterioration and salt output loads with runoff indicate the salinity threats posed by treated wastewater irrigation on the salinization of the Salinas River. Similarly, crops respond to the salinity in the root zone over the entire growing season [18]. Thus, we used seasonal-averaged root zone EC of the saturated extract or EC_{eS} (dS/m), deep percolation S_d (kg/ha), and surface runoff S_r (kg/ha) salt output loads as key output state variables to describe long-term (decadal) impacts of irrigation with treated wastewater and farm management practices (i.e., applied water quality and depths, irrigation technology, and crop rotations). Annual rainfall mitigates impacts of irrigating with saline water as such accounting for rainfall leaching is important for evaluating long-term dynamics.

Daily EC of the saturated extract (EC_e) in the root zone layers (k) is calculated as:

$$EC_e(k)_t = \frac{ECsw(k)_t \times \theta(k)_t}{SP(k)} \tag{8}$$

where SP(k) is the saturation percentage (water content of the saturated soil paste expressed on a dry weight basis) for layer k. Traditionally, for most mineral soils it is assumed that field capacity is half of SP, so EC_{et} is the mean EC_e of the 4 rootzone layers; daily values EC_{et} are then averaged over the entire growing season yielding the seasonal-averaged root zone EC_e (EC_{eS}):

$$EC_{eS} = \frac{\sum_{\text{Growing season}} EC_{et}}{\text{Days in the growing season}} \tag{9}$$

We applied the model to both meteorological series and management practices for 13 years of cropping practices in the study area. The objective was to simulate a 13-year continuous cropping and provide a multiple-year record of seasonal EC_{eS} and related parameters. Following [15], we assumed a factor of 640 to convert EC (dS/m) into TDS (mg/L) for EC $\leq$ 5 dS/m and a factor of 800 for EC $>$ 5 dS/m.

The model estimates daily lower boundary water flux (D) along with an estimate of soil–water EC in the bottom layer to determine the daily salt output load associated with deep percolation water calculated as:

$$S_d = D \times EC_{sw}(4) \times 6.4 \tag{10}$$

where S_d is the daily drainage salt output load in kg/ha; $EC_{sw}(4)$ is the EC of soil water in the bottom root zone layer in dS/m and 6.4 is the conversion factor assuming a factor of 640 to convert EC (dS/m into TDS (mg/L) and flux in mm. Additionally, the model estimates daily runoff volumes (SR) from the soil surface along with the runoff water EC such that the daily salt output load associated with runoff is calculated as:

$$S_r = SR \times EC_{sr} \times 6.4 \tag{11}$$

where S_r is the daily runoff salt output load in kg/ha; EC_{sr} is the EC of surface runoff in dS/m and 6.4 is the conversion factor.

2.5. Calibration and Validation

The primary objective of model calibration was to capture the long-term soil salinity dynamics in the fields irrigated with treated wastewater in the study region by satisfactorily reproducing the 2000 to 2012 soil-water salinity (EC_{sw}) data set described by [11]. The study consisted of six test sites and one control site randomly distributed across the Castroville region that were chosen to provide a typical range of soil characteristics, drainage systems, types of crops grown, irrigation methods, and farming practices found in the region. Average annual water quality delivered to each site was determined as well as soil samples collected from depths of 0.03–0.30 m, 0.30–0.61 m, and 0.61 to 0.91 m at four different locations within 1 m of a designated global positioning system (GPS) point.

Soil samples were collected following winter rains before the spring planting, during and at the end of the summer growing season prior to winter rains. Saturated paste extracts from these soil samples were analyzed for EC and solute concentrations. The control site received only well (2000–2009) or surface (2010–2012) water and this site was used for calibration, while site 3 received 94–98% recycled water and this site was used for validation (see Table 2). Annual crop rotations that included lettuce, broccoli, cauliflower, cabbage, and strawberry are shown in Figure A1. Control site vegetables were established with sprinklers for 20–30 days and drip irrigated, while at site 3, vegetables were sprinkler irrigated and then drip irrigation was used for strawberries.

Sensitivity analysis, calibration and validation were performed for soil–water EC (EC_{sw}) at three depth intervals. The goodness-of-fit measures of the model predictions were evaluated using a set of statistical indices including root-mean-squared-error (RMSE), Nash-Sutcliffe efficiency (NSE), coefficient of determination (R^2), coefficient of regression (b), and mean relative error (MRE). A perfect fit between observations and model predictions yields a RMSE = 0.0, NSE = 1.0, R^2 = 1.0, b = 1.0 and MRE = 0.0.

3. Results

3.1. Sensitivity Analysis

The sensitivity analyses were performed on the 33 model parameters as indicated in Table 3. We measure the sensitivity of the root mean squared error (RMSE) metric, calculated using the sampled soil water salinity (EC_{sw}) at the three depth intervals to ensure that our sensitivity indices are grounded relative to the observed soil salinity. The sample sizes and corresponding number of model evaluations required for both the Elementary Effect (EE) and Sobol' methods are listed in Table 4. For the EE method, a sample size of n = 1000 with 10 optimal trajectories were used resulting in 340 model evaluations. Since the Sobol analysis was performed with the reduced parameter space results determined from the prior EE approach, the number of samples selected ranged from 10 to 12 and iterations were discontinued at the point where the sensitivity results converged. The Latin Hypercube sampling design was used with n = 50 sampling points to generate a near-random sample of parameter values as proposed by Reference [56]. The open-source implementations were used for both methods [48]. Convergence was considered acceptable when within the 95% confidence interval.

Table 4. Sampling sizes and number of model runs performed for each sensitivity analysis method.

Method	Sample Size	Model Evaluations
Elementary Effect (EE) w/10 trajectories	1000	340
Sobol's	10	200
	11	220
	12	240

Among the 33 rootzone model parameters (Table 3), the EE method revealed that only nine strongly influenced the output soil–water EC (EC_{sw}) as summarized in Table 5. Table 5 indicates parameter sensitivity values μ, μ*, μ*_conf and Σ/EE from the elementary effect analysis. The mean of the distribution of EEi (Σ/EE) is a proxy for a total sensitivity index of ith input factor and μ*_conf is the confidence interval of the μ* at 95%. We chose these nine parameters based on the combination of small μ* values with corresponding large σ values (Figure 5). High σ value indicates that the EE are strongly affected by the choice of the sample points at which they are computed, therefore a non-negligible interaction with other factor values as illustrated in Figures 5 and 6.

Next, we applied Sobol's sensitivity analysis to determine factor fixing using the nine influential factors from the EE results as summarized in Table 6, we reported the first-order (Si) and total order sensitivity (S_{Ti}). Of the nine factors, only two were non-influential by this analysis, that is, they resulted in $S_T = 0$. These non-influential parameters towards average root zone salinity included the capillary fringe height (Hd) and depth to groundwater (Hwt), so these were fixed at average values for the model calibration analysis. Interestingly, S_T values for Tsrz for layer 2 and layer 3 indicate that saturated soil moisture content is more influential for these layers than Ks, and especially that Tsrz in layer 2 is the most influential parameter in the model. Given that we used the control site for model calibration and irrigation of vegetables for this site is managed with sprinklers for establishment and then drip for the rest of the plant development stages, we suspect that plant water uptake is mainly from the bottom layers. However, in our model plant water uptake even with drip is assumed to be mainly from the top layer: 60–30–7–3% uptake pattern. Root water uptake (RWU) although included in the sensitivity analysis was not influential. Mostly likely, upward flow of soil water from the second layer provides water required for uptake in the top layer. Soil moisture is directly related to unsaturated conductivity (a driving force for soil water from wet to dry soil layer). As such the water saturated water content in the second layer ended up being most influential in our calibration. It is important to note that with respect to the crop rooting depth, it is likely that the lettuce rootzone depth is important for the control site specifically as it was the main crop grown for the majority of the experiment period. For other sites, it may be important to include variability of the crop rooting depth.

Table 5. Parameter sensitivity based on Morris indices.

Code	Parameter	μ*	μ	μ*_Conf	Σ/EE
	Root zone Sampling Depth of 0.03–0.3 m				
k1	Rate of dissolution at 0.03–0.30 m depth	0.16	−0.14	0.07	0.15
ZrL	Rooting depth of lettuce	0.1	−0.10	0.06	0.1
Tsrz	Root zone saturated water content	0.09	−0.07	0.08	0.14
Ksrz	Root zone saturated hydraulic conductivity	0.08	0.03	0.06	0.12
Twprz	Root zone water content at wilting point	0.05	−0.05	0.03	0.05
Hd	Capillary fringe height	0.04	0.04	0.08	0.12
pL	Fraction of depletable moisture for lettuce	0.04	0.03	0.02	0.04
	Root zone Sampling Depth of 0.30–0.61 m				
k1	Rate of dissolution at 0.03–0.30 m depth	0.12	−0.1	0.07	0.13
Ksrz	Root zone saturated hydraulic conductivity	0.07	−0.03	0.04	0.11
Tsrz	Root zone saturated water content	0.07	−0.05	0.06	0.12
Tfc5	Unsaturated zone saturated water content	0.05	−0.02	0.04	0.08
k2	Rate of dissolution at 0.3–0.61 m depth	0.05	−0.05	0.02	0.04
ZrL	Rooting depth of lettuce	0.05	−0.04	0.03	0.06
Twprz	Root zone water content at wilting point	0.04	−0.02	0.02	0.05
Hd	Capillary fringe height	0.04	0.03	0.06	0.1
Tr5	Unsaturated zone residual water content	0.03	−0.03	0.02	0.04
	Root zone Sampling Depth of 0.61–0.91 m				
k1	Rate of dissolution at 0.03–0.30 m depth	0.13	−0.08	0.06	0.16
Tsrz	Root zone saturated water content	0.09	−0.05	0.06	0.14
Ksrz	Root zone saturated hydraulic conductivity	0.08	−0.01	0.06	0.13
Tfc5	Unsaturated zone saturated water content	0.06	−0.03	0.06	0.11
Twprz	Root zone water content at wilting point	0.05	−0.03	0.03	0.07
Hwt	Depth to groundwater table	0.05	0.05	0.06	0.1
ZrL	Rooting depth of lettuce	0.05	−0.04	0.04	0.07
k2	Rate of dissolution at 0.3–0.61 m depth	0.05	−0.04	0.03	0.04
Hd	Capillary fringe height	0.04	0.04	0.07	0.13
Tr5	Unsaturated zone residual water content	0.03	−0.03	0.02	0.04

μ* is a good proxy of the total sensitivity index [53].

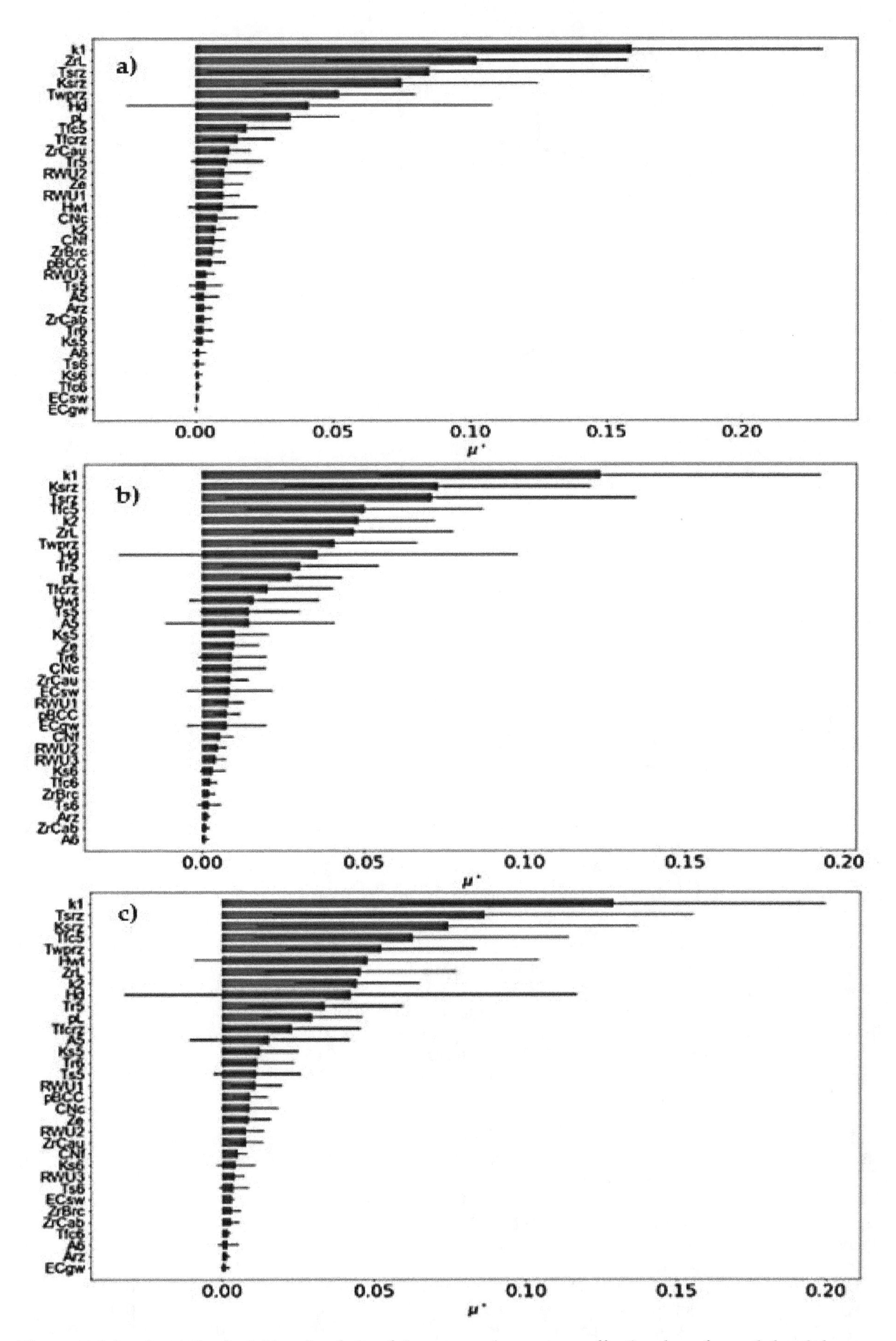

Figure 5. Morris method μ* (the absolute of the mean elementary effect) values for soil depth layers. (**a**) 0.03–0.30 m; (**b**) 0.3–61 m; (**c**) 0.61–0.91 m depths.

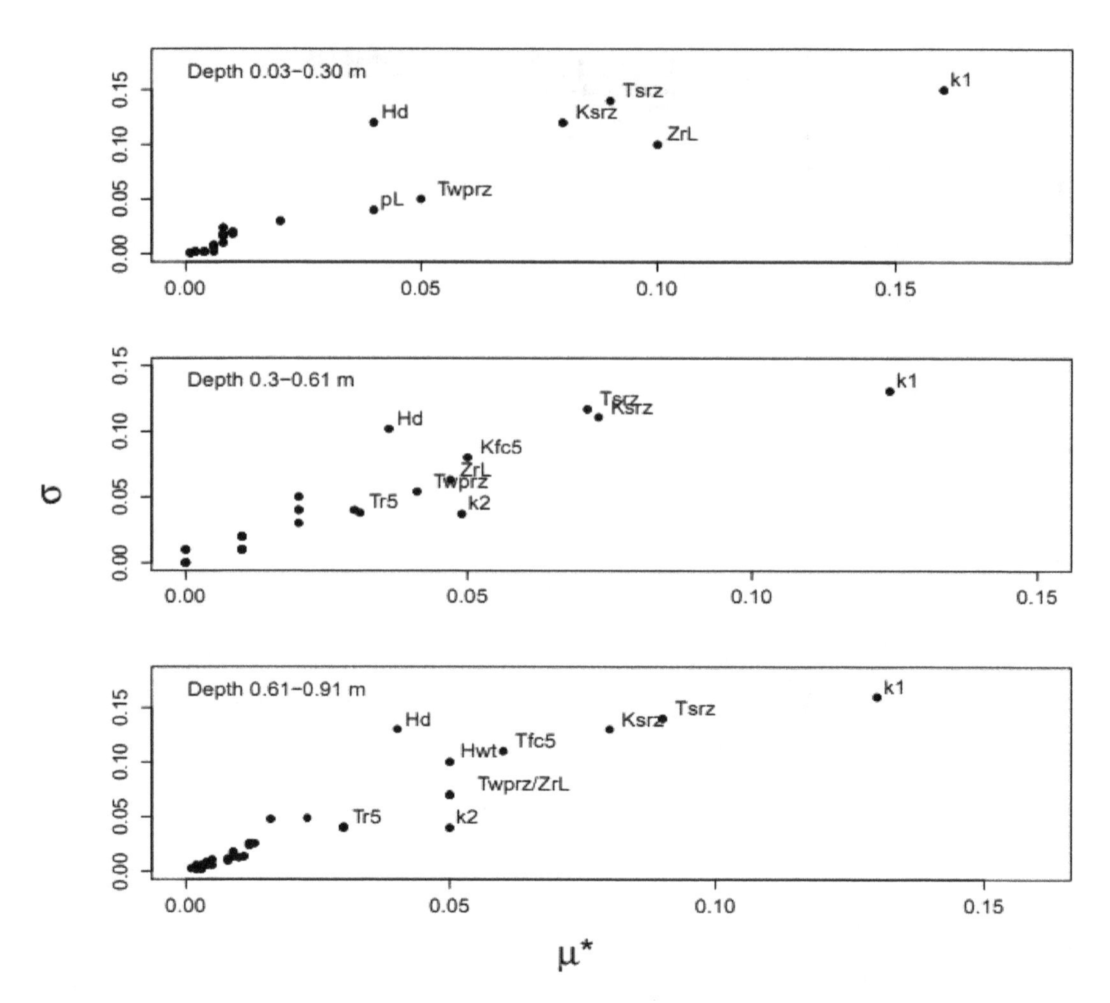

Figure 6. Morris method μ* (the absolute of the mean elementary effect) vs. σ (the standard deviation of the elementary effect) for the different soil depths.

Table 6. Sobol's sensitivity indices.

Code	Parameter	S_T	S_T_conf	Si	Si_conf
	Root zone Sampling Depth of 0.03–0.3 m				
k1	Rate of dissolution at 0.03–0.30 m depth	0.65	0.42	0.42	0.33
ZrL	Rooting depth of lettuce	0.43	0.47	−0.10	0.29
Tfcrz	Unsaturated zone saturated water content	0.07	0.06	−0.04	0.13
Ksrz	Root zone saturated hydraulic conductivity	0.05	0.08	−0.01	0.06
Tsrz	Root zone saturated water content	0.04	0.05	−0.07	0.11
Twprz	Root zone water content at wilting point	0.02	0.02	−0.08	0.07
k2	Rate of dissolution at 0.3–0.61 m depth	0.01	0.01	0.03	0.05
Hwt	Depth to groundwater table	0	0	0	0.02
H_d	Capillary fringe height	0	0	0	0.02
	Root zone Sampling Depth of 0.30–0.61 m				
Tsrz	Root zone saturated water content	3.38	9.1	0.01	0.11
k1	Rate of dissolution at 0.03–0.30 m depth	0.86	0.53	0.31	0.5
k2	Rate of dissolution at 0.3–0.61 m depth	0.27	0.19	0.28	0.28
ZrL	Rooting depth of lettuce	0.24	0.49	0.03	0.24
Tfcrz	Unsaturated zone saturated water content	0.05	0.06	0	0.14
Ksrz	Root zone saturated hydraulic conductivity	0.04	0.07	0.01	0.13
Twprz	Root zone water content at wilting point	0.02	0.02	−0.03	0.07
Hwt	Depth to groundwater table	0	0.01	−0.02	0.05
H_d	Capillary fringe height	0	0.01	−0.02	0.05

Table 6. *Cont.*

Code	Parameter	S_T	S_T_conf	Si	Si_conf
	Root zone Sampling Depth of 0.61–0.91 m				
k1	Rate of dissolution at 0.03–0.30 m depth	0.89	0.68	0.05	0.44
ZrL	Rooting depth of lettuce	0.36	0.48	0.05	0.21
k2	Rate of dissolution at 0.3–0.61 m depth	0.29	0.17	0.09	0.24
Tsrz	Root zone saturated water content	0.12	0.22	0.02	0.08
Ksrz	Root zone saturated hydraulic conductivity	0.07	0.1	−0.03	0.08
Tfcrz	Unsaturated zone saturated water content	0.06	0.07	0.03	0.1
Twprz	Root zone water content at wilting point	0.03	0.03	−0.01	0.04
H_d	Capillary fringe height	0	0.01	0	0.01
Hwt	Depth to groundwater table	0	0.01	0	0.01

3.2. Calibration and Validation

Model calibration was performed allowing the parameters k1, k2, ZrL, Tsrz, Ksrz, Tfcrz, and Twprz identified as most sensitive to model determination of soil–water EC (EC_{sw}) to vary within their ranges and output compared to the measured soil salinity at the control site. Model validation was completed by simulating EC_{sw} for site 3. The inclusion of the saturated soil water content and saturated hydraulic conductivity in the calibration was crucial as infiltration rates were expected to vary with changes in exchangeable sodium in the soil. O'Geen [57] provides classification of salt-affected soils based on trends of soil water EC, exchangeable sodium percentage (ESP), and SAR. We assessed the potential infiltration problems caused by irrigation water quality following Reference [6] (p. 44) and found that slight-to-moderate reduction in infiltration rates due to irrigation water salinity were expected for the control site and site 4 (Figure A2). It is however interesting to note that blending well water and recycled water alleviated the possible adverse effects of well water on soil infiltration rates.

The latin hypercube sampling design was used with $N = 50$ sampling points as proposed by Reference [56]. Intervals were sampled without replacement to ensure even distribution of points with respect to each variable. We executed the model 50 times and computed the corresponding RMSE associated with model predictions. An open-source global optimization code DEoptim written in R was used to find a global minimum RMSE [58]. DEoptim implements the differential evolution algorithm for global optimization. The estimated best fits with the least RMSE values are listed in Table 7.

Table 7. Best-fit parameter values estimated with calibration for EC_{sw}.

Property	Code	Units	Value
Soil Hydraulic Parameters			
Root zone saturated water content	Tsrz	m^3/m^3	0.467
Root zone water content at field capacity	Tfcrz	m^3/m^3	0.361
Root zone water content at wilting point	Twprz	m^3/m^3	0.172
Root zone saturated hydraulic conductivity	Ksrz	mm/day	347
Plant Parameters			
Rooting depth of lettuce	ZrL	m	0.5
Soil–Water Chemistry Parameters			
Rate of dissolution at 0.03–0.30 m	k1	dSm^{-1}/day	0.014
Rate of dissolution at 0.30–0.61 m	k2	dSm^{-1}/day	0.022

Summarized in Table 8 are the indices associated with comparisons between EC_{sw} measured in the field and model predicted values at different soil depths. For all depth intervals, the sum of first-order effects and sum of total order indices is greater than one, indicating that there are interactions among model factors. Moreover, for factors that have total indices greater than their first-order values, other factors are taking part in the interaction such that throughout the soil profile the root zone hydraulic parameters, rooting depth and dissolution rate are taking part in determination of the soil-water EC, with the dissolution rate accounting for the largest fraction of output variance. This observation provides insight about how well water flow and salinity is modeled.

Table 8. Statistical indices for simulated vs. observed soil water EC (EC_{SW}).

Index	RMSE	MRE	NSE	R^2	b
Optimal value	0	0	1	1	1
Control Site (Calibration)					
0.03–0.30 m	0.45	0.23	0.72	0.73	0.89
0.30–0.61 m	0.41	0.25	0.52	0.52	0.89
0.61 – 0.91 m	0.34	0.26	0.16	0.27	0.95
Site 3 (Validation)					
0.03–0.30 m	0.95	0.24	0.56	0.60	0.87
0.30–0.61 m	0.70	0.21	0.24	0.39	0.88
0.61–0.91 m	0.60	0.20	0.48	0.51	0.94

Model realizations for the calibration and validation runs were compared with measured EC_{SW} and shown in Figure 7. Model predictions did not capture large values of observed EC_{SW} though model performance improved with increased soil depth. The RMSE index indicated large discrepancies between predicted and measured values, hence the greater values at site 3. Whereas R^2 values reflect the combined dispersion against the single dispersion of the observed and predicted values. The mean relative error (MRE < 30%) for all layers indicates satisfactory model performance, while the larger NSE values for the top soil layer indicate that the modeling effort is worthwhile in predicting near surface salinity to depths of 0.3 m.

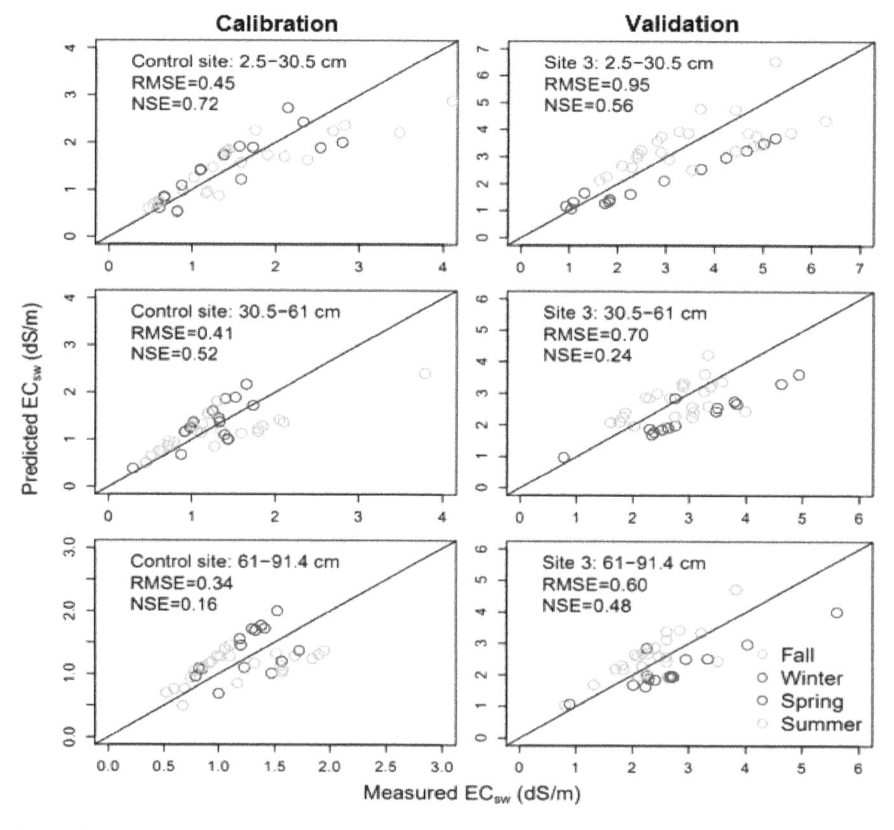

Figure 7. Predicted vs. measured EC_{SW} at 2.5–30.5 cm, 30.5–61 cm and 61–91.4 cm soil depths for control site (calibration) and site 3 (validation).

For all calibrated model runs, regressions of the predicted vs. observed values resulted in a non-zero intercept, b of nearly 1 dS/m. Taken together with the low R^2-values, we conclude that the model persistently underestimates EC_{sw}, especially those observed EC_{sw} values greater than ~2 dS/m. Based on the relatively small MRE values that provide an indication of the magnitude of the error relative to observed values without considering the error direction, the model captures salinity dynamics for all layers in the root zone. However, the Figure 8 plot of residuals vs. predicted EC_{sw} for the calibration and validation results exhibits heteroscedasticity, that is, residuals grow as the predicted EC_{sw} values increase. Overall, this latter observation suggests that although the NSE, RMSE, and MRE statistics show that the model has some predictive capacity, it does not capture some processes apparently involved in the soil salinity dynamics.

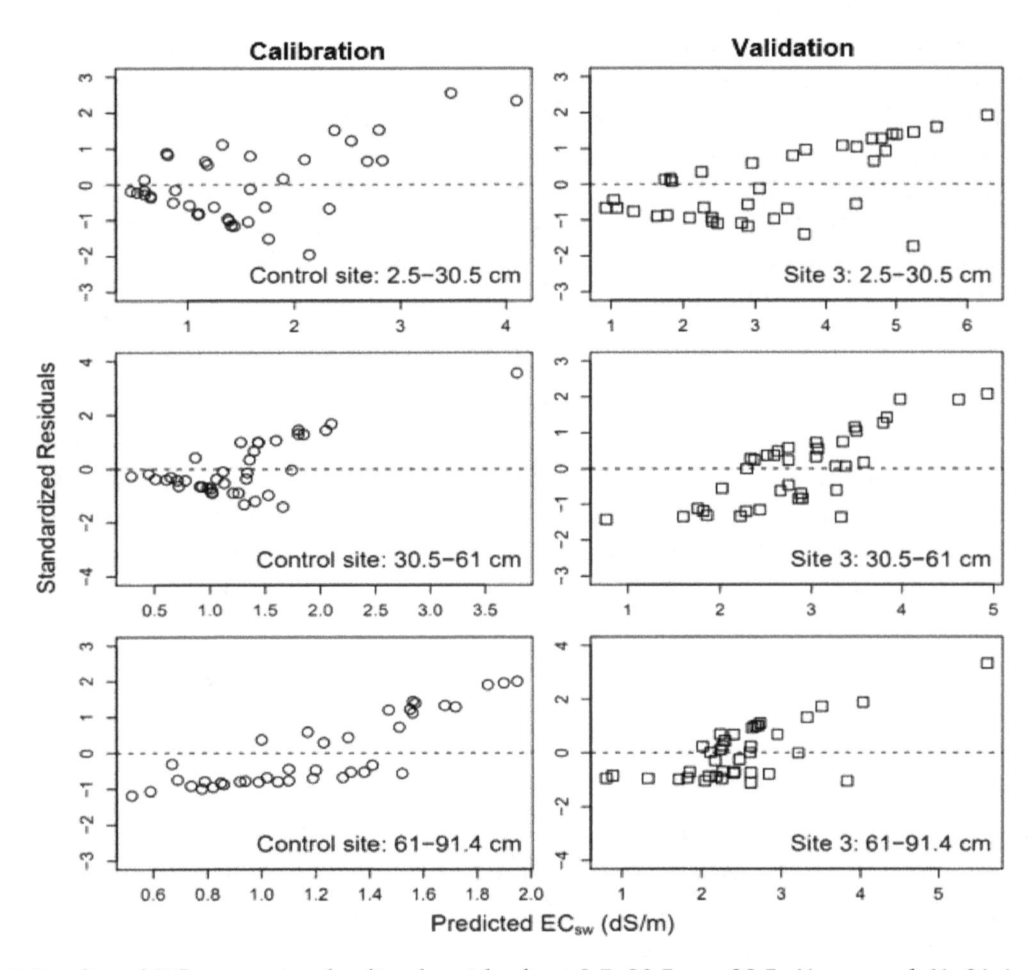

Figure 8. Predicted EC_{sw} vs. standardized residuals at 2.5–30.5 cm, 30.5–61 cm, and 61–91.4 cm soil depths for control site (calibration) and site 3 (validation).

The possible explanation for the differences in measured and predicted soil salinity is that the model does not account for fertilizer and soil amendment management, or plant root uptake of solutes and fertilizer. Generally, transformation (e.g., dissolution) in the soil of different chemicals added during fertigation will increase soil salinity. For example, urea is converted to ammonium that is adsorbed in the soil depending on sol temperature; ammonium is converted to nitrate by nitrification that depends on soil temperature, soil moisture, pH and oxygen content, and nitrate is highly mobile but ammonium, potassium, and phosphorus remain relatively immobile in the root zone. Although the salt index (SI) based on equivalent units of sodium nitrate (developed in 1943 to evaluate the salt hazard of fertilizers) alone cannot be used to evaluate the effect of increased soil salinity from fertilizer applications, it can be used as indicator for the long-term effects on soil salinity. The most commonly used fertilizers in the study region (from California Department of Food and Agriculture annual reports) were nitrogen fertilizers that included urea ammonium nitrate solution (SI = 95), ammonium nitrate (SI =102) and calcium nitrate (SI = 53); phosphorus fertilizer (SI = 7.8-29), potassium sulfate (SI = 46), gypsum, and lime. Sodium nitrate was arbitrarily set at 100, where EC of 0.5 to 40 mass percentage of sodium nitrate is 0.54 to 17.8 dS/m and for a mixture of materials it is reasonable to assume EC is additive for horticulture. As such, the lower the index value the smaller the contribution the fertilizer makes to the level of soluble salts. Thus, fertilizer applications likely add to soil salinities exacerbating the problem over time.

Although the model underestimates EC_{sw}, it adequately captures salinity trends in the leaching of salt during winter months and an increase in salinity water applications and ET during the crop season. In an effort to evaluate performance of transient vs. steady state models, Reference [12] concluded that the transient models better predict the dynamics of the chemical–physical–biological interactions in an agricultural system. However, since we account for irrigation water salinity, rainfall salinity and dissolution of salts in the soil and exclude additions of fertilizer and soil amendment our simulated EC_{sw} values can be viewed as a likely lower bound of soil salinity associated with the irrigation and farm management practices considered in the model description.

Another complexity possible affecting model prediction is the spatial distribution of salinity with drip irrigation as noted by Reference [59]. They used the transient Hydrus-2D model to compare results between field experiments having both drip and sprinkler irrigated processing tomatoes under shallow water table conditions for a wide range of irrigation water salinities. Both field and model results showed that soil-wetting patterns occurring under drip irrigation caused localized leaching which was concentrated near the drip line. In addition, a high-salinity soil volume was found near the soil surface that increased with increasing applied water EC. Overall, localized leaching occurred near the drip line while soil salinity increased with increasing distance from the emitter and with increasing soil depth. Such localized non-uniformities in leaching are not captured in the one-dimensional model but may have affected field soil sampling in the drip–irrigated fields of our study region. That is, soil samples collected some distance from where dripline emitters were previously operating would likely have greater salinities than would otherwise occur under the uniform leaching and dissolution conditions assumed in the model. Nonetheless, it is important to note that this model is user-friendly and less data intensive and it can be very useful for setting reference benchmarks of long-term salinity impacts of using saline water for irrigation.

3.3. Long-Term Salinity with Treated Wastewater Irrigation

We used the calibrated and validated model to simulate the long-term (50-year periods) soil salinity in the fields irrigated with varying fractions of treated wastewater, that is, we applied the model to control site and sites 2 to 7 (Tables 2 and 3). Fifty-year simulations assumed randomly selected rainfall and ET_o data from historical records (1983 to 2014) from 2013 to 2049 and the 13-year cropping patterns and irrigation management. In the model calibration, the groundwater table height (Hwt) and groundwater salinity (ECgw) were found to be non-influential parameters with respect to the measured soil water EC (EC_{sw}). As such average values of measured Hwt and ECgw from the monitoring well located ~5 km west of the study site were used, that is, 0.95 m and 0.97 dS/m, respectively. For each simulated case, the three output variables of interest were averaged root zone salinity over the growing season expressed as EC_{eS} (dS/m), annual drainage salt output load as S_d (kg/ha/year), and annual runoff salt output load S_r (kg/ha/year).

Values for the annual average water-balance terms over the 13-water year simulation at each site are summarized in Table 9a and for the 50-water year simulation in Table 9b. Somewhat greater actual crop water uptake (ET) is achieved under drip and sprinkler irrigated fields with similar crop rotations (sites 2 and 3 as compared to 6 and 7). Leaching fractions were generally very low for all sites at less than 2%. The greatest leaching occurred in the vegetable–strawberry rotation that was first sprinkler than drip irrigated, while the other irrigation and cropping practices yielded similar leaching volumes with the exception of no leaching from drip-irrigated artichokes. Similarly, surface runoff is smaller from drip or sprinkler irrigated fields as compared to sprinkler-furrow irrigated fields.

Table 9. Summary of annual average water balance and salinity variables for the different field management scenarios.

a. 13 Years Simulations							
Site No.	**Control Site**	**2**	**3**	**4**	**5**	**6**	**7**
Crop management	Vegetables	Vegetables	Vegetables and strawberry	Perennial artichoke		Vegetables and strawberry	Vegetables
Irrigation management	Sprinkler then drip	Sprinkler or drip	Sprinkler then drip	Sprinkler or drip		Sprinkler then furrow	Sprinkler then furrow
Irrigation (mm/yr)	652	597	682	648		694	656
Seasonal Evapotranspiration ET (mm/year)	254	200	252	392		204	177
Surface runoff (mm/year)	299	232	236	158		315	265
Leaching (mm/yr)	7.1	4.4	16.3	0		23.7	12.7
Irrigation water electrical conductivity EC_W (dS/m)	0.71	0.94	1.36	1.06	1.36	1.1	1.37
Root zone EC_{SW} (dS/m)	1.48	1.74	2.45	2.4	2.89	1.76	1.95
Root zone seasonal-averaged electrical conductivity of the saturated extract EC_{eS} (dS/m)	0.09	0.13	0.19	0.15	0.18	0.15	0.17
Salt output load with deep percolation S_d (kg/ha/year)	61	42	207	0	0	153	104
Salt output load with runoff S_r (kg/ha/year)	304	851	1860	648	1012	1073	1465

b. 50 Years Simulations							
Site No.	**Control Site**	**2**	**3**	**4**	**5**	**6**	**7**
Seasonal ET (mm/year)	187	128	188	395		141	131
Surface runoff (mm/year)	323	295	236	163		275	250
Leaching (mm/year)	8.0	7.2	9.4	1.0		13.0	8.8
EC_W (dS/m)	0.71	0.94	1.36	1.06	1.36	1.1	1.37
Root zone EC_{SW} (dS/m)	1.12	1.20	1.86	2.05	2.52	1.34	1.50
Root zone EC_{eS} (dS/m)	0.14	0.19	0.27	0.20	0.23	0.24	0.26
S_d (kg/ha/year)	27	20	64	0.8	1.0	56	38
S_r (kg/ha/year)	305	491	1028	339	518	598	764

Simulated average (50-year) annual root zone soil water salinity for all sites is shown in Figure 9. Sites managed with sprinkler or drip irrigation and with higher salinity water (EC_W) had higher estimated annual average EC_{SW}, for example, at sites 3, 4, and 5, EC_{SW} was 2.94 dS/m, 3.15 dS/m, and 3.44 dS/m, respectively. For sites 6 and 7 that used sprinklers for germination then furrow irrigation, the latter site received water with higher salinity than the former 1.37 dS/m compared to 1.1 dS/m on vegetable and strawberry rotation, but the resulting average annual soil water salinity differed little, that is, 2.05 versus 2.89 dS/m, respectively. The control site irrigated with well water had the least annual average rootzone soil–water salinity. Furthermore, soil–water salinity equilibrium $EC_{SW} \leq 2.0$ dS/m was reached throughout the 50-year horizon for the control site irrigated with well water and after 12 years of irrigated with blended wastewater for sites 2, 6, and 7, whereas for sites 3, 4, and 5 soil water EC increased above 2.0 dS/m in the simulation period. Using Mann–Kendall analysis we found that actual ET had a positive and significant association whereas irrigation amounts had a negative and significant association with $EC_{SW} \leq 2.0$ dS/m (Tau = 0.321, p-value = 0.016 and

Tau = −0.268, p-value = 0.046 respectively). The Mann–Kendall Tau values indicate the strength and direction of monotonic trends, with −1 and 1 representing perfectly negative and positive monotonic trends, respectively, while the p-value indicates relative significance [60].

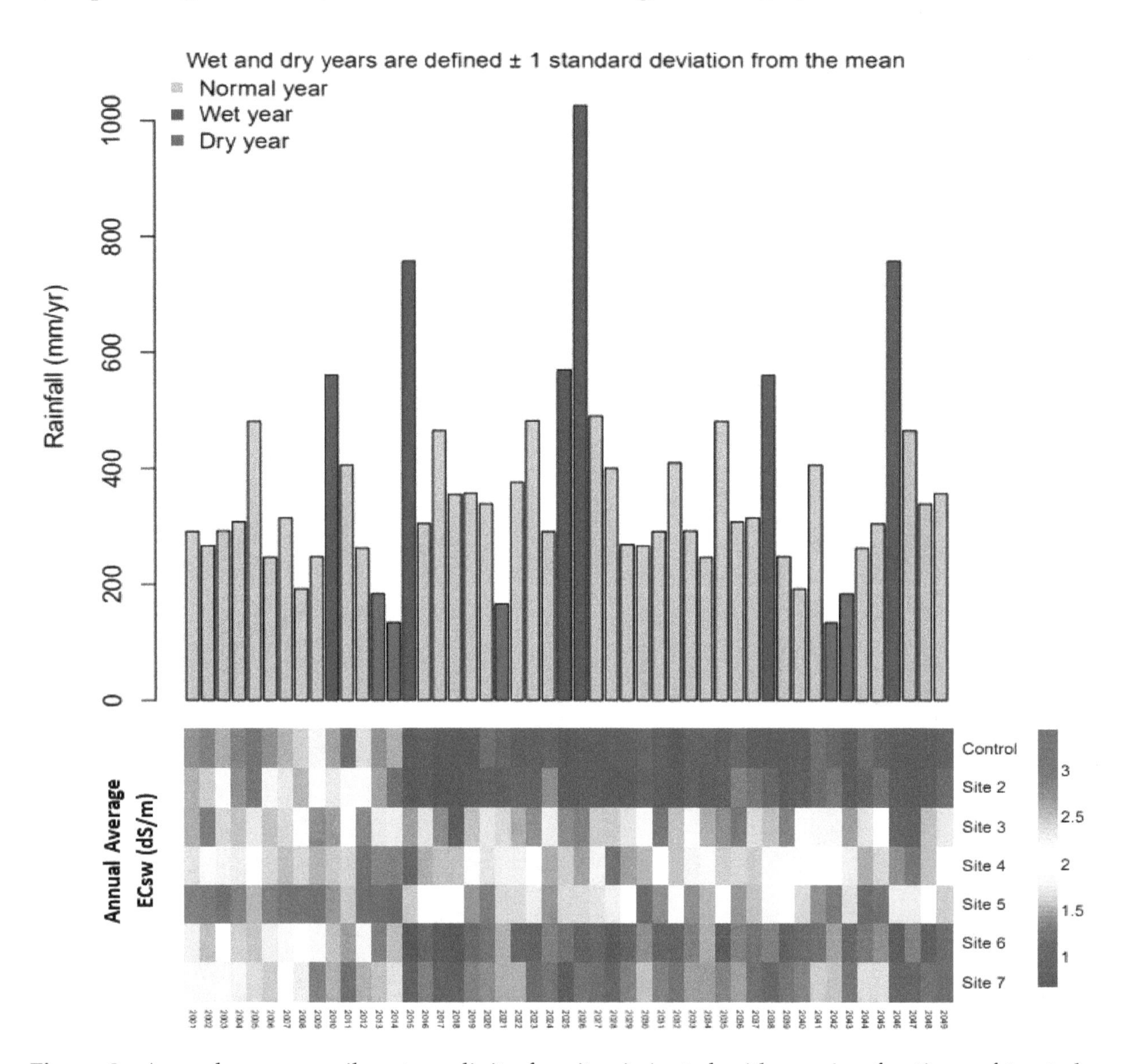

Figure 9. Annual average soil–water salinity for sites irrigated with varying fractions of treated wastewater (2000 to 2049).

As Platts and Grismer [11] found that salt leaching to deeper soil layers occurred during the rainy season (October–March), while during the growing season soil–water EC increases in soil layers near the surface due to evapo-concentration, on the other hand, applied water salinity causes soil–water EC spikes during the growing season. Rainfall was important towards salt leaching from the root zone at all sites as evident during the wet years 2010, 2016, 2025, 2026, 2038, and 2046 when average annual soil water EC decreased. With the exception of sites 3, 4, and 5, there were no upward trends in soil water salinity over the 50-year period. Overall, relatively constant soil–water EC after 50 years simulation of EC_{sw} < initial EC_{sw} of 2.19 dS/m for all sites except site 5 suggest that there was adequate soil leaching in the region for sustained use of the treated wastewater for irrigation. However, the question remained as to what level of soil salinity would be acceptable especially for annual strawberry production.

Crops are generally assumed to respond to seasonal-averaged root zone salinity of the saturated paste (EC_{eS}) and yield loss thresholds and rates of decline with increasing salinity have been determined based on salinity thresholds in [61]. We calculated EC_{eS} for all the sites and plot the range of EC_{eS} in Figure 10. The maximum seasonal-averaged saturated paste EC for each site was 0.19, 0.27, 0.20, 0.23, 0.24, and 0.26 dS/m for sites 2, 3, 4, 5, 6, and 7, respectively. These values are less than half that of the lowest Mass–Hoffman threshold value of $EC_e^* = 1.0$ dS/m associated the most sensitive crop (strawberry) in the rotations considered. As such, it is unlikely that long-term irrigation with treated wastewater in the region will adversely affect crop yields significantly.

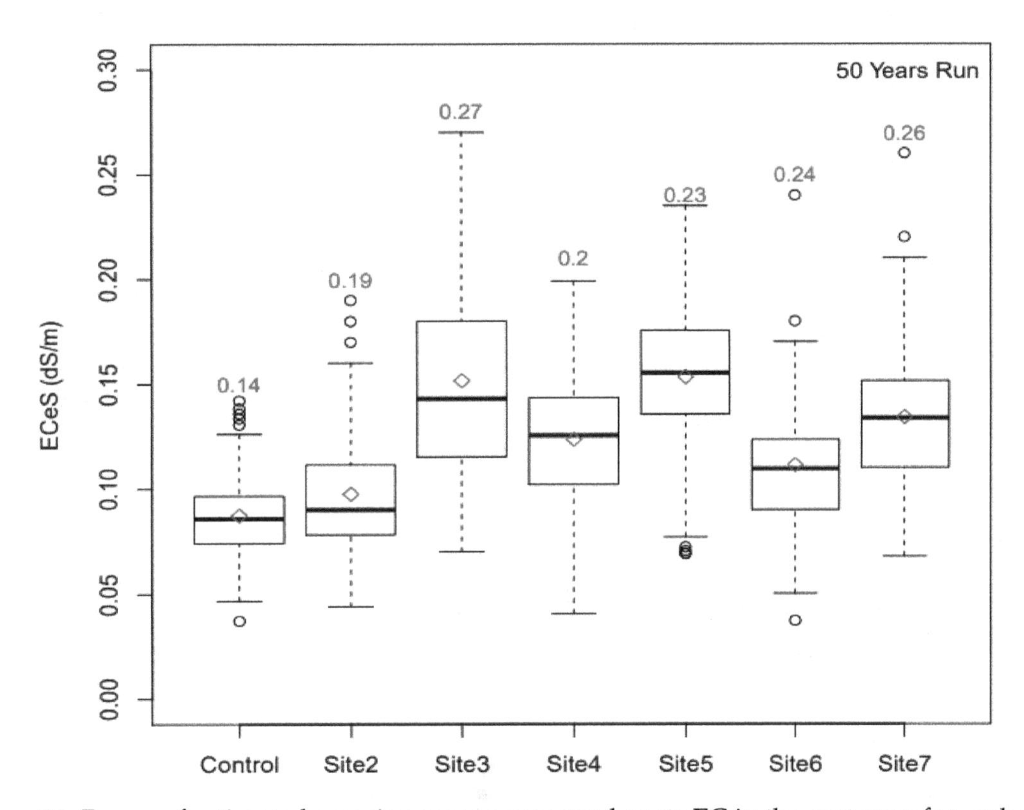

Figure 10. Range of estimated growing season saturated paste EC in the root zone for each site.

In terms of possible adverse environmental effects associated with salinization of surface and ground waters in the region, we determined the cumulative salt output load with deep percolation (S_d) and salt output load with runoff (S_r) during the 50-water year simulation period for the different sites as shown in Figure 11. Salts accompanying surface runoff pose a larger threat in the watershed as these are an order-of-magnitude greater than the cumulative salt output loads to groundwater. Salt loading with deep percolation for the 50-year simulation range up to 3,377 kg/ha with the greatest loads from site 3 and minimal loading for sites 4 and 5 (40 kg/ha and 49 kg/ha respectively) on which annual artichoke crops were grown. Cumulative salt loads accompanying runoff ranged from 19,918 to 59,552 kg/ha, with the greatest loading emanating from site 3 and least from site 4. In comparison to the control site irrigated with well water cumulative salt output loading with deep percolation was 1325 kg/ha and salt output load with runoff 21,505 kg/ha.

To clarify what factors were key to affecting adverse environmental salinization within the region, we tested the effect of applied water EC (EC_w) and depths, rainfall depths, actual crop ET and potential crop ET, and number of days fallow on soil water EC (EC_{sw}) and salt output loads with runoff (S_r) or deep percolation (S_d) using the non-parametric Mann–Kendall trend analysis (the 'Kendall' test in the R package [60]. With respect to the Mann–Kendall Tau values (Table 10), the applied water EC (EC_w) and rainfall depths had positive and significant effects on annual average soil water EC (EC_{sw}) and annual salt output loads with runoff (S_r) and deep percolation (S_d). Calculated actual crop ET had a positive and significant effect on EC_{sw}; this is expected as water uptake by the plant and evaporation leave salts behind. On the other hand, actual crop ET had a negative and significant effect on S_d, and had no significant effect on S_r. Applied water depths had positive and significant effect on S_d and S_r. The number of days fields were fallowed had a negative effect on EC_{sw} and a positive effect on S_d and S_r. Overall, these observations were consistent with that expected from the field observations and described above.

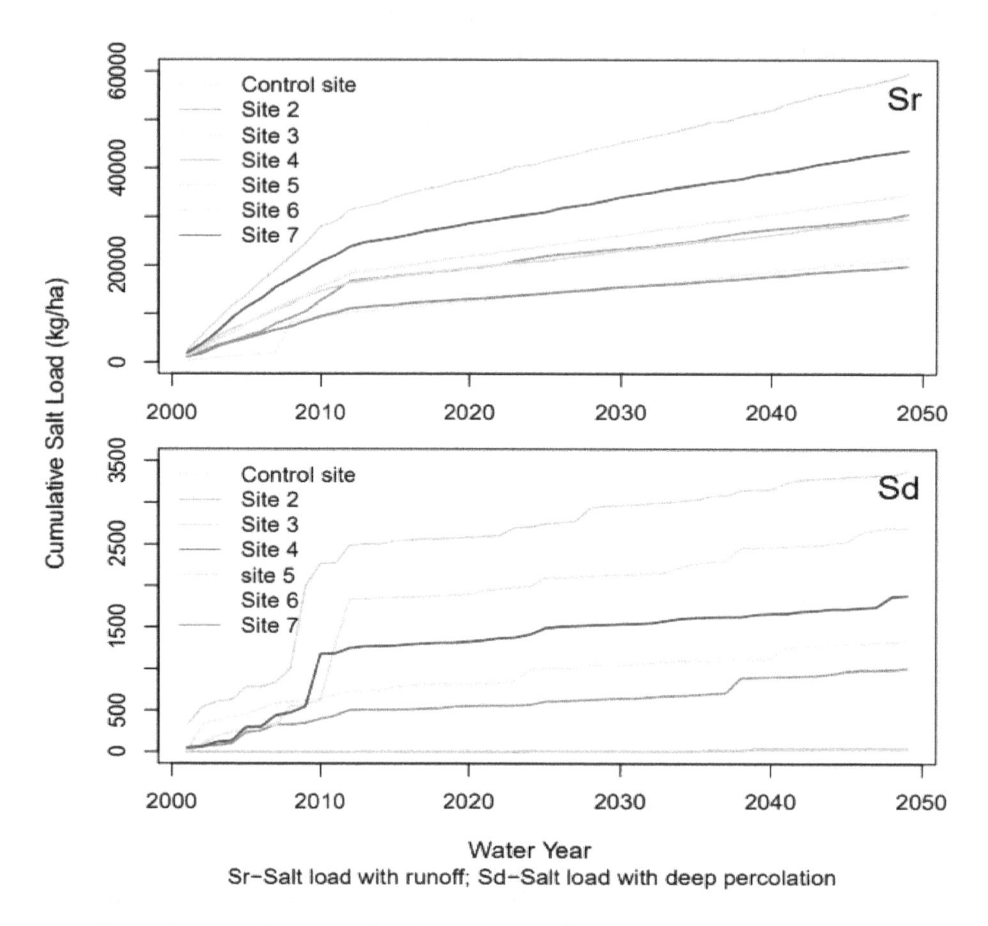

Figure 11. Cumulative salt output loads with runoff and deep percolation from each site.

Table 10. Mann–Kendall trend analysis for average annual soil water EC (EC_{sw}), annual salt output load with runoff (S_r), and with deep percolation (S_d).

Parameter	EC_{sw} (dS/m)		S_r (kg/ha/year)		S_d (kg/ha/year)	
	Tau	*p*-Value [a]	Tau	*p*-Value [a]	Tau	*p*-Value [a]
EC_w (dS/m)	0.33	***	0.49	***	0.04	ns
Rainfall (mm/year)	−0.18	*	0.03	ns	0.06	ns
Potential crop ET (mm/year)	0.03	ns	0.24	**	0.17	*
Actual crop ET (mm/year)	0.44	***	−0.11	ns	−0.33	***
Irrigation (mm/year)	0	ns	0.41	***	0.24	**
Days fallow (days)	−0.44	***	0.26	**	0.39	***

[a] Two-sided *p*-value ranges: 0 ≤ *** ≤ 0.001; 0.001 < ** ≤ 0.01; 0.01 < * ≤ 0.05; ns > 0.05.

4. Discussion

We calibrated and validated a modified root zone salinity model originally developed by Isidoro and Grattan [17], which was then applied to estimate long-term soil salinity in fields irrigated with treated wastewater. We conducted a global sensitivity analysis using the elementary effect/Morris and Sobol's methods to first reduce the number of influential model parameters important to calibration and that need to be acquired from the field. Seven of the thirty-three model parameters were found to be critical to root zone soil salinity dynamics. These were parameters accounting for salt dissolution in the soil, root zone hydraulic parameters, and crop rooting depth. Model calibration resulted in a satisfactory fit to the observed field data; however, the model underestimated soil water salinity (EC_{SW}), especially for large $EC_{SW} > 2$ dS/m measured during the growing season. We attributed this error to the model's failure to account for fertilizer and soil amendment applications and transformation thereof (e.g., gypsum dissolution) in the soil. In addition, drip irrigation leads to very localized variations in soil salinity that depend on the distance from the emitter that are not considered in the model and may have affected field soil sampling between plantings and harvests. Nonetheless, the model adequately captured soil–water EC trends that were congruent with observed data.

Sites irrigated with greater salinity water (EC_W) combined with sprinkler or drip had greater estimated annual average soil–water salinity (EC_{SW}). Sites that combined sprinkler irrigation for germination with furrow for the remaining development stages resulted in lower annual average EC_{SW} in the root zone even when more saline irrigation water was applied. Rainfall played an important role in the leaching of salts from the root zone as during wet years average annual soil water EC decreased at all sites. Moreover, rainfall had a negative and significant effect on annual average root zone EC_{SW}. We found that for all sites use of treated wastewater for irrigation over the 50-year period does not affect strawberry yields, the most salt sensitive of the crops in the rotations encountered. Overall irrigation water EC (EC_W), rainfall amounts, actual calculated crop ET and the number of days fields were fallowed had significant effects on annual average soil water EC (EC_{SW}). EC_W, rainfall and actual crop ET effects were positive and fallowing decreased average root zone EC_{SW}. On the other hand, irrigation amounts and number of days fallowed had positive and significant effects on salt output loads associated with runoff and deep percolation. Moreover, soil water salinity equilibrium $EC_{SW} \leq 2.0$ dS/m is reached throughout the 50-year horizon for the control site irrigated with well water and after 8, 9 and 14 years of irrigated with blended wastewater for sites 2, 6, and 7 respectively. For sites 3, 4, and 5 soil water EC increased above 2.0 dS/m in the simulation period. Actual ET had a positive and significant association whereas irrigation amounts had a negative and significant association with $ECsw \leq 2.0$ dS/m.

While we believe that the modeling results can inform recommendations about irrigation management practices and for estimating salt output loading resulting from use of saline waters for irrigation, difficulties in linking field observations of soil salinity and model predictions remain troubling. However, since we account for irrigation water salinity, rainfall salinity, and dissolution of salts in the soil and exclude additions of fertilizer and soil amendment our simulated EC_{SW} values are likely a lower bound of soil salinity associated with the irrigation and farm management practices considered in the modeling. Nonetheless, it is important to note that this model is user-friendly and less data intensive and it can be very useful for setting reference benchmarks of long-term salinity impacts of using saline water for irrigation.

Author Contributions: Conceptualization, P.Z. and M.G.; Data curation, P.Z. and M.G.; Formal analysis, P.Z.; Funding acquisition, M.G.; Investigation, M.G.; Methodology, P.Z., I.K. and M.G.; Project administration, M.G.; Resources, M.G.; Software, P.Z.; Supervision, I.K. and M.G.; Validation, P.Z. and M.G.; Visualization, P.Z.; Writing—Original draft, P.Z. and M.G.; Writing—Review & editing, I.K. and M.G.

Acknowledgments: We thank Stephen Grattan and Daniel Isidoro for providing notes and equations for the root zone salinity model and Belinda Platts and the Monterey County Water Resources Agency for water and soil sampling and crop data.

Appendix A

Table A1. Cropping patterns for the control site and sites irrigated with treated wastewater.

Site #	Cropping Pattern	Crop	Planting Month	Harvest Month	Average Growing Days
Control	Lettuce, Broccoli, Cauliflower, Cabbage	Lettuce	Mar–Aug	Jun–Nov	72
		Broccoli	Jul–Aug	Oct–Dec	101
		Cauliflower	Jul–Nov	Apr–Oct	118
		Cabbage	Apr	Jul	98
2	Lettuce, Broccoli, Cauliflower, Spinach, Celery	Lettuce	Jan–Sep	Apr–Nov	74
		Broccoli	Jan–Jun	May–Oct	104
		Cauliflower	May–Aug	Aug–Nov	94
		Spinach	Sep	Oct	49
		Celery	Jul	Oct	93
3	Lettuce, Broccoli, Cauliflower, Strawberry	Lettuce	Mar–Jul	May–Oct	73
		Broccoli	Feb–Jul	Jun–Oct	104
		Cauliflower	Feb–Apr	May–Jul	94
		Strawberry	Nov	Oct–Nov	344
4 & 5	Artichoke	1st crop		May	Annual
		2nd crop		Oct–Nov	
6	Lettuce, Broccoli, Cauliflower, Strawberry, Celery	Lettuce	Jan–Jul	Apr– Sep	74
		Broccoli	Apr–Jul	Jul–Oct	92
		Cauliflower	Jan–Jul	May–Nov	99
		Strawberry	Nov	Nov	344
		Celery	May–Jul	Aug–Oct	95
7	Lettuce, Cauliflower, Broccoli	Lettuce	Mar–Aug	May–Oct	72
		Cauliflower	Apr–Aug	Jul–Aug	92
		Broccoli	Jul	Oct	97

Table A2. Time-averaged crop coefficients and maximum rooting depth.

Crop	K_c Values			Rooting Depth (cm)
	Initial	Midseason	Late	
Artichoke	0.5	1	0.95	90
Broccoli	0.7	1.05	0.95	60
Cauliflower	0.7	1.05	0.95	70
Celery	0.7	1.05	1	50
Lettuce	0.7	1	0.95	50
Spinach	0.7	0.9	0.95	50
Strawberry	0.4	0.9	0.85	30

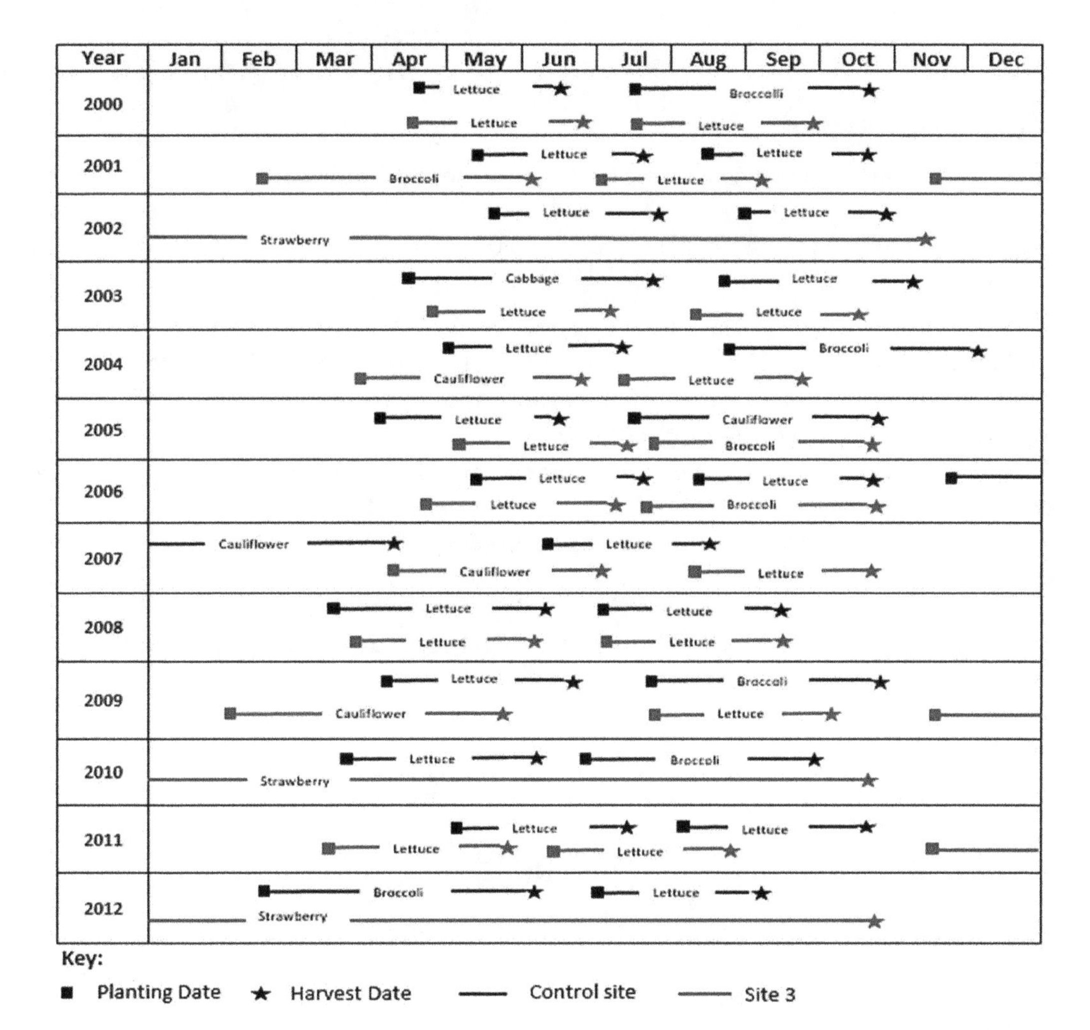

Figure A1. Crop rotation schedule for the control site and site 3.

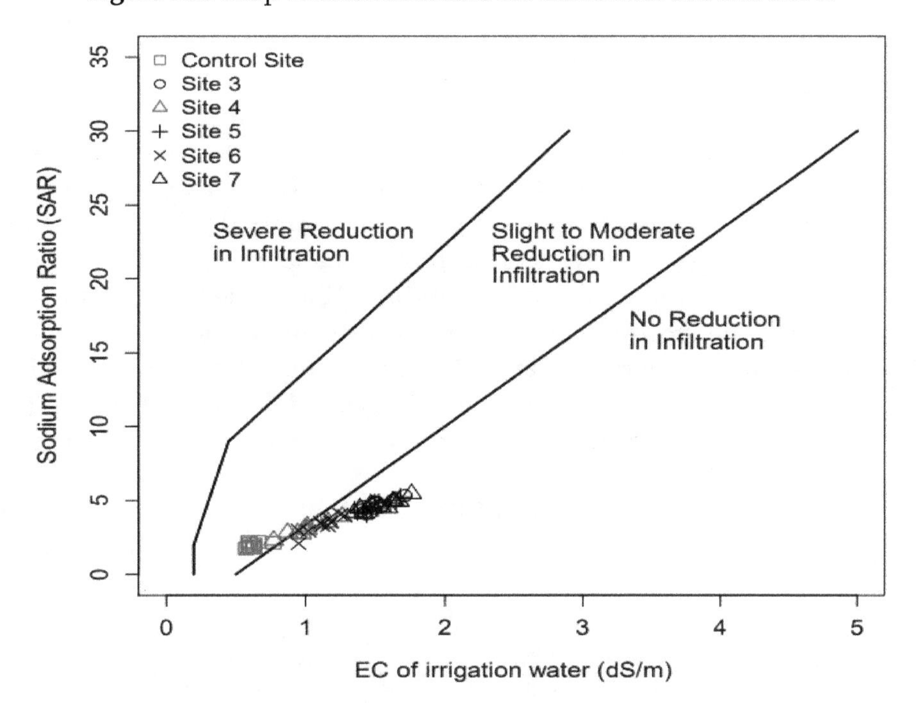

Figure A2. Effect of salinity and sodium adsorption ratio of irrigation water on infiltration rate. (Modified from Reference [6] (p. 44)).

Appendix B

Unsaturated Soil Water Movement

The soil matric potential (ψ) was related to the volumetric water content (θ) by means of Equation (A1) [32]:

$$\varphi = \varphi_s \times \left(\frac{\theta}{\theta_s}\right)^{-b} \quad (A1)$$

Where θ_s was the volumetric water content at saturation, ψ_s is the water entry potential or "saturation" water potential and b is the slope of the water retention curve on a logarithmic plot. For each soil type, b and ψ_s were calculated from the volumetric water content at field capacity and wilting point and their respective potentials in absolute value (ψ_{FC} = 316 cm and ψ_{WP} = 15,849 cm; so that pF (FC) = 2.5 and pF (WP) = 4.2). Taking logarithms, the expression of the potentials for FC and WP become linear equations:

$$\begin{aligned} \log(\varphi_{FC}) &= 2.5 = \log\varphi_s - b.\log\left(\tfrac{\theta_{FC}}{\theta_s}\right) \\ \log(\varphi_{WP}) &= 4.2 = \log\varphi_s - b.\log\left(\tfrac{\theta_{WP}}{\theta_s}\right) \end{aligned} \quad (A2)$$

from which, b and ψ_s are estimated. The unsaturated hydraulic conductivity for a given θ was given by:

$$K = K_s \times \left(\frac{\theta}{\theta_s}\right)^{2b+3} \quad (A3)$$

where K_s is the saturated hydraulic conductivity. Thus, unsaturated flow between layers (U) can be calculated as:

$$U = K\left(\frac{\Delta\varphi}{\Delta Z}\right) \quad (A4)$$

where ΔZ is the center to center simulation distance selected between layers and neglecting the gravitational gradient.

Crop Water Uptake

Non-stressed crop ET is calculated as:

$$ET_c = K_c \times ET_o \quad (A5)$$

where K_c is the crop coefficient and varies with the crop development stages (Table A2) and ET_o is the reference ET. Between cropping seasons, all ET or evaporation E was assumed to take place from the upper layer. For this period Kc was calculated from the mean interval between precipitation events of each month and the mean precipitation event in each month and ET_o [38].

In each layer (k), the actual crop ET can be lower than $ET_c(k)$ due to water stress, which depends on the soil water content and the sensitivity of the crop to low water contents, accounted for through the crop-specific parameter p: the ratio of readily available soil water (RAW) to total available water (TAW) (p = RAW/TAW) [38]. When the soil water content (W(k)) in a layer fell below We(k) = WP + (1 − p) TAW, the ET from that layer actual crop ET(k) dropped below the ETc(k), and the actual ET of the layer was calculated as:

$$\text{actual crop ET} = K_s \times ET_c \quad (A6)$$

where K_s is a stress coefficient [38]:

$$K_s = \begin{cases} 1 & \text{if } W(k) > W_e(k) \\ \dfrac{W(k) - W_e(k)}{W_e(k) - WP(k)} & \text{if } WP(k) < W(k) < W_e(k) \\ 0, & \text{if } W(k) < WP(k) \end{cases} \quad (A7)$$

when one layer was stressed during the growing season ($W(k) < W_e(k)$), the model allowed increase in the extraction coefficient of the lower layer to supply the ET demand of the day. The root zone water uptake pattern depends on irrigation frequency. Root uptake patterns were taken from [18,42,43].

Root length increase a function time is calculated as [45]:

$$L_z = L_o + (L_{max} - L_0) \times \sqrt{\frac{(t - \frac{t_0}{2})}{(t_{L_{max}} - \frac{t_0}{2})}} \quad \text{(A8)}$$

where L_z is the rooting depth at time t, L_o is the starting root depth, L_{max} is the maximum root length, $t_{L_{max}}$ time after planting when L_{max} is reached and t_o is time to reach 90% crop emergence. This is a linear root expansion; the method assumes that once half of the time required for crop emergence is passed by $\frac{t_0}{2}$, the rooting depth starts to increase from an initial depth L_o till L_{max} is reached.

Water Balance

Surface runoff for winter rainfall and a fraction of applied water is modelled using the SCS method. We define curve number (CN) associated with row crop cover for the growing season and bare soil for non-growing season from the SCS tables and calculate precipitation runoff as:

$$SR(p) = \begin{matrix} 0 & \text{if } P \leq 0.2S \\ \frac{(P - 0.2S)^2}{P + 0.8S} & \text{if } P > 0.2S \end{matrix} \quad \text{(A9)}$$

where P is runoff producing precipitation and S is the potential maximum retention after runoff begins related to CN by:

$$S = 254 \times (\frac{100}{CN} - 1) \quad \text{(A10)}$$

Daily water balance for the 4-layered root zone and 2-layered vadose zone is performed. To account for the slow water movement between layers for low water content below field capacity, a slow upward or downward flow U is calculated dependent upon the difference in matric potential between soil layers (Equations B4). In the first quarter of the root zone inflows and outflows include applied water (I) and rainfall (P), the drainage above field capacity (D (1)) to layer 2, actual crop ET and U. For the underlying root zone layers, inflows and outflows include drainage (D) from the overlying layer, U and actual crop ET and finally for the unsaturated layers below the root zone inflows and outflows include D and U.

When the soil water content in layer "k" is above field capacity, the excess water drains to the lower layer over a two-day period, the higher flow in the first day than the second. The fraction α of the excess water that drains the first day is calculated from the soil texture in the layer through an empirical relation obtained to match results presented by approximately 0.9 for sand, 0.85 for loam and 0.7 for clay [49]. Two arbitrary water contents were defined from field capacity to saturation for each layer W_a and W_b defined as:

$$\begin{matrix} W_a = (1 - \alpha) \times (W_s - W_{FC}) + W_{FC} \\ W_b = (1 - \alpha) \times (W_s - W_{FC}) + W_{FC} \end{matrix} \quad \text{(A11)}$$

Drainage (D) is calculated as:

$$D = \begin{matrix} \alpha \times (W_s - W_{FC}) & \text{if } W > W_b \\ (1 - \alpha) \times (W_s - W_{FC}) & \text{if } W_a < W < W_b \\ W - W_{FC} & \text{if } W < W_a \end{matrix} \quad \text{(A12)}$$

where W_{FC} and W_s define field capacity and saturation of a soil layer. After taking out actual crop ET and D outflows from the layer, we also need to account for upwards or downward movement of water (U) dependent upon the difference in matric potential between soil layers (Equation (A4)).

Salt Balance

Salt balance was performed in conjunction with the water balance assuming complete mixing of water entering each layer with that already stored in that layer. The electrical conductivity of water (EC) was used as an indicator of salinity, assuming implicitly that there was a unique relationship between EC and total dissolved solids (TDS) in these dilute solutions and that the EC behaves like a non-reactive solute. Salinity of the input waters (irrigation water (EC_w) and precipitation (EC_p)) must be known. The mass of salts in layer k (Z(k)) is estimated from the product $EC_{sw}(k)\ W(k)$, where EC_{sw} is the electrical conductivity of the soil water in that layer. The mass of salts in layer k in day 1 (t + 1) results from the salinity in day 0 (or t) and the salt fluxes in day 1 that are added sequentially. Accounting for layer 1 for example is as follows: salts in I, P, and mineral dissolution (k_d) are added to the salt mass in layer 1 to obtain $Z^a(1)_1$:

$$Z^a(1)_1 = Z(1)_0 + EC_w \times I_1 + EC_p \times P_1 + k_d \quad \text{(A13)}$$

This results in a soil water concentration of:

$$EC^a_{sw}(1)_1 = Z^a(1)_1/(W(1)_0 + I_1 + P_1) \quad \text{(A14)}$$

Drainage takes place with concentration EC^a_{sw} so that the new mass of salts is:

$$Z^b(1)_1 = Z^a(1)_1 - EC^a_{sw}(1)_1 \times D(1)_1 \quad \text{(A15)}$$

and the new soil water concentration is:

$$EC^b_{sw}(1)_1 = Z^b(1)_1/(W(1)_t + I_1 + P_1 - D(1)_1) \quad \text{(A16)}$$

The soil at this state is evapo-concentrated by crop water uptake (actual crop ET):

$$EC^c_{sw}(1)_1 = Z^b(1)_1/\left(W(1)_t + I_1 + P_1 - D(1)_1 - \text{actual crop ET }(1)_1\right) \quad \text{(A17)}$$

The mass of salts in the slow flow U are then added or removed to obtain the final mass of salts in the layer:

$$Z(1)_1 = \begin{array}{ll} Z^b(1)_1 - U_{1-2} \times EC^c_{sw}(2)_1 & \text{if } U_{1-2} < 0 \\ Z^b(1)_1 - U_{1-2} \times EC^c_{sw}(1)_1 & \text{if } U_{1-2} > 0 \end{array} \quad \text{(A18)}$$

which allows for calculating the final soil water concentration:

$$EC_{sw}(1)_1 = Z(1)_1/W(1)_1 \quad \text{(A19)}$$

References

1. Hillel, D. *Salinity Managment for Sustainable Irrigation: Integrating Science, Environment, and Economics*; The World Bank: Washington, DC, USA, 2000.
2. USBR. *Quality of Water—Colorado River Progress Report No. 24*; U.S. Bureau of Reclamation: Salt Lake City, UT, USA, 2013.
3. Rose, D.A.; Konukcu, F.; Gowing, J.W. Effect of water table depth on evaporation and salt accumulation from saline groundwater. *Aust. J. Soil Res.* **2005**, *43*, 565–573. [CrossRef]
4. Grismer, M.E.; Gates, T.K. Estimating saline water table contributions to crop water use. *Calif. Agric.* **1988**, *42*, 23–24.

5. Ragab, R.A.; Amer, F. Estimating water table contribution to the water supply of maize. *Agric. Water Manag.* **1986**, *11*, 221–230. [CrossRef]
6. Hanson, B.R.; Grattan, R.R.; Fulton, A. *Agricultural Salinity and Drainage*; University of California, Davis: Davis, CA, USA, 2006.
7. Konikow, L.F.; Rielly, T.E. Seawater intrusion in the United States. In *Seawater Intrution in Coastal Aquifers*; Bear, J., Cheng, A.H.D., Sorek, S., Ouazar, D., Herrera, I., Eds.; Springer: dordrecht, The Netherlands, 1999; pp. 463–506.
8. Platts, B.E.; Grismer, M.E. Chloride levels increase after 13 years of recycled water use in the Salinas Valley. *Calif. Agric.* **2014**, *68*, 7. [CrossRef]
9. MCWRA. *State of the Salinas River Groundwater Basin*; Monterey County Water Resources Agency: Salinas, CA, USA, 2015.
10. Vengosh, A.; Gill, J.; Davisson, L.M.; Hudson, G.B. A multi-isotope (B, Sr, O, H, and C) and age dating (^{3}H–^{3}He and ^{14}C) study of groundwater from Salinas Valley, California: Hydrochemistry, dynamics, and contamination processes. *Water Resour. Res.* **2002**, *38*, 9–1–9–17. [CrossRef]
11. Platts, B.; Grismer, M.E. Rainfall leaching is critical for long-term use of recycled water in the Salinas Valley. *Calif. Agric.* **2014**, *68*, 75–78. [CrossRef]
12. Letey, J.; Hoffman, G.J.; Hopmans, J.W.; Grattan, S.R.; Suarez, D.; Corwin, D.L.; Oster, J.D.; Wu, L.; Amrhein, C. Evaluation of soil salinity leaching requirement guidelines. *Agric. Water Manag.* **2011**, *98*, 502–506. [CrossRef]
13. SHI National Soil Health Measurements to Accelerate Agricultural Transformation. Available online: http://soilhealthinstitute.org/national-soil-health-measurements-accelerate-agricultural-transformation/ (accessed on 23 August 2017).
14. Maas, E.V.; Hoffman, G.J. Crop Salt Tolerance. *J. Irrig. Drain.* **1977**, *103*, 20.
15. Rhoades, J.D.; Kandiah, A.; Mashali, A.M. *The Use of Saline Waters for Crop Production*; FAO: Rome, Italy, 1992; Volume FAO Irrigation & Drainage, p. 48.
16. Shelia, V.; Simunek, J.; Boote, K.; Hoogenboom, G. Coupling DSSAT and HYDRUS-1D for simulations of soil water dynamics in the soil-plant-atmosphere system. *J. Hydrol. Hydromech.* **2018**, *66*, 232–245. [CrossRef]
17. Isidoro, D.; Grattan, S.R. Predicting soil salinity in response to different irrigation practices, soil types and rainfall scenarios. *Irrig. Sci.* **2011**, *29*, 197–211. [CrossRef]
18. Ayers, R.S.; Westcot, D.W. *Water Quality for Agriculture*; Food and Agriculture Organization of the United Nations: Rome, Italy, 1985.
19. Gates, T.K.; Grismer, M.E. Stochastic approximation applied to optimal irrigation and drainage planning. *J. Irrig. Drain.* **1989**, *115*, 255–283. [CrossRef]
20. Grismer, M.E. Pore-size distribution and infiltration. *Soil Sci.* **1986**, *141*, 249–260. [CrossRef]
21. Ayars, J.; Christen, E.; Soppe, R.; Meyer, W. The resource potential of in-situ shallow groundwater use in irrigated agriculture. *Irrig. Sci.* **2006**, *24*, 147–160. [CrossRef]
22. Grismer, M.E. *Use of Shallow Groundater for Crop Production*; UC Agriculture & Natural Resources: Davis, CA, USA, 2015; pp. 1–6.
23. Grismer, M.E.; Bali, K.M. Subsurface drainage systems have little impact on water tables salinity of clay soils. *Calif. Agric.* **1998**, *52*, 18–22. [CrossRef]
24. Grismer, M.E.; Gates, T.K.; Hanson, B.R. Irrigation and drainage strategies in saline problem areas. *Calif. Agric.* **1988**, *42*, 23–24.
25. Talsma, T. *The Control of Saline Groundwater*; Meded, Landbouwhogeschool: Wageningen, The Netherlands, 1963; Volume 63, pp. 1–68.
26. Durbin, T.J.; Kapple, G.W.; Freckleton, J.R. *Two-Dimensional and Three-Dimensional Digital Flow Models of the Salinas Valley grouNd-Water Basin, California*; Water Resources Division, US Geological Survey: Reston, VA, USA, 1978.
27. Hall, P. *Selected Geological Cross Sections in the Salinas Valley Using GeoBASE*; Monterey County Water Resources Agency: Salinas, CA, USA, 1992.
28. Fogg, G.E.; Labolle, E.M.; Weissmann, G.S. Groundwater Vulnerability Assessment: Hydrogeologic Perspective and Example from Salinas Valley, California. In *Assessment of Non-Point Source Pollution in the Vadose Zone*; American Geophysical Union: Washington, DC, USA, 2013; pp. 45–61.
29. University of California, Division of Agriculture and Natural Resources. SoilWeb. Available online: https://casoilresource.lawr.ucdavis.edu/gmap/ (accessed on 7 October 2018).

30. Šimůnek, J.; Šejna, M.; Saito, H.; Sakai, M.; van Genuchten, M.T. *The Hydrus-1D Software Package for Simulating the Movement of Water, Heat, and Multiple Solutes in Variably Saturated Media*; 4.14; Department of Environmental Sciences, University of California Riverside: Riverside, CA, USA, 2013.
31. DWR CIMIS: California Irrigation Managemetn Information System. Available online: https://cimis.water.ca.gov/Resources.aspx (accessed on 6 April 2017).
32. Clapp, R.B.; Hornberger, G.M. Empirical equations for some soil hydraulic properties. *Water Resour. Res.* **1978**, *14*, 601–604. [CrossRef]
33. Brooks, R.H.; Corey, A.T. Hydraulic properties of porous media. In *Hydrology Papers*; Colorado State University: Fort Collins, CO, USA, 1964.
34. Gardner, W.R. Some steady state solutions of unsaturated moisture flow equations with application to evaporation from a water table. *Soil Sci.* **1958**, *84*, 228–232. [CrossRef]
35. Haverkamp, R.; Vaclin, M.; Touma, J.; Wierenga, P.J.; Vachaud, G. A comparison of numerical simulation models for one-dimensional infiltration. *Soil Sci. Soc. Am. J.* **1977**, *41*, 285–294. [CrossRef]
36. Mualem, Y. A new model for predicting the hydraulic conductivity of unsaturated porous media. *Water Resour. Res.* **1976**, *12*, 513–522. [CrossRef]
37. van Genuchten, M.T. A closed-form equation for predicting the hydraulic conductivity of unsaturated soils. *Soil Sci. Soc. Am. J.* **1980**, *44*, 892–898. [CrossRef]
38. Allen, R.G.; Pereira, L.S.; Raes, D.; Smith, M. Crop evapotranspiration-Guidelines for computing crop water requirements-FAO Irrigation and drainage paper 56. *FAO Rome* **1998**, *300*, D05109.
39. Hjelmfelt, A.T. Investigation of curve number procedure. *J. Hydraul. Eng.* **1991**, *117*, 725–737. [CrossRef]
40. USDA. *Natural Resources Conservation Service National Engineering Handbook*; USDA: Washington, DC, USA, 2008; Volume Part 623 Section 15.
41. Raats, P.A.C. Distribution of salts in the root zone. *J. Hydrol.* **1975**, *27*, 237–248. [CrossRef]
42. Rhoades, J.D. Use of saline drainage water for irrigation. In *Agricultural Drainage*; Skaggs, R.W., van Schilfgaarde, J., Eds.; Agronomy Monograph, ASA-CSSA-SSSA: Madison, WI, USA, 1999; Volume 38, pp. 615–657.
43. Skaggs, T.H.; Anderson, R.G.; Corwin, D.L.; Suarez, D.L. Analytical steady-state solutions for water-limited cropping systems using saline irrigation water. *Water Resour. Res.* **2014**, *50*, 9656–9674. [CrossRef]
44. Grattan, S.R.; Grieve, C.M. Salinity–mineral nutrient relations in horticultural crops. *Sci. Horticult.* **1998**, *78*, 127–157. [CrossRef]
45. Steduto, P.; Hsiao, T.C.; Raes, D.; Fereres, E. *Crop yield Response to Water*; Food and Agriculture Organization of the United Nations Rome: Rome, Italy, 2012.
46. Orang, M.N.; Snyder, R. *Consumptive Use Program-CUP+*; California Department of Water Resources: Sacramento, CA, USA, 2013.
47. Cahn, M. Estimated Crop Coefficients for Strawberry. In *Salinas Valley Agriculture*; UCANR: Salinas Valley, CA, USA, 2012; Volume 2017.
48. Herman, J.; Usher, W. SALib: An open-source Python library for sensitivity analysis. *J. Open Source Softw.* **2017**, *2*, 97. [CrossRef]
49. Hillel, D.; van Bavel, C.H.M. Simulation of Profile Water Storage as Related to Soil Hydraulic Properties. *Soil Sci. Soc. Am. J.* **1976**, *40*, 807–815. [CrossRef]
50. DWR. *California's Groundwater Bulletin 118*; California Department of Water Resources: Sacramento, CA, USA, 2004.
51. Saltelli, A.; Ratto, M.; Andres, T.; Campolongo, F.; Cariboni, J.; Gatelli, D.; Saisana, M.; Tarantola, S. *Global Sensitivity Analysis: The Primer*; Wiley: Hoboken, NJ, USA, 2008.
52. Morris, M.D. Factorial sampling plans for preliminary computational experiments. *Technometrics* **1991**, *33*, 161–174. [CrossRef]
53. Campolongo, F.; Cariboni, J.; Saltelli, A. An effective screening design for sensitivity analysis of large models. *Environ. Model. Softw.* **2007**, *22*, 1509–1518. [CrossRef]
54. Sobol′, I.M. Global sensitivity indices for nonlinear mathematical models and their Monte Carlo estimates. *Math. Comput. Simul.* **2001**, *55*, 271–280. [CrossRef]
55. Saltelli, A. Making best use of model evaluations to compute sensitivity indices. *Comput. Phys. Commun.* **2002**, *145*, 280–297. [CrossRef]

56. McKay, M.; Beckman, R.; Conover, W. A comparison of three methods for selecting values of input variables in the analyss of output from a computer code. *Technometrics* **1979**, *21*, 239–245.
57. O'Geen, A. Reclaiming Saline, Sodic, and Saline-Sodic Soils. In *Drought Tip*; University of California, Agriculture and Natural Resources: Richmond, CA, USA, 2015.
58. Mullen, K.; Ardia, D.; Gil, D.; Windover, D.; Cline, J. DEoptim: An R Package for Global Optimization by Differential Evolution. *J. Stat. Softw.* **2011**, *40*, 1–26. [CrossRef]
59. Hanson, B.; Hopmans, J.W.; Simunek, J. Leaching with Subsurface Drip Irrigation under Saline, Shallow Groundwater Conditions. *Vadose Zone J.* **2008**, *7*, 810–818. [CrossRef]
60. McLeod, A.I. *Kendall*, version 2.2; Univerity of Western Ontario: London, ON, Canada, 2015.
61. Grieve, C.M.; Grattan, S.R.; Maas, E.V. Plant salt tolerence. In *Angricultural Salinity Assessment and Management*, 2nd ed.; Wallender, W.W., Tanji, K.K., Eds.; American Society of Civil Engineers: Reston, VA, USA, 2012; pp. 405–459.

15

Adjustment of Irrigation Schedules as a Strategy to Mitigate Climate Change Impacts on Agriculture in Cyprus

Panagiotis Dalias *, Anastasis Christou and Damianos Neocleous

Agricultural Research Institute, Ministry of Agriculture, Rural Development and Environment, P.O. Box 22016, 1516 Nicosia, Cyprus; Anastasis.Christou@ari.gov.cy (A.C.); d.neocleous@ari.gov.cy (D.N.)
* Correspondence: dalias@ari.gov.cy.

Abstract: The study aimed at investigating eventual deviations from typical recommendations of irrigation water application to crops in Cyprus given the undeniable changes in recent weather conditions. It focused on the seasonal or monthly changes in crop evapotranspiration (ETc) and net irrigation requirements (NIR) of a number of permanent and annual crops over two consecutive overlapping periods (1976–2000 and 1990–2014). While the differences in the seasonal ETc and NIR estimates were not statistically significant between the studied periods, differences were identified via a month-by-month comparison. In March, the water demands of crops appeared to be significantly greater during the recent past in relation to 1976–2000, while for NIR, March showed statistically significant increases and September showed significant decreases. Consequently, the adjustment of irrigation schedules to climate change by farmers should not rely on annual trends as an eventual mismatch of monthly crop water needs with irrigation water supply might affect the critical growth stages of crops with a disproportionately greater negative impact on yields and quality. The clear increase in irrigation needs in March coincides with the most sensitive growth stage of irrigated potato crops in Cyprus. Therefore, the results may serve as a useful tool for current and future adaptation measures.

Keywords: climate change adaptation; irrigated crops; net irrigation requirements; crop evapotranspiration; monthly changes

---

1. Introduction

Theoretical considerations, climate simulation models and empirical evidence indicate that global warming is leading to increased water vapor and to increased land precipitation at higher latitudes, notably over North America and Eurasia [1]. However, contrary to many mid-to-upper latitude regions of the world, several regional studies have shown a dominant decreasing trend over the Mediterranean Basin [2], although changes will not be equivalent across all Mediterranean regions or seasons [3].

In Cyprus, this change is already being manifested by a decrease in mean annual rainfall and an increase in annual mean temperature. Model projections agree on its future warming and drying, with a likely increase of heatwaves and dry spells; a prospect that will worsen the already existing water scarcity [4,5].

The consequences of such temperature and precipitation changes on a number of aspects of human life and agriculture might be considerable. In agriculture, increased temperatures or the extension of dryness may have a negative impact on crop yields [6] and in turn on food security [7] and may influence crops dynamics, e.g., the exclusion of some crops, or their replacement by others more adapted to the new conditions [8]. Changes in climatic conditions might also affect the proliferation and spread of invasive species, weeds, or diseases [9].

Crop production in Cyprus is covered by annual (e.g., potatoes and vegetables) and permanent (e.g., citrus, olives and grapevines) crops summing over 100×10^3 ha, of which 30% is irrigable land. Citrus and potatoes are the most widely grown crops in the country and consume over 30% of the total agricultural water (150×10^6 m^3). Crop production is constrained by a highly variable climate, limited precipitation and high temperatures from mid-May to mid-September [10]. However, crop water needs may be fully or partly met by rainfall mainly from October to March. Given the projected lower precipitation it can reasonably be assumed that irrigation water availability and crop yields will be affected [11]. Nonetheless, previous work showed that, considering the changes over recent years in mean rainfall and pan evaporation data, the total irrigation needs of crops in Cyprus have not been modified, at least until now [12].

While extreme weather events, which are predicted to increase under future climate scenarios, are already considered a significant challenge for producers [13], little work has been done so far on the current seasonal or monthly changes of temperature or the distribution of precipitation throughout the year and the consequences that these modifications may bring upon crops. Such changes may have an impact on some critical stages of the biological cycle of plants and disproportionately affect productivity and yields [14]. For example, irrigation experiments showed different effects on wheat yield, quality, and water-use efficiencies depending on the plant-growth or phenological stage at which water deficits were applied [15,16]. Therefore, an eventual mismatch of crop water needs with irrigation water supply might be critical, and adaptation measures related to irrigation schedules or the adjustment of planting/seeding dates might be necessary.

This study aims to investigate whether one of the characteristics of the ongoing climate change in Cyprus is a significant modification of the seasonal or monthly water needs and irrigation demands of crops, and discusses the consequences for agricultural production of an eventual deviation of the prevailing irrigation schedules to the current climatic conditions. It also investigates the possibilities for adaptation to climate change challenges using planting period shifts or irrigation schedule modifications.

2. Materials and Methods

The analysis of water and irrigation needs in this study was applied to 35 irrigated crops cultivated in Cyprus. Some of these crops require their water needs to be met fully or partially by irrigation, while some require irrigation only occasionally.

Mean monthly precipitation (mm) and pan evaporation (screened USWB Class A pan) data from 16 weather stations provided by the Cyprus Meteorological Department were used. These stations were situated in the main agricultural areas of Cyprus.

Crop evapotranspiration (ETc) was calculated from potential evapotranspiration (ETo) and pan evaporation (Epan) data obtained from the weather stations using the methodology proposed by the Food and Agriculture Organization of the United Nations (FAO) [17]. More precisely, Epan measurements were converted to reference evapotranspiration (ETo) using the equation:

$$\mathrm{ETo} = \mathrm{Kp} \times \mathrm{Epan} \quad (1)$$

where Kp is the pan coefficient, which takes into account the type of pan, its environment and climate. Potential crop evapotranspiration (ETc) was calculated from ETo according to:

$$\mathrm{ETc} = \mathrm{Kc} \times \mathrm{ETo} \quad (2)$$

where Kc is the crop coefficient, which depends on the kind of crop and its stage of development. Combining the previous two equations, ETc can be expressed as:

$$\mathrm{ETc} = \mathrm{Kc} \times \mathrm{Kp} \times \mathrm{Epan} \quad (3)$$

Substituting Kc × Kp with a coefficient C equation, (3) becomes:

$$ETc = C \times Epan \quad (4)$$

i.e., crop evapotranspiration is calculated directly from pan evaporation using a single coefficient. Values of the coefficients can be derived from the literature [18,19] but they were extensively studied and adjusted to local conditions by the Agricultural Research Institute of Cyprus (e.g., [20,21]).

The net irrigation requirements (NIR) of the crops were calculated by subtracting from the actual water requirements of crops (ETc values) the effective rainfall (Pe), i.e., rain water that is not percolated below the root zone or run-off, but is stored in the root zone and can be used by the plants [17].

ETc and NIR values were estimated for the 35 crops cultivated in Cyprus under two overlapping periods, 1976–2000 and 1990–2014, for each of the 16 weather stations. The mean values obtained for these two periods for each season (winter: December–January–February, spring: March–April–May, summer: June–July–August, and autumn: September–October–November) and each month were compared for each crop.

Statistical Analysis

The mean seasonal or monthly ETc and NIR values for each studied crop and mean seasonal or monthly precipitation and Epan values obtained from the 16 weather stations were compared between the two studied periods (1976–2000 and 1990–2014) using a paired *t*-test (16 double samples for each crop over the 16 weather stations). A p-value < 0.05 in these tests was considered as statistically significant (GraphPad Software, Inc., San Diego, CA, USA).

3. Results

Differences of seasonal ETc and NIR estimates were not statistically significant between the studied periods. The results of the comparison of seasonal NIR between the two studied periods for selected crops cultivated in Cyprus are shown in Table 1. All crops showed decreased average irrigation needs in all seasons apart from spring, where some crops appeared to have greater irrigation water demands in the recent past. The p values for the NIR of many crops were close to statistical significance in autumn.

Table 1. p-Values for the comparison of seasonal net irrigation requirements (NIR) between the two periods (1976–2000 and 1990–2014) for selected crops cultivated in Cyprus. Months that are not included in the irrigation period of a crop are indicated by n/a (non-applicable).

	Winter	Spring	Summer	Autumn
Fruit trees (mountains)	n/a	0.7157	0.1989	0.0680
Green beans: greenhouse	0.1132	0.6404	n/a	0.1973
Haricot beans	n/a	n/a	n/a	0.0728
Lettuce	0.6292	n/a	n/a	0.1639
Marrows: outside grown	n/a	0.6714	0.1924	n/a
Melons: outside grown	n/a	0.6714	0.1879	n/a
Monkey nuts	n/a	0.6080	0.1667	0.0680
Okra (lady´s fingers)	n/a	0.9315	0.1750	n/a
Onions dried	n/a	0.9814	0.5567	n/a
Peas general	n/a	0.8110	n/a	n/a
Peppers: outside grown	n/a	0.6533	0.1773	0.0680
Pistachio	n/a	n/a	0.1903	0.0680
Potatoes (spring crop)	n/a	0.8070	n/a	n/a
Radish	n/a	0.0054	n/a	0.0959
Spinach	n/a	n/a	n/a	0.0959
Table grapes	n/a	0.6484	0.5567	n/a
Table olives	n/a	0.6271	0.1953	0.0712

Table 2, showing the results for only the most water-consuming crops, indicates that March was the only month in which there was a statistically significant difference (increase) in ETc between the two periods ($p < 0.05$). The water demand of all crops in this month was significantly greater in recent years than in the distant past. Void cells in Table 2 indicate months that are not included in the irrigation period of crops in Cyprus. For NIR, apart from March, statistically significant differences between the two periods were also found for September, as seen in Table 3. In March, the irrigation requirements were greater for 1990–2014 than for 1976–2000, in contrast to what was found for September.

Table 2. Crop evapotranspiration (ETc) values in mm of the most water-consuming crops in Cyprus. The upper number for each month indicates the average value for the 1976–2000 period and the lower number indicates the average value for the period 1990–2014. Non-significant differences between these two averages are indicated by n.s. (paired *t*-test) and significant differences ($p < 0.05$) by *. Months that are not included in the irrigation period of a crop are indicated by n/a (non-applicable).

	Bananas (*Musa* spp.)		Citrus (*Citrus* spp.)		Taro (*Colocasia esculenta*)		Potatoes (*Solanum tuberosum*)	
January	n/a		n/a		n/a		n/a	
February	n/a		n/a		n/a		n/a	
March	22.9 23.9	*	18.3 19.1	*	32.9 34.4	*	54.9 57.4	*
April	70.0 69.3	n.s.	65.2 64.6	n.s.	157.3 155.7	n.s.	95.9 94.9	n.s.
May	116.4 115.9	n.s.	99.7 99.2	n.s.	186.3 185.5	n.s.	130.4 129.8	n.s.
June	174.1 173.1	n.s.	132.3 131.6	n.s.	378.0 375.9	n.s.	n/a	
July	219.4 214.6	n.s.	138.3 135.3	n.s.	448.3 438.5	n.s.	n/a	
August	231.8 227.2	n.s.	176.0 172.5	n.s.	452.0 443.1	n.s.	n/a	
September	195.5 189.0	n.s.	119.4 115.5	n.s.	365.9 353.8	n.s.	n/a	
October	123.5 119.7	n.s.	52.7 51.0	n.s.	153.2 148.5	n.s.	n/a	
November	49.2 47.9	n.s.	9.6 9.4	n.s.	135.0 131.5	n.s.	n/a	
December	n/a		n/a		n/a		n/a	

Meteorological precipitation and evaporation data was analyzed to gain insight into the causes of this change. Climate charts of the distribution of rainfall and Epan over the months of the year (Figure 1a,b) illustrate the differences in these meteorological variables between the two 24-year periods. The mean March Epan value for 1990–2014 was significantly increased in relation to the 1976–2000 interval. March and September were the only months that this statistically significant difference was observed. An increase in Epan in March was recorded at 13 of 16 weather stations.

All 16 stations showed decreased average precipitation in more recent years, with an average reduction of 36%. In all other months, the stations showed both increases and decreases in average rainfall when the two periods were compared. The statistically strong tendency to decreased precipitation during March was not followed by a respective decrease during the following two months of spring. In May, for example, at 12 out of 16 stations a tendency towards an increase in precipitation was noted. For September, which also showed a statistically significant change in the irrigation needs of crops, the opposite trend was manifested, with only 1 station out of 16 recording a decrease in rainfall.

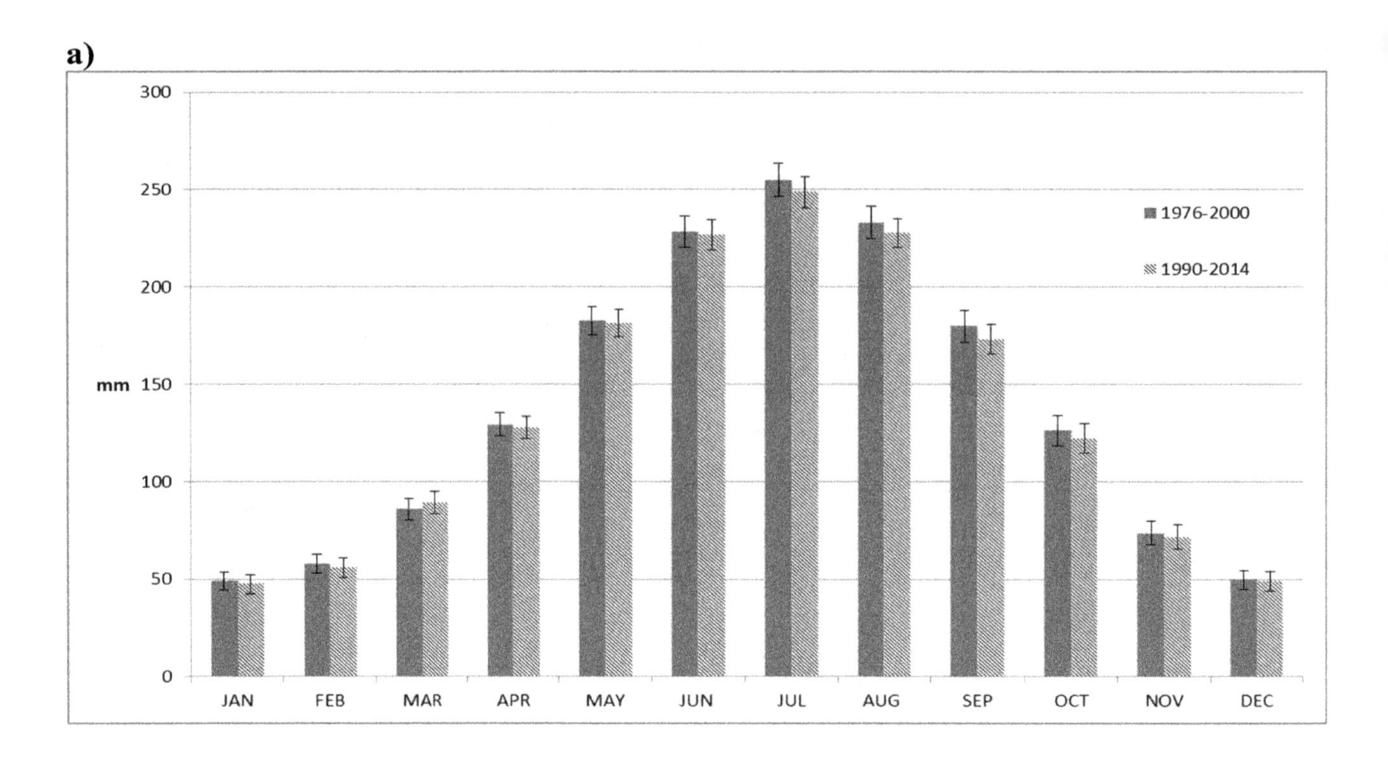

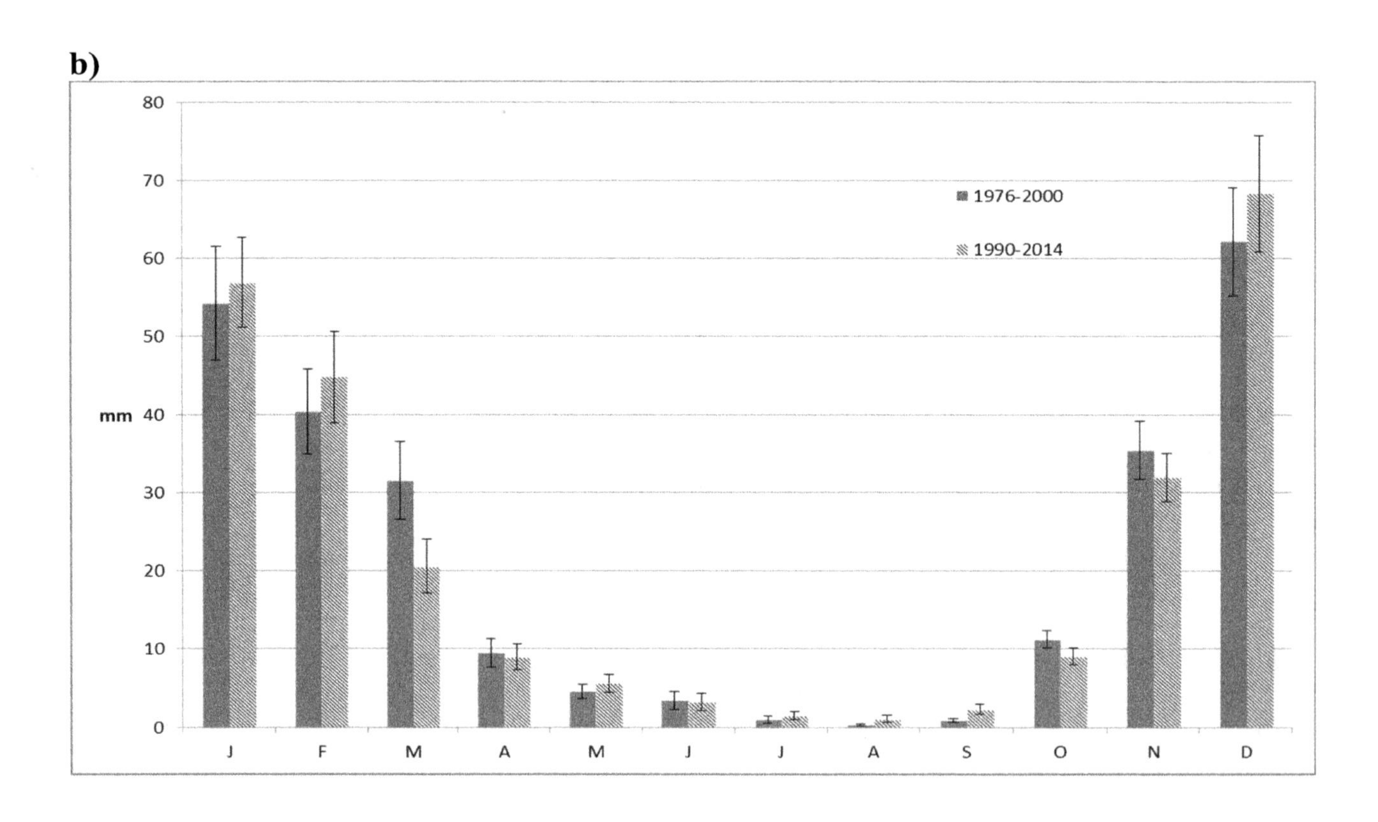

Figure 1. Monthly averages for (**a**) Class A pan evaporation (mm) and (**b**) effective rainfall over two 24-year periods (1976–2000 and 1990–2014) in Cyprus. Data from 16 meteorological weather stations.

Table 3. Net irrigation requirements (NIR) values in mm of the most water-consuming crops in Cyprus. The upper number for each month indicates the average value for the 1976–2000 period and the lower number the average value for the period 1990–2014. Non-significant differences between these two averages are indicated by n.s. (paired *t*-test) and significant differences ($p < 0.05$) by *. Months that are not included in the irrigation period of a crop are indicated by n/a (non-applicable).

	Bananas (*Musa* spp.)		Citrus (*Citrus* spp.)		Taro (*Colocasia esculenta*)		Potatoes (*Solanum tuberosum*)	
January	n/a		n/a		n/a		n/a	
February	n/a		n/a		n/a		n/a	
March	8.1 13.2	*	5.0 8.1	*	15.4 21.3	*	34.5 39.9	*
April	64.7 64.5	n.s.	59.5 59.4	n.s.	157.7 156.6	n.s.	92.3 91.8	n.s.
May	119.5 117.9	n.s.	101.6 100.0	n.s.	194.0 192.0	n.s.	134.4 132.7	n.s.
June	182.1 181.2	n.s.	137.5 136.9	n.s.	399.6 397.6	n.s.	n/a	
July	218.4 213.0	n.s.	137.3 133.7	n.s.	447.3 436.9	n.s.	n/a	
August	231.5 226.1	n.s.	175.7 171.4	n.s.	451.7 442.0	n.s.	n/a	
September	194.6 186.7	*	118.5 113.2	*	365.0 351.5	*	n/a	
October	120.0 118.2	n.s.	44.4 45.0	n.s.	151.7 148.9	n.s.	n/a	
November	20.1 20.3	n.s.	n/a	n.s.	107.9 107.5	n.s.	n/a	
December	n/a		n/a		n/a		n/a	

4. Discussion

The re-estimation of irrigation required two successive past periods in order to evaluate the effect of the ongoing changes in precipitation and evaporative demand of the atmosphere on the water demand of crops. The results revealed some interesting effects of ongoing climate change, which usually do not receive the deserved attention, and which could prove to be a useful guide for farmers, policy makers, government officers and agricultural advisors.

The trends in the change of mean annual precipitation and mean annual temperature in Cyprus are not reflected equally or proportionally at different times of the year. Consequently, the adjustment of irrigation schedules to climate change by farmers should not rely on annual trends as practiced by local growers. Focusing on month-by-month changes revealed strong trends towards an increase in evaporation during March at all meteorological stations, which in combination with a respective decrease in precipitation attests that an adjustment of irrigation water provision to crops is needed. Irrigation programs that are based on "old" meteorological data would result in water deficiencies, which may affect critical growth stages of plants. Moreover, in many cases, farmers now need to irrigate their crops during March, whereas previously, irrigation in March was negligible. In March, the precipitation dropped by 36% and the amount of water that would be needed to compensate for this reduction was estimated to be also 36% on average, as rainfall covered a large part of the total water demand of crops. Climate change effects on irrigation scheduling parameters were also found in Calabria, Italy. From an analysis of reference evapotranspiration (ETo) during the last decades, it was shown that a positive trend in summer precipitation also caused an advance of the last watering, resulting in a slight decrease of the length of the irrigation season [22].

The example of potatoes is probably indicative of the necessary adaptation measures. Potatoes are one of the most exportable products of Cyprus and one of the most water-consuming crops. The "spring crop" or the "main crop" is planted in November/February and harvested in March/June mainly for export, but also for local consumption. Stolonization and tuber initiation are the stages that are most sensitive to water shortage [23], mainly because they are the stages of the highest crop water demand. If water shortages occur during the mid-season stage, which in Cyprus coincides with March, the negative effect on the yield will be pronounced. Karafyllidis et al. [24] showed that limited soil moisture availability affected yield and the number and size of tubers. In the following year, seed produced under conditions of moisture stress produced plants with 20% fewer stems, 24–33% less yield, 18–22% fewer tubers and 19–22% fewer large tubers than plants from seed produced under abundant water supply.

Hence, irrigation should be applied as an adaptation measure to safeguard yields if meteorological trends continue as they are today. An earlier shift of plantation dates could alternatively also be envisaged as an adaptation measure of potato cultivation, as crops would have completed their water-sensitive stage before the less favorable conditions of March. An analysis of the optimum adjustment of planting dates for corn and soybean was also suggested by Woznicki et al. [25] as one of the best adaptation strategies to cope with future climate change scenarios.

However, in contrast, precipitation increased from 0.9 to 2.3 mm in September, affecting irrigation water demand for this month by only 5%. This is because the contribution of rainfall to the total amount of water that needed to be applied to crops in September was nevertheless very small. In this case, following current irrigation guidelines would result in supplying crops with an excess of water. This would not have a negative effect on productivity and yields but it would result in wasting water. The results, therefore, support the notion that in changing climatic conditions, the irrigation adaptation actions required are different in each case depending on specific conditions. Using a modeling approach to simulate the impact of various climate change scenarios on crop water and downscaling climatic parameters derived from global circulation models, Doria and Madramootoo [26] similarly suggested that in order to sustain crop production in the future, efficient irrigation scheduling for producers should be used as an adaptation measure.

The monthly changes in weather conditions that were highlighted in this study and their significant effects on agricultural production constitute a very subtle aspect of climate change, as they are not obvious even as seasonal changes. As a result, we advocate for further examination and verification in other places with a similar climate. However, if the shown precipitation and evaporation trends continue in the future, rainfed crops could also be affected and emphasis should be placed on supplementary irrigation during March. The addition of small amounts of water in this month could improve and stabilize yields, providing the missing moisture for normal plant growth.

5. Conclusions

Irrigation schedules that are based on the average evaporation and rainfall records of an area have to be adjusted to recent changes of climatic parameters even if the year-round changes are not significantly affected. Shifts in rainfall and temperature "allocation" across the months of the year call for a corresponding adjustment of the irrigation water applied to crops, as an eventual mismatch with plant needs could significantly affect some of their critical growth stages. The adjustment of irrigation schedules should be based on more local studies, even if they are in opposition to trends found in wider areas.

Author Contributions: Each author made substantial contributions to this publication. P.D. collected and analyzed the data and wrote the first draft. D.N. and A.C. had a significant contribution to the improvement and revision of the manuscript.

Acknowledgments: This work was supported by the Agricultural Research Institute of Cyprus (ARI) and authors did not receive any specific grant from funding agencies in the public, commercial, or not-for-profit sectors. The authors would like to thank the staff of the Natural Resources and Environment Section of ARI for their assistance.

References

1. Trenberth, K.E.; Jones, P.D.; Ambenje, P.; Bojariu, R.; Easterling, D.; Klein Tank, A.; Parker, D.; Rahimzadeh, F.; Renwick, J.A.; Rusticucci, M.; et al. Observations: Surface and atmospheric climate change. In *Climate Change 2007: The Physical Science Basis*; Solomon, S., Qin, D., Manning, M., Chen, Z., Marquis, M., Averyt, K.B., Tignor, M., Miller, H.L., Eds.; Intergovernmental Panel on Climate Change 4th Assessment Report; Cambridge University Press: Cambridge, UK, 2007; pp. 235–336.
2. Alpert, P.; Ben-Gai, T.; Baharad, A.; Benjamini, Y.; Yekutieli, D.; Colacino, M.; Diodato, L.; Ramis, C.; Homar, V.; Romero, R.; et al. The paradoxical increase of Mediterranean extreme daily rainfall in spite of decrease in total values. *Geophys. Res. Lett.* **2002**, *29*, 1536. [CrossRef]
3. Misra, A.K. Climate change and challenges of water and food security. *Int. J. Sustain. Built Environ.* **2014**, *3*, 153–165. [CrossRef]
4. Lelieveld, J.; Hadjinicolaou, P.; Kostopoulou, E.; Chenoweth, J.; El Maayar, M.; Giannakopoulos, C.; Hannides, C.; Lange, M.A.; Tanarhte, M.; Tyrlis, E.; et al. Climate change and impacts in the eastern Mediterranean and the Middle East. *Clim. Chang.* **2012**, *114*, 667–687. [CrossRef] [PubMed]
5. Lionello, P.; Abrantes, F.; Gacic, M.; Planton, S.; Trigo, R.; Lbrich, U. The climate of the Mediterranean region: Research progress and climate change impacts. *Reg. Environ. Chang.* **2014**, *14*, 1679–1684. [CrossRef]
6. Lobell, D.B.; Field, C.B. Global scale climate–crop yield relationships and the impacts of recent warming. *Environ. Res. Lett.* **2007**, *2*, 014002. [CrossRef]
7. Schmidhuber, J.; Tubiello, F.N. Global food security under climate change. *Proc. Natl. Acad. Sci. USA* **2007**, *104*, 19703–19708. [CrossRef]
8. Olesen, J.E.; Bindi, M. Consequences of climate change for European agricultural productivity, land use and policy. *Eur. J. Agron.* **2002**, *16*, 239–262. [CrossRef]
9. Rosenzweig, C.; Iglesias, A.; Yang, X.B.; Epstein, P.R.; Chivian, E. Climate change and extreme weather events—Implications for food production, plant diseases, and pests. *Glob. Chang. Hum. Health* **2001**, *2*, 90–104. [CrossRef]
10. Papadavid, G.; Neocleous, D.; Kountios, G.; Markou, M.; Michailidis, A.; Ragkos, A.; Hadjimitsis, D. Using SEBAL to Investigate How Variations in Climate Impact on Crop Evapotranspiration. *J. Imaging* **2017**, *3*, 30. [CrossRef]
11. Fraga, H.; Carcia de Cortázar Atauri, I.; Santos, J.A. Viticulture irrigation demands under climate change scenarios in Portugal. *Agric. Water Manag.* **2018**, *196*, 66–74. [CrossRef]
12. Christou, A.; Dalias, P.; Neocleous, D. Spatial and temporal variations in evapotranspiration and net water requirements of typical Mediterranean crops on the island of Cyprus. *J. Agric. Sci.* **2017**, *155*, 1311–1323. [CrossRef]
13. Barlow, K.M.; Christy, B.P.; OLeary, G.J.; Riffkin, P.A.; Nuttall, J.G. Simulating the impact of extreme heat and frost events on wheat crop production: A review. *Field Crops Res.* **2015**, *171*, 109–119. [CrossRef]
14. Hatfield, J.L.; Prueger, J.H. Temperature extremes: Effect on plant growth and development. *Weather Clim. Extremes* **2015**, *10*, 4–10. [CrossRef]
15. Ali, M.H.; Hoque, M.R.; Hassan, A.A.; Khair, A. Effects of deficit irrigation on yield, water productivity, and economic returns of wheat. *Agric. Water Manag.* **2007**, *92*, 151–161. [CrossRef]
16. Tari, A.F. The effects of different deficit irrigation strategies on yield, quality, and water-use efficiencies of wheat under semi-arid conditions. *Agric. Water Manag.* **2016**, *167*, 1–10. [CrossRef]
17. Allen, R.G.; Pereira, L.S.; Raes, D.; Smith, M. *Crop Evapotranspiration—Guidelines for Computing Crop Water Requirements*; FAO Irrigation and Drainage Paper No. 56; FAO: Rome, Italy, 1998.
18. Doorenbos, J.; Pruitt, W.O. *Guidelines for Predicting Crop Water Requirements*; FAO Irrigation and Drainage Paper No. 24; FAO: Rome, Italy, 1975.
19. Doorenbos, J.; Kassam, A.H. *Yield Response to Water*; FAO Irrigation and Drainage Paper No. 33; FAO: Rome, Italy, 1979.
20. Metochis, C. Irrigation of lucerne under semi-arid conditions in Cyprus. *Irrig. Sci.* **1980**, *1*, 247–252. [CrossRef]
21. Metochis, C. Water requirement, yield and fruit quality of grapefruit irrigated with high-sulphate water. *J. Hortic. Sci.* **1989**, *64*, 733–737. [CrossRef]

22. Capra, A.; Mannino, R. Effects of climate change on the irrigation scheduling parameters in Calabria (south Italy) during 1925–2013. *Irrig. Drain. Syst. Eng.* **2015**, *S1*, 003.
23. Brouwer, C.; Prins, K.; Heibloem, M. *Irrigation Water Management: Irrigation Scheduling*; Training Manual No. 4; FAO: Rome, Italy, 1989.
24. Karafyllidis, D.I.; Stavropoulos, N.; Georgakis, D. The effect of water stress on the yielding capacity of potato crops and subsequent performance of seed tubers. *Potato Res.* **1996**, *39*, 153–163. [CrossRef]
25. Woznicki, S.A.; Pouyan Nejadhashemi, A.; Parsinejad, M. Climate change and irrigation demand: Uncertainty and adaptation. *J. Hydrol. Reg. Stud.* **2015**, *3*, 247–264. [CrossRef]
26. Doria, R.O.; Madramootoo, C.A. Estimation of irrigation requirements for some crops in southern Quebec using CROPWAT. *Irrig. Drain.* **2009**, *64*, 1–11.

Permissions

All chapters in this book were first published by MDPI; hereby published with permission under the Creative Commons Attribution License or equivalent. Every chapter published in this book has been scrutinized by our experts. Their significance has been extensively debated. The topics covered herein carry significant findings which will fuel the growth of the discipline. They may even be implemented as practical applications or may be referred to as a beginning point for another development.

The contributors of this book come from diverse backgrounds, making this book a truly international effort. This book will bring forth new frontiers with its revolutionizing research information and detailed analysis of the nascent developments around the world.

We would like to thank all the contributing authors for lending their expertise to make the book truly unique. They have played a crucial role in the development of this book. Without their invaluable contributions this book wouldn't have been possible. They have made vital efforts to compile up to date information on the varied aspects of this subject to make this book a valuable addition to the collection of many professionals and students.

This book was conceptualized with the vision of imparting up-to-date information and advanced data in this field. To ensure the same, a matchless editorial board was set up. Every individual on the board went through rigorous rounds of assessment to prove their worth. After which they invested a large part of their time researching and compiling the most relevant data for our readers.

The editorial board has been involved in producing this book since its inception. They have spent rigorous hours researching and exploring the diverse topics which have resulted in the successful publishing of this book. They have passed on their knowledge of decades through this book. To expedite this challenging task, the publisher supported the team at every step. A small team of assistant editors was also appointed to further simplify the editing procedure and attain best results for the readers.

Apart from the editorial board, the designing team has also invested a significant amount of their time in understanding the subject and creating the most relevant covers. They scrutinized every image to scout for the most suitable representation of the subject and create an appropriate cover for the book.

The publishing team has been an ardent support to the editorial, designing and production team. Their endless efforts to recruit the best for this project, has resulted in the accomplishment of this book. They are a veteran in the field of academics and their pool of knowledge is as vast as their experience in printing. Their expertise and guidance has proved useful at every step. Their uncompromising quality standards have made this book an exceptional effort. Their encouragement from time to time has been an inspiration for everyone.

The publisher and the editorial board hope that this book will prove to be a valuable piece of knowledge for researchers, students, practitioners and scholars across the globe.

List of Contributors

Rosa Francaviglia and Claudia Di Bene
Council for Agricultural Research and Economics, Research Centre for Agriculture and Environment (CREA-AA), 00184 Rome, Italy

Simone Bergonzoli, Massimo Brambilla, Elio Romano, Maurizio Cutini, Pietro Toscano and Carlo Bisaglia
Council for Agricultural Research and Economics-Research Centre for Engineering and Agro-Food Processing, CREA-IT, Treviglio, 24047 Bergamo, Italy

Sergio Saia, Paola Cetera and Luigi Pari
Council for Agricultural Research and Economics—Research Centre for Engineering and Agro-Food Processing (CREA-IT), Via della Pascolare, 16, Monterotondo, 00015 Roma, Italy

Gulom Bekmirzaev
Tashkent Institute of Irrigation and Agricultural Mechanization Engineers, Department of Irrigation and Melioration, Kori-Niyoziy street 39, 100000 Tashkent, Uzbekistan

Jose Beltrao
Research Centre for Spatial and Organizational Dynamics, University of Algarve, Campus de Gambelas, 8005-139 Faro, Portugal

Baghdad Ouddane
Physico-Chemistry Team of the Environment, Sciences and Technologies, University of Lille, LASIR UMR-CNRS 8516, Building C8, 59655 Villeneuve d'Ascq, CEDEX, France

Amir Haghverdi and Somayeh Ghodsi
Department of Environmental Sciences, University of California, Riverside, 900 University Avenue, Riverside, CA 92521, USA

Wesley Wright, Paul Ayers, Douglas Hayes, Brian Leib, Wesley C. Wright, Timothy Grant, Muzi Zheng and Phue Vanchiasong
Department of Biosystems Engineering & Soil Science, University of Tennessee, 2506 E.J. Chapman Drive, Knoxville, TN 37996-4531, USA

Robert Washington-Allen
Department of Agriculture, Nutrition, and Veterinary Science (ANVS), University of Nevada, Reno, Mail Stop 202, Reno, NV 89557, USA

Ali Montazar and Oli Bachie
University of California Division of Agriculture and Natural Resources, UC Cooperative Extension Imperial County, 1050 East Holton Road, Holtville, CA 92250, USA

Dennis Corwin
USDA-ARS, United States Salinity Laboratory, 450 West Big Springs Road, Riverside, CA 92507, USA

Daniel Putnam
Department of Plant Sciences, University of California Davis, One Shields Ave., Davis, CA 95616, USA

Michael Cahn
Division of Agriculture and Natural Resources, University of California, UCCE Monterey County, 1432 Abbott Street, Salinas, CA 93901, USA

Alexander Putman
Department of Microbiology and Plant Pathology, University of California, Riverside, 900 University Avenue, Riverside, CA 92521, USA

Zied Hammami and Asad S. Qureshi
International Center for Biosaline Agriculture (ICBA), Dubai, UAE

Ali Sahli, Fatma Ezzahra Ben Azaiez, Sawsen Ayadi and Youssef Trifa
Laboratory of Genetics and Cereal Breeding, National Agronomic Institute of Tunisia, Carthage University, 43 Avenue Charles Nicole, 1082 Tunis, Tunisia

Arnaud Gauffreteau
INRA–INA-PG–AgroParisTech, UMR 0211, Avenue Lucien Brétignières, F-78850 Thiverval Grignon, France

Zoubeir Chamekh
Laboratory of Genetics and Cereal Breeding, National Agronomic Institute of Tunisia, Carthage University, 43 Avenue Charles Nicole, 1082 Tunis, Tunisia
Carthage University, National Agronomic Research Institute of Tunisia, LR16INRAT02, Hédi Karray, 1082 Tunis, Tunisia

Peng-fei Liu and Yong-guang Hu
Key Laboratory of Modern Agricultural Equipment and Technology, Ministry of Education Jiangsu Province, Jiangsu University, Zhenjiang 212013, China

Aliasghar Montazar
Division of Agriculture and Natural Resources, UC Cooperative Extension, University of California, Imperial County, Holtville, CA 92250, USA

Kyaw-Tha Paw U
Department of Land, Air and Water Resources, University of California, Davis, CA 95616, USA

Yong-zong Lu
Key Laboratory of Modern Agricultural Equipment and Technology, Ministry of Education Jiangsu Province, Jiangsu University, Zhenjiang 212013, China
Division of Agriculture and Natural Resources, UC Cooperative Extension, University of California, Imperial County, Holtville, CA 92250, USA
Department of Land, Air and Water Resources, University of California, Davis, CA 95616, USA

Nicolas Quintana-Ashwell and Drew M. Gholson
National Center for Alluvial Aquifer Research, Mississippi State University, 4006 Old Leland Rd, Leland, MS 38756, USA

L. Jason Krutz
Mississippi Water Resources Research Institute, 885 Stone Blvd, Ballew Hall, Mississippi State University, Leland, MS 38756, USA

Christopher G. Henry
Rice Research and Extension Center, University of Arkansas Cooperative Extension Service, Stuttgart, AR 72160, USA

Trey Cooke
Delta Farmers Advocating Resource Management, Delta F.A.R.M., Stoneville, MS 38776, USA

David Butler
Department of Plant Sciences, University of Tennessee, 2431 Joe Johnson Dr., Knoxville, TN 37996-4531, USA

Bradley Crookston
Department of Plants, Soils and Climate, Utah State University, Logan, UT 84322, USA

Brock Blaser and Marty Rhoades
Department of Agricultural Sciences, West Texas A&M University, Canyon, TX 79016, USA

Murali Darapuneni
Department of Plant and Environmental Sciences, New Mexico State University, Tucumcari, NM 88401, USA

Divya Handa, Robert S. Frazier and Saleh Taghvaeian
Department of Biosystems and Agricultural Engineering, Oklahoma State University, Stillwater, OK 74078, USA

Jason G. Warren
Department of Plant and Soil Sciences, Oklahoma State University, Stillwater, OK 74078, USA

Ahmed A. Hashem
College of Agriculture, Arkansas State University, 422 University Loop W, Jonesboro, AR 72401, USA
Agricultural Engineering Department, Suez Canal University, Kilo 4.5 Ring Road, Ismailia 41522, Egypt
Agricultural & Biological Engineering Department, Purdue University, 225 South University Street, West Lafayette, IN 47907, USA

Sherif A. Radwan
Agricultural Engineering Department, Suez Canal University, Kilo 4.5 Ring Road, Ismailia 41522, Egypt

Bernard A. Engel and Vincent F. Bralts
Agricultural & Biological Engineering Department, Purdue University, 225 South University Street, West Lafayette, IN 47907, USA

Gary W. Marek
USDA-ARS Conservation and Production Research Laboratory, 300 Simmons Road, Unit 10, Bushland, TX 79012, USA

Jerry E. Moorhead
Lindsay Corporation, 8948 Centerport Blvd, Amarillo, TX 79108, USA

Prasanna H. Gowda
USDA, ARS Southeast Area, 141 Experimental Station Road, Stoneville, MS 38776, USA

Prudentia Zikalala
Department of Land, Air, and Water Resources, University of California, One Shield Avenue, Davis, CA 95616, USA

Isaya Kisekka and Mark Grismer
Department of Land, Air, and Water Resources & Biological and Agricultural Engineering, University of California, One Shield Avenue, Davis, CA 95616, USA

Panagiotis Dalias, Anastasis Christou and Damianos Neocleous
Agricultural Research Institute, Ministry of Agriculture, Rural Development and Environment, 1516 Nicosia, Cyprus

Index

Printed in the USA
CPSIA information can be obtained
at www.ICGtesting.com
JSHW051413091023
49903JS00006B/409